ELECTRONICS ENGINEERING

Edited by
Rongie N. Bolanos

Delve Publishing

Electronics Engineering

Edited by Rongie N. Bolanos

Edition 2018

Delve Publishing

Published by

Delve Publishing LLC

708 3rd Avenue, 6th Floor

New York

NY 10801

United States of America

www.delvepublishing.com

ISBN: 978-1-68095-685-6 (Hardback)

Printed in United States of America

Library of Congress Control Number: 2016920458

About the Editor

Rongie N. Bolanos

Rongie N. Bolanos finished Bachelor of Science in Electronics and Communication Engineering at the Rizal Technological University and is currently working as drafter.

Table of Contents

2 HISTORY OF ELECTRONIC ENGINEERING

3 SEMICONDUCTOR DEVICES

4 DIODES

5 MICROCONTROLLER:TYPES AND DESIGN

6 AMPLIFIERS

7 BIPOLAR JUNCTION TRANSISTOR THEORY AND MODELING

8 JFET AND MOSFET

List of Abbreviations

FET	A field-effect transistor
AC	Alternating current,
ADC	Analog-to-digital converter
BJT	Bipolar junction transistor
BJTs	Bipolar junction transistors
COD	Catastrophic optical damage
CE	Common emitter
dB	Decibels
DSP	Digital signal processor
DAC	Digital-to-analog converter
DTL	Diode-transistor logic
DC	Direct current,
DFB	Distributed feedback laser
ECL	Emitter-coupled logic
FETs	Field-effect transistors
GaAs	Gallium arsenide, ,
GPIO	General purpose input/output pins
ICs	Integrated circuits
JFET	Junction field-effect transistor
LEDs	Light-emitting diodes,
LPE	Liquid phase epitaxy
MOSFET	Metal–oxide–semiconductor field-effect transistor
MOSFETs	Metal oxide semiconductor field-effect transistors
MOS	Metal–oxide–semiconductor

MADT	Micro-alloy diffused transistor
MAT	Micro-alloy transistor
μA	Micro-ampere
OAI	OR-AND-Invert
PADT	Post-alloy diffused transistor
PCB	Printed circuit board
PIT	Programmable Interval Timer
PWM	Pulse Width Modulation
RHBD	Radiation-hardened-by-design
RF	Radio frequency
RSD	Raised source/drain
RTL	Resistor-transistor logic
SEMOI	Semiconductor on insulator
SPI	Serial Peripheral Interface
SOI	Silicon on insulator
TED	Transferred electron device
TTL	Transistor-transistor logic
TWTAs	Traveling wave tube amplifiers
UTC	Ultrathin channel
UART	Universal Asynchronous Receiver/Transmitter
USB	Universal Serial Bus
VCSELs	Vertical-cavity surface-emitting lasers

Preface

Electronics means study of flow of electrons in electrical circuits. The word Electronics comes from electron mechanics which means learning the way how an electron behaves under different conditions of externally applied fields. Fundamentals of electronics are the core subject in all branches of engineering nowadays. It deals with electrical circuits that involve active electrical components such as vacuum tubes, transistors, diodes, integrated circuits, associated passive. Electronics has made tremendous advancement during last few decades and our day to day life involves the use of electronic devices. Availability of economical and fast means of communication paves the way for progress of a country. Few decades ago, the main application of electronics was in the field of telephony and telegraphy. Now, with the aid of radio waves we can transmit any message from one place to another, without the use of wires. Radio and TV broadcasting offers a means of both entertainment as well as communication. Today, Electronics gadgets are widely used for entertainment. Defense applications are completely controlled by electronic circuits. RADAR that is Radio Detection and Ranging is the most important development in electronics field. With the help of radar it is possible to detect and find the exact location of enemy aircraft. Radar and anti-craft guns can be linked by an automatic control system to make a complete unit. Electronics circuits are widely being used in industrial applications such as control of thickness, quality, weight and moisture content of a material. Electronic amplifier circuits are used to amplify signals and thus control the operations of automatic door openers, power systems and safety devices. Electronically controlled systems are used for heating and welding in the industry. Electronics systems are being used by Doctors and scientists in the diagnosis and treatment of various diseases. X-rays, ECG, Short eave diathermy units and oscillographs are some instruments which have been used so far in medical science. The use of electronics in medical science has grown so extremely and is useful in saving the life of mankind from a lot of sufferings. Electronics engineering is an electrical engineering discipline which utilizes non-linear and active electrical components to design electronic circuits, devices, microprocessors, microcontrollers and other systems. The discipline typically also designs passive electrical components, usually based on printed circuit boards. This text introduces reader to the basics of electronic engineering.

Content Coverage

Chapters One and *Two* reveal an introduction of electronics and historical overview of electronic engineering, respectively. Electronic engineering as a profession sprang from technological improvements in the telegraph industry during the late 19th century and in the radio and telephone industries during the early 20th century.

Chapters Three and *Four* are intended to effort on semiconductor devices and diodes, separately. Semiconductor devices exploit the electronic properties of semiconductor materials, principally silicon, germanium, and gallium arsenide, as well as organic semiconductors, while diode is like one-way valve which allows electric current to flow in one direction but generally does not allow it to flow in the opposite direction.

Chapters Five and *Six* present a coverage of microcontrollers and power amplifiers, correspondingly. Microcontrollers are used in automatically controlled products and devices, such as automobile engine control systems, implantable medical devices, power tools, toys and other embedded systems, while amplifier is an electronic device that receives an electrical signal and reprocesses it to amplify, or increase, its power.

Chapters Seven and *Eight* elucidate the details of bipolar junction transistor (BJT) and JFET & MOSFET, separately. The primary purpose of BJT is for use in electrical devices to amplify or change electrical signals. However, JFETs are exclusively voltage-controlled in that they do not need a biasing current and MOSFET are made exhibit both insulation like properties and conduction like properties.

CHAPTER 1

An Introduction to Electronics

Chapter Contents

- Introduction
- Electronic Devices and Components
- History of Electronic Components
- Types of Circuits
- Logic Gate
- Adder (Electronics)
- Counter (Digital)
- Processor Register
- Multiplexer
- Heat Dissipation and Thermal Management
- Noise
- Electronics Theory
- Electronics Lab
- Computer Aided Design (CAD)
- Construction Methods
- Degradation

Electronics is the branch of science that deals with the study of flow and control of electrons (electricity) and the study of their behavior and effects in vacuums, gases, and semiconductors, and with devices using such electrons. This control of electrons is accomplished by devices that resist, carry, select, steer, switch, store, manipulate, and exploit the electron.

Some of the basic electrical units and definitions are mentioned below:

Passive: Capable of operating without an external power source. Typical passive components are resistors, capacitors, inductors and diodes (although the latter are a special case).

Active: Requiring a source of power to operate. Includes transistors (all types), integrated circuits (all types), TRIACs, SCRs, LEDs, etc.

INTRODUCTION

Electronics is the science of controlling electrical energy electrically, in which the electrons have a fundamental role. Electronics deals with electrical circuits that involve active electrical components such as vacuum tubes, transistors, diodes, integrated circuits, associated passive electrical components, and interconnection technologies. Commonly, electronic devices contain circuitry consisting primarily or exclusively of active semiconductors supplemented with passive elements; such a circuit is described as an electronic circuit.

The science of Electronics is also considered to be a branch of Physics and Electrical Engineering.

Surface-mount electronic components

The nonlinear behaviour of active components and their ability to control electron flows makes amplification of weak signals possible, and electronics is widely used in information processing, telecommunication, and signal processing. The ability of electronic devices to act as switches makes digital information processing possible. Interconnection technologies such as circuit boards, electronics packaging technology, and other varied forms of communication infrastructure complete circuit functionality and transform the mixed components into a regular working system.

Electronics is distinct from electrical and electro-mechanical science and technology, which deal with the generation, distribution, switching, storage, and conversion of electrical energy to and from other energy forms using wires, motors, generators, batteries, switches, relays, transformers, resistors, and other passive components. This distinction started around 1906 with the invention by Lee De Forest of the triode, which made electrical amplification of weak radio signals and audio signals possible with a non-mechanical device. Until 1950 this field was called "radio technology" because its principal application was the design and theory of radio transmitters, receivers, and vacuum tubes.

Today, most electronic devices use semiconductor components to perform electron control. The study of semiconductor devices and related technology is considered a branch of solid-state physics, whereas the design and construction of electronic circuits to solve practical problems come under electronics engineering. This article focuses on engineering aspects of electronics.

Branches of Electronics

Electronics has branches as follows:

- Digital electronics
- Analogue electronics
- Microelectronics
- Circuit design
- Integrated circuits
- Optoelectronics
- Semiconductor devices
- Embedded systems

ELECTRONIC DEVICES AND COMPONENTS

Electronics Technician performing a voltage check on a power circuit card in the air navigation equipment room aboard the aircraft carrier USS *Abraham Lincoln* (CVN 72).

An electronic component is any physical entity in an electronic system used to affect the electrons or their associated fields in a manner consistent with the intended function of the electronic system. Components are generally intended to be connected together, usually by being soldered to a printed circuit board (PCB), to create an electronic circuit with a particular function (for example an amplifier, radio receiver, or oscillator). Components may be packaged singly, or in more complex groups as integrated circuits. Some common electronic components are capacitors, inductors, resistors, diodes, transistors, etc. Components are often categorized as active (e.g. transistors and thyristors) or passive (e.g. resistors, diodes, inductors and capacitors).

HISTORY OF ELECTRONIC COMPONENTS

Vacuum tubes (Thermionic valves) were among the earliest electronic components. They were almost solely responsible for the electronics revolution of the first half of the Twentieth Century. They took electronics from parlor tricks and gave us radio, television, phonographs, radar, long distance telephony and much more. They played a leading role in the field of microwave and high power transmission as well as television receivers until the middle of the 1980s. Since that time, solid state devices have all but completely taken over. Vacuum tubes are still used in some specialist applications such as high power RF amplifiers, cathode ray tubes, specialist audio equipment, guitar amplifiers and some microwave devices.

In April 1955 the IBM 608 was the first IBM product to use transistor circuits without any vacuum tubes and is believed to be the world's first all-transistorized calculator to be manufactured for the commercial market. The 608 contained more than 3,000 germanium transistors. Thomas J. Watson Jr. ordered all future IBM products to use transistors in their design. From that time on transistors were almost exclusively used for computer logic and peripherals.

TYPES OF CIRCUITS

Circuits and components can be divided into two groups: analog and digital. A particular device may consist of circuitry that has one or the other or a mix of the two types.

Analog circuits

Hitachi J100 adjustable frequency drive chassis

Most analog electronic appliances, such as radio receivers, are constructed from combinations of a few types of basic circuits. Analog circuits use a continuous range of voltage or current as opposed to discrete levels as in digital circuits.

The number of different analog circuits so far devised is huge, especially because a ‹circuit' can be defined as anything from a single component, to systems containing thousands of components.

Analog circuits are sometimes called linear circuits although many non-linear effects are used in analog circuits such as mixers, modulators, etc. Good examples of analog circuits include vacuum tube and transistor amplifiers, operational amplifiers and oscillators.

One rarely finds modern circuits that are entirely analog. These days analog circuitry may use digital or even microprocessor techniques to improve performance. This type of circuit is usually called "mixed signal" rather than analog or digital.

Sometimes it may be difficult to differentiate between analog and digital circuits as they have elements of both linear and non-linear operation. An example is the comparator which takes in a continuous range of voltage but only outputs one of two levels as in a digital circuit. Similarly, an overdriven transistor amplifier can take on the characteristics of a controlled switch having essentially two levels of output. In fact, many digital circuits are actually implemented as variations of analog circuits similar to this example—after all, all aspects of the real physical world are essentially analog, so digital effects are only realized by constraining analog behavior.

Digital Circuits

Digital circuits are electric circuits based on a number of discrete voltage levels. Digital circuits are the most common physical representation of Boolean algebra, and are the basis of all digital computers. To most engineers, the terms "digital circuit", "digital system" and "logic" are interchangeable in the context of digital circuits. Most digital circuits use a binary system with two voltage levels labeled "0" and "1". Often logic "0" will be a lower voltage and referred to as "Low" while logic "1" is referred to as "High". However, some systems use the reverse definition ("0" is "High") or are current based. Quite often the logic designer may reverse these definitions from one circuit to the next as he sees fit to facilitate his design. The definition of the levels as "0" or "1" is arbitrary.

Ternary (with three states) logic has been studied, and some prototype computers made.

Computers, electronic clocks, and programmable logic controllers (used to control industrial processes) are constructed of digital circuits. Digital signal processors are another example.

Building blocks:

- Logic gates
- Adders
- Flip-flops
- Counters
- Registers
- Multiplexers
- Schmitt triggers

Highly integrated devices:

- Microprocessors
- Microcontrollers
- Application-specific integrated circuit (ASIC)
- Digital signal processor (DSP)
- Field-programmable gate array (FPGA)

LOGIC GATE

In electronics, a **logic gate** is an idealized or physical device implementing a Boolean function; that is, it performs a logical operation on one or more binary inputs, and produces a single binary output. Depending on the context, the term may refer to an **ideal logic gate**, one that has for instance zero rise time and unlimited fan-out, or it may refer to a non-ideal physical device (see Ideal and real op-amps for comparison).

Logic gates are primarily implemented using diodes or transistors acting as electronic switches, but can also be constructed using vacuum tubes, electromagnetic relays (relay logic), fluidic logic, pneumatic logic, optics, molecules, or even mechanical elements. With amplification, logic gates can be cascaded in the same way that Boolean functions can be composed, allowing the construction of a physical model of all of Boolean logic, and therefore, all of the algorithms and mathematics that can be described with Boolean logic.

Logic circuits include such devices as multiplexers, registers, arithmetic logic units (ALUs), and computer memory, all the way up through complete microprocessors, which may contain more than 100 million gates. In modern practice, most gates are made from field-effect transistors (FETs), particularly MOSFETs (metal–oxide–semiconductor field-effect transistors).

Compound logic gates AND-OR-Invert (AOI) and OR-AND-Invert (OAI) are often employed in circuit design because their construction using MOSFETs is simpler and more efficient than the sum of the individual gates. In reversible logic, Toffoli gates are used.

Electronic gates

To build a functionally complete logic system, relays, valves (vacuum tubes), or transistors can be used. The simplest family of logic gates using bipolar transistors is called resistor-transistor logic (RTL). Unlike simple diode logic gates (which do not have a gain element), RTL gates can be cascaded indefinitely to produce more complex logic functions. RTL gates were used in early integrated circuits. For higher speed and better density, the resistors used in RTL were replaced by diodes resulting in diode-transistor logic (DTL). Transistor-transistor logic (TTL) then supplanted DTL. As integrated circuits became more complex, bipolar transistors were replaced with smaller field-effect transistors (MOSFETs); see PMOS and NMOS. To reduce power consumption still further, most contemporary chip implementations of digital systems now use CMOS logic. CMOS uses complementary (both n-channel and p-channel) MOSFET devices to achieve a high speed with low power dissipation.

For small-scale logic, designers now use prefabricated logic gates from families of devices such as the TTL 7400 series by Texas Instruments, the CMOS 4000 series by RCA, and their more recent descendants. Increasingly, these fixed-function logic gates are being replaced by programmable logic devices, which allow designers to pack a large number of mixed logic gates into a single integrated circuit. The field-programmable nature of programmable logic devices such as FPGAs has reduced the 'hard' property of hardware; it is now possible to change the logic design of a hardware system by reprogramming some of its components, thus allowing the features or function of a hardware implementation of a logic system to be changed.

Other types of logic gates include, but are not limited to

Logic family	Abbreviation	Description
Tunnel diode logic	TDL	Exactly the same as Diode Logic but can perform at a faster speed.
Neon logic	NL	Uses neon bulbs or 3 element neon trigger tubes to perform logic.
Core diode logic	CDL	Performed by semiconductor diodes and small ferrite toroidal cores for moderate speed and moderate power level.

4Layer Device Logic	4LDL	Uses thyristors and SCR's to perform logic operations where high current and or high voltages are required.
Direct-coupled transistor logic	DCTL	Uses transistors switching between bottom saturated and tunnel states to perform logic, depends upon use of transistors with carefully controlled parameters. Economical but tends to be susceptible to noise because of the lower voltage levels employed. Often considered to be the father to modern TTL logic.
Current-mode logic	CML	Uses transistors to perform logic but biasing is from constant current sources to prevent saturation and allow extremely fast switching. Has high noise immunity despite fairly low logic levels.
Diode logic	DL	

Electronic logic gates differ significantly from their relay-and-switch equivalents. They are much faster, consume much less power, and are much smaller (all by a factor of a million or more in most cases). Also, there is a fundamental structural difference. The switch circuit creates a continuous metallic path for current to flow (in either direction) between its input and its output. The semiconductor logic gate, on the other hand, acts as a high-gain voltage amplifier, which sinks a tiny current at its input and produces a low-impedance voltage at its output. It is not possible for current to flow between the output and the input of a semiconductor logic gate.

Another important advantage of standardized integrated circuit logic families, such as the 7400 and 4000 families, is that they can be cascaded. This means that the output of one gate can be wired to the inputs of one or several other gates, and so on. Systems with varying degrees of complexity can be built without great concern of the designer for the internal workings of the gates, provided the limitations of each integrated circuit are considered.

The output of one gate can only drive a finite number of inputs to other gates, a number called the ‹fanout limit'. Also, there is always a delay, called the 'propagation delay', from a change in input of a gate to the corresponding change in its output. When gates are cascaded, the total propagation delay is approximately the sum of the individual delays, an effect which can become a problem in high-speed circuits. Additional delay can be caused when a large number of inputs are connected to an output, due to the distributed capacitance of all the inputs and wiring and the finite amount of current that each output can provide.

History and Development

The binary number system was refined by Gottfried Wilhelm Leibniz (published in 1705) and he also established that by using the binary system, the principles of arithmetic and logic could be combined.

In an 1886 letter, Charles Sanders Peirce described how logical operations could be carried out by electrical switching circuits. Eventually, vacuum tubes replaced relays for logic operations. Lee De Forest's modification, in 1907, of the Fleming valve can be used as an AND logic gate. Ludwig Wittgenstein introduced a version of the 16-row truth table as proposition 5.101 of *Tractatus Logico-Philosophicus* (1921). Walther Bothe, inventor of the coincidence circuit, got part of the 1954 Nobel Prize in physics, for the first modern electronic AND gate in 1924. Konrad Zuse designed and built electromechanical logic gates for his computer Z1 (from 1935–38).

Claude E. Shannon introduced the use of Boolean algebra in the analysis and design of switching circuits in 1937. Using this property of electrical switches to implement logic is the fundamental concept that underlies all electronic digital computers. Shannon's work became the foundation of digital circuit design, as it became widely known in the electrical engineering community during and after World War II. The theoretical rigor of Shannon's work superseded the *ad hoc* methods that had prevailed previously.

Active research is taking place in molecular logic gates.

Symbols

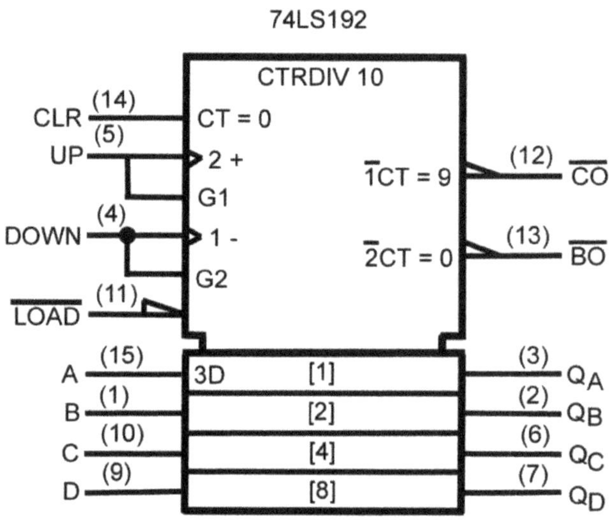

A synchronous 4-bit up/down decade counter symbol (74LS192) in accordance with ANSI/IEEE Std. 91-1984 and IEC Publication 60617-12.

There are two sets of symbols for elementary logic gates in common use, both defined in ANSI/IEEE Std 91-1984 and its supplement ANSI/IEEE Std 91a-1991. The "distinctive shape" set, based on traditional schematics, is used for simple drawings, and derives from MIL-STD-806 of the 1950s and 1960s. It is sometimes unofficially described as "military", reflecting its origin. The "rectangular shape" set, based on ANSI Y32.14 and other early industry standards, as later refined by IEEE and IEC, has rectangular outlines for all types of gate and allows representation of a much wider range of devices than is possible with the traditional symbols. The IEC standard, IEC 60617-12, has been adopted by other standards, such as EN 60617-12:1999 in Europe, BS EN 60617-12:1999 in the United Kingdom, and DIN EN 60617-12:1998 in Germany.

The mutual goal of IEEE Std 91-1984 and IEC 60617-12 was to provide a uniform method of describing the complex logic functions of digital circuits with schematic symbols. These functions were more complex than simple AND and OR gates. They could be medium scale circuits such as a 4-bit counter to a large scale circuit such as a microprocessor.

IEC 617-12 and its successor IEC 60617-12 do not explicitly show the "distinctive shape" symbols, but do not prohibit them. These are, however, shown in ANSI/IEEE 91 (and 91a) with this note: "The distinctive-shape symbol is, according to IEC Publication 617, Part 12, not preferred, but is not considered to be in contradiction to that standard." IEC 60617-12 correspondingly contains the note (Section 2.1) "Although non-preferred, the use of other symbols recognized by official national standards, that is distinctive shapes in place of symbols [list of basic gates], shall not be considered to be in contradiction with this standard. Usage of these other symbols in combination to form complex symbols (for example, use as embedded symbols) is discouraged." This compromise was reached between the respective IEEE and IEC working groups to permit the IEEE and IEC standards to be in mutual compliance with one another.

A third style of symbols was in use in Europe and is still widely used in European academia. See the column "DIN 40700" in the table in the German Wikipedia.

In the 1980s, schematics were the predominant method to design both circuit boards and custom ICs known as gate arrays. Today custom ICs and the field-programmable gate array are typically designed with Hardware Description Languages (HDL) such as Verilog or VHDL.

Type	**Distinctive shape (IEEE Std 91/91a-1991)**	**Rectangular shape (IEEE Std 91/91a-1991 IEC 60617-12 : 1997)**	**Boolean algebra between A & B**	**Truth table**
AND	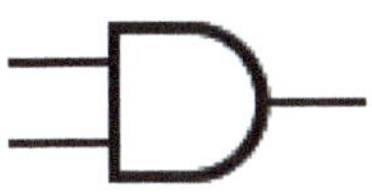		{\displaystyle A\ cdot B{\text{ or }}A\;\&\;B}	**INPUT** | **OUTPUT** A | B | A AND B 0 | 0 | 0 0 | 1 | 0 1 | 0 | 0 1 | 1 | 1
OR			{\displaystyle A+B}	**INPUT** | **OUTPUT** A | B | A OR B 0 | 0 | 0 0 | 1 | 1 1 | 0 | 1 1 | 1 | 1

<table>
<tr><td>NOT</td><td>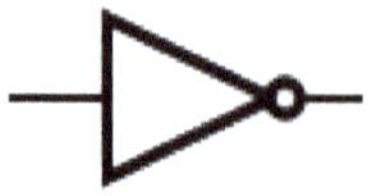</td><td></td><td>{\displaystyle {\ overline {A}}{\ text{ or }}{\sim }A}</td><td><table><tr><th>INPUT</th><th>OUTPUT</th></tr><tr><td>A</td><td>NOT A</td></tr><tr><td>0</td><td>1</td></tr><tr><td>1</td><td>0</td></tr></table></td></tr>
<tr><td colspan="5"></td></tr>
<tr><td>NAND</td><td>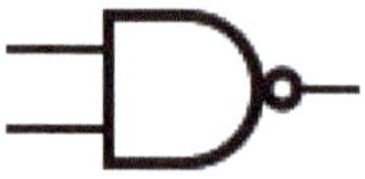</td><td></td><td>{\displaystyle {\ overline {A\cdot B}}{\text{ or }} A\uparrow B}</td><td><table><tr><th colspan="2">INPUT</th><th>OUTPUT</th></tr><tr><td>A</td><td>B</td><td>A NAND B</td></tr><tr><td>0</td><td>0</td><td>1</td></tr><tr><td>0</td><td>1</td><td>1</td></tr><tr><td>1</td><td>0</td><td>1</td></tr><tr><td>1</td><td>1</td><td>0</td></tr></table></td></tr>
<tr><td>NOR</td><td>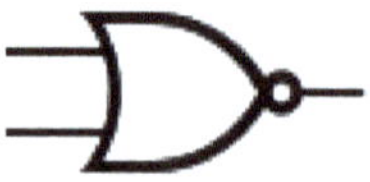</td><td></td><td>{\displaystyle {\ overline {A+B}} {\text{ or }} A-B}</td><td><table><tr><th colspan="2">INPUT</th><th>OUTPUT</th></tr><tr><td>A</td><td>B</td><td>A NOR B</td></tr><tr><td>0</td><td>0</td><td>1</td></tr><tr><td>0</td><td>1</td><td>0</td></tr><tr><td>1</td><td>0</td><td>0</td></tr><tr><td>1</td><td>1</td><td>0</td></tr></table></td></tr>
<tr><td>XOR</td><td>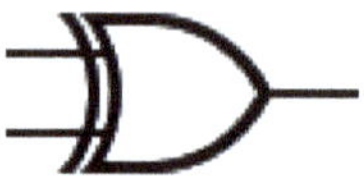</td><td></td><td>{\displaystyle A\ oplus B}</td><td><table><tr><th colspan="2">INPUT</th><th>OUTPUT</th></tr><tr><td>A</td><td>B</td><td>A XOR B</td></tr><tr><td>0</td><td>0</td><td>0</td></tr><tr><td>0</td><td>1</td><td>1</td></tr><tr><td>1</td><td>0</td><td>1</td></tr><tr><td>1</td><td>1</td><td>0</td></tr></table></td></tr>
<tr><td colspan="5"></td></tr>
<tr><td>XNOR</td><td>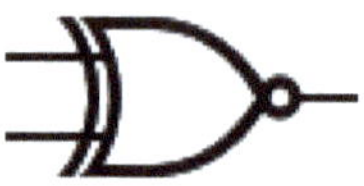</td><td>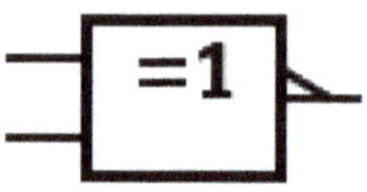</td><td>{\displaystyle {\overline {A\ oplus B}}{\text{ or }}{A\odot B}}</td><td><table><tr><th colspan="2">INPUT</th><th>OUTPUT</th></tr><tr><td>A</td><td>B</td><td>A XNOR B</td></tr><tr><td>0</td><td>0</td><td>1</td></tr><tr><td>0</td><td>1</td><td>0</td></tr><tr><td>1</td><td>0</td><td>0</td></tr><tr><td>1</td><td>1</td><td>1</td></tr></table></td></tr>
</table>

In electronics a NOT gate is more commonly called an inverter. The circle on the symbol is called a *bubble*, and is used in logic diagrams to indicate a logic negation between the external logic state and the internal logic state (1 to 0 or vice versa). On a circuit diagram it must be accompanied by a statement asserting that the *positive logic convention* or *negative logic convention* is being used (high voltage level = 1 or high voltage level = 0, respectively). The *wedge* is used in circuit diagrams to directly indicate an active-low (high voltage level = 0) input or output without requiring a uniform convention throughout the circuit diagram. This is called *Direct Polarity Indication*. See IEEE Std 91/91A and IEC 60617-12. Both the *bubble* and the *wedge* can be used on distinctive-shape and rectangular-shape symbols on circuit diagrams, depending on the logic convention used. On pure logic diagrams, only the *bubble* is meaningful.

The output of a two input exclusive-OR is true only when the two input values are *different*, and false if they are equal, regardless of the value. If there are more than two inputs, the output of the distinctive-shape symbol is undefined. The output of the rectangular-shaped symbol is true if the number of true inputs is exactly one or exactly the number following the "=" in the qualifying symbol. In practice, these gates are built from combinations of simpler logic gates.

Universal Logic Gates

The 7400 chip, containing four NANDs. The two additional pins supply power (+5 V) and connect the ground.

Charles Sanders Peirce (during 1880–81) showed that NOR gates alone (or alternatively NAND gates alone) can be used to reproduce the functions of all the other logic gates, but his work on it was unpublished until 1933. The first published proof was by Henry M. Sheffer in 1913, so the NAND logical operation is sometimes called Sheffer stroke; the logical NOR is sometimes called *Peirce's arrow*.Consequently, these gates are sometimes called *universal logic gates*.

De Morgan Equivalent Symbols

By use of De Morgan's laws, an *AND* function is identical to an *OR* function with negated inputs and outputs. Likewise, an *OR* function is identical to an *AND* function with negated inputs and outputs. A NAND gate is equivalent to an OR gate with negated inputs, and a NOR gate is equivalent to an AND gate with negated inputs.

This leads to an alternative set of symbols for basic gates that use the opposite core symbol (*AND* or *OR*) but with the inputs and outputs negated. Use of these alternative symbols can make logic circuit diagrams much clearer and help to show accidental connection of an active high output to an active low input or vice versa. Any connection that has logic negations at both ends can be replaced by a negationless connection and a suitable change of gate or vice versa. Any connection that has a negation at one end and no negation at the other can be made easier to interpret by instead using the De Morgan equivalent symbol at either of the two ends. When negation or polarity indicators on both ends of a connection match, there is no logic negation in that path (effectively, bubbles “cancel”), making it easier to follow logic states from one symbol to the next. This is commonly seen in real logic diagrams - thus the reader must not get into the habit of associating the shapes exclusively as OR or AND shapes, but also take into account the bubbles at both inputs and outputs in order to determine the “true” logic function indicated.

A De Morgan symbol can show more clearly a gate’s primary logical purpose and the polarity of its nodes that are considered in the “signaled” (active, on) state. Consider the simplified case where a two-input NAND gate is used to drive a motor when either of its inputs are brought low by a switch. The “signaled” state (motor on) occurs when either one OR the other switch is on. Unlike a regular NAND symbol, which suggests AND logic, the De Morgan version, a two negative-input OR gate, correctly shows that OR is of interest. The regular NAND symbol has a bubble at the output and none at the inputs (the opposite of the states that will turn the motor on), but the De Morgan symbol shows both inputs and output in the polarity that will drive the motor.

De Morgan’s theorem is most commonly used to implement logic gates as combinations of only NAND gates, or as combinations of only NOR gates, for economic reasons.

Data Storage

Logic gates can also be used to store data. A storage element can be constructed by connecting several gates in a “latch” circuit. More complicated designs that use clock signals and that change only on a rising or falling edge of the clock are called edge-triggered “flip-flops”. Formally, a flip-flop is called a bistable circuit, because it has two stable states which it can maintain indefinitely. The combination of multiple flip-flops in parallel, to store a multiple-bit value, is known as a register. When using any of these gate setups the overall system has memory; it is then called a sequential logic system since its output can be influenced by its previous state(s), i.e. by the *sequence* of input states. In contrast, the output from combinatorial logic is purely a combination of its present inputs, unaffected by the previous input and output states.

These logic circuits are known as computer memory. They vary in performance, based on factors of speed, complexity, and reliability of storage, and many different types of designs are used based on the application.

Three-State Logic Gates

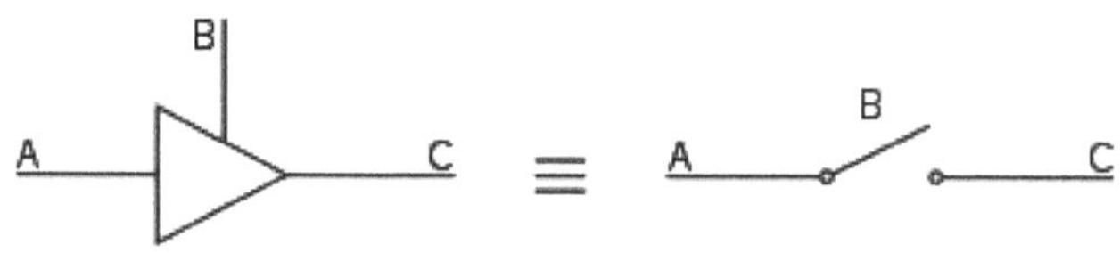

A tristate buffer can be thought of as a switch. If *B* is on, the switch is closed. If B is off, the switch is open.

A three-state logic gate is a type of logic gate that can have three different outputs: high (H), low (L) and high-impedance (Z). The high-impedance state plays no role in the logic, which is strictly binary. These devices are used on buses of the CPU to allow multiple chips to send data. A group of three-states driving a line with a suitable control circuit is basically equivalent to a multiplexer, which may be physically distributed over separate devices or plug-in cards.

In electronics, a high output would mean the output is sourcing current from the positive power terminal (positive voltage). A low output would mean the output is sinking current to the negative power terminal (zero voltage). High impedance would mean that the output is effectively disconnected from the circuit.

Implementations

Since the 1990s, most logic gates are made in CMOS technology (i.e. NMOS and PMOS transistors are used). Often millions of logic gates are packaged in a single integrated circuit.

There are several logic families with different characteristics (power consumption, speed, cost, size) such as: RDL (resistor-diode logic), RTL (resistor-transistor logic), DTL (diode-transistor logic), TTL (transistor-transistor logic) and CMOS (complementary metal oxide semiconductor). There are also sub-variants, e.g. standard CMOS logic vs. advanced types using still CMOS technology, but with some optimizations for avoiding loss of speed due to slower PMOS transistors.

Non-electronic implementations are varied, though few of them are used in practical applications. Many early electromechanical digital computers, such as the Harvard Mark I, were built from relay logic gates, using electro-mechanical relays. Logic gates can be made using pneumatic devices, such as the Sorteberg relay or mechanical logic gates, including on a molecular scale. Logic gates have been made out of DNA. Logic gates can be made from quantum mechanical effects (though quantum computing usually diverges from boolean design). Photonic logic gates use non-linear optical effects.

In principle any method that leads to a gate that is functionally complete (for example, either a NOR or a NAND gate) can be used to make any kind of digital logic circuit. Note that the use of 3-state logic for bus systems is not needed, and can be replaced by digital multiplexers, which can be built using only simple logic gates (such as NAND gates, NOR gates, or AND and OR gates).

ADDER (ELECTRONICS)

An **adder**, also called **summer**, is a digital circuit that performs addition of numbers. In many computers and other kinds of processors, adders are used not only in the arithmetic logic units, but also in other parts of the processor, where they are used to calculate addresses, table indices, increment and decrement operators, and similar operations.

Although adders can be constructed for many numerical representations, such as binary-coded decimal or excess-3, the most common adders operate on binary numbers. In cases where two's complement or ones' complement is being used to represent negative numbers, it is trivial to modify an adder into an adder–subtractor. Other signed number representations require more logic around the basic adder.

Half adder

Half adder logic diagram

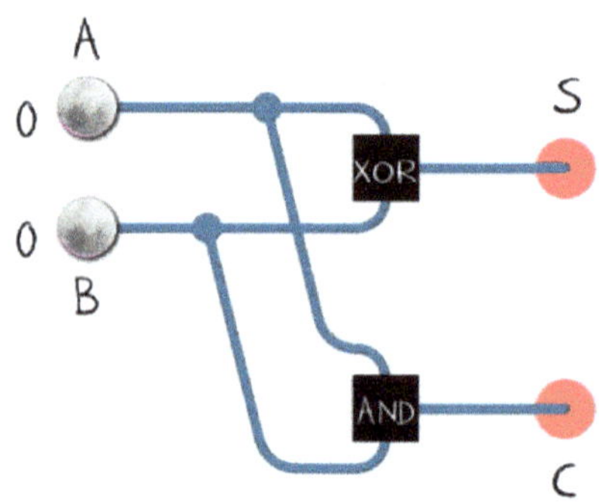

Half adder in action

The **half adder** adds two single binary digits *A* and *B*. It has two outputs, sum (*S*) and carry (*C*). The carry signal represents an overflow into the next digit of a multi-digit addition. The value of the sum is $2C + S$. The simplest half-adder design, pictured on the right, incorporates an XOR gate for *S* and an AND gate for *C*. With the addition of an OR gate to combine their carry outputs, two half adders can be combined to make a full adder. The half adder adds two input bits and generates a carry and sum, which are the two outputs of a half adder. The input variables of a half adder are called the augend and addend bits. The output variables are the sum and carry. The truth table for the half adder is:

Inputs		Outputs	
A	***B***	***C***	***S***
0	0	0	0

1	0	0	1
0	1	0	1
1	1	1	0

Full adder

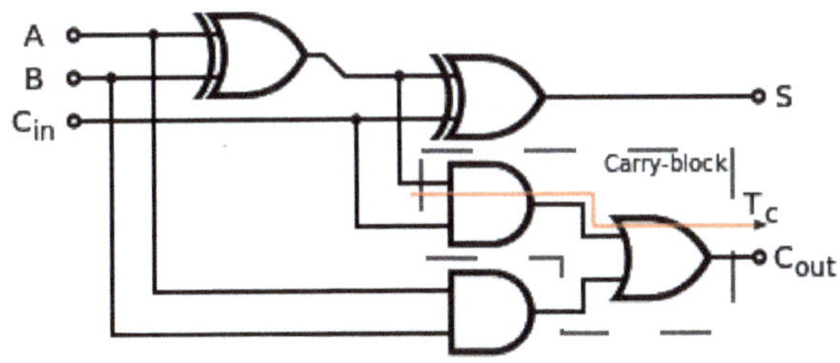

Logic diagram for a full adder.

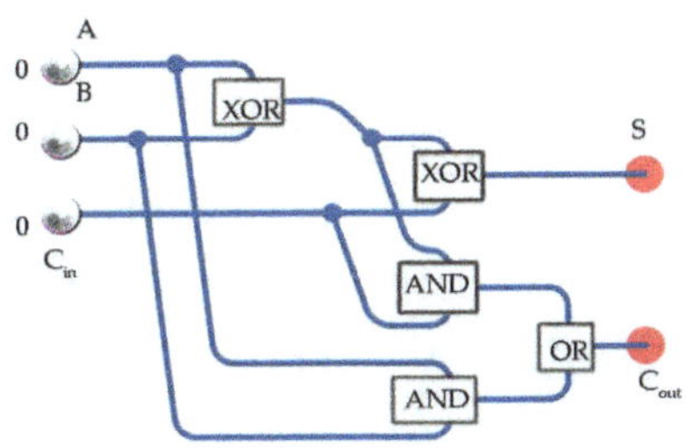

Full adder in action. A full adder gives the number of 1s in the input in binary representation.

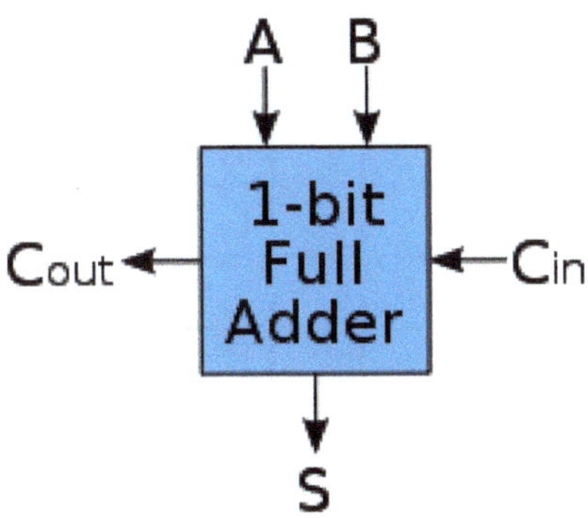

Schematic symbol for a 1-bit full adder with C_{in} and C_{out} drawn on sides of block to emphasize their use in a multi-bit adder.

A **full adder** adds binary numbers and accounts for values carried in as well as out. A one-bit full adder adds three one-bit numbers, often written as A, B, and C_{in}; A and B are the operands, and C_{in} is a bit carried in from the previous less-significant stage. The full adder is usually a component in a cascade of adders, which add 8, 16, 32, etc. bit binary numbers. The circuit produces a two-bit output, output carry and sum typically represented by the signals C_{out} and S, where $\mathrm{sum} = 2 \times C_{out} + S$.

A full adder can be implemented in many different ways such as with a custom transistor-level circuit or composed of other gates. One example implementation is with $S = A \oplus B \oplus C_{in}$ and $C_{out} = (A \cdot B) + (C_{in} \cdot (A \oplus B))$.

In this implementation, the final OR gate before the carry-out output may be replaced by an XOR gate without altering the resulting logic. Using only two types of gates is convenient if the circuit is being implemented using simple IC chips which contain only one gate type per chip.

A full adder can be constructed from two half adders by connecting A and B to the input of one half adder, connecting the sum from that to an input to the second adder, connecting C_i to the other input and OR the two carry outputs. The critical path of a full adder runs through both XOR-gates and ends at the sum bit S. Assumed that an XOR-gate takes 1 delays to complete, the delay imposed by the critical path of a full adder is equal to

$$T_{\mathrm{FA}} = 2 \cdot T_{\mathrm{XOR}} = 2D.$$

The critical path of a carry runs through 1 XOR-gate in adder and through 2 gates (AND and OR) in carry-block and therefore, if AND- or OR-gates takes 1 delay to complete, has a delay of

$$T_c = T_{\mathrm{XOR}} + T_{\mathrm{AND}} + T_{\mathrm{OR}} = D + D + D = 3D.$$

The truth table for the full adder is:

Inputs			**Outputs**	
A	B	C_{in}	C_{out}	S
0	0	0	0	0
0	0	1	0	1
0	1	0	0	1
0	1	1	1	0
1	0	0	0	1
1	0	1	1	0
1	1	0	1	0
1	1	1	1	1

More Complex Adders

Ripple-carry Adder

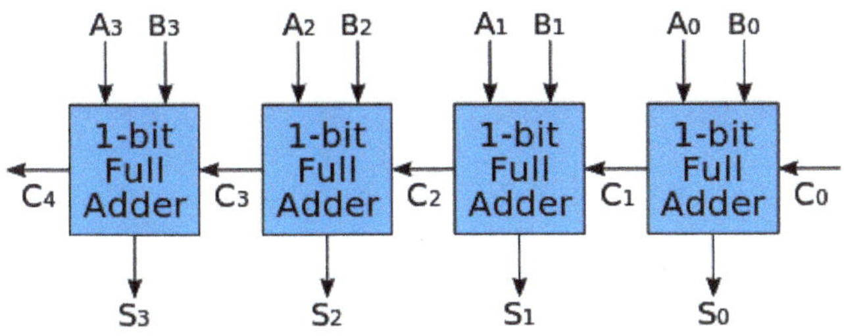

4-bit adder with logic gates shown.

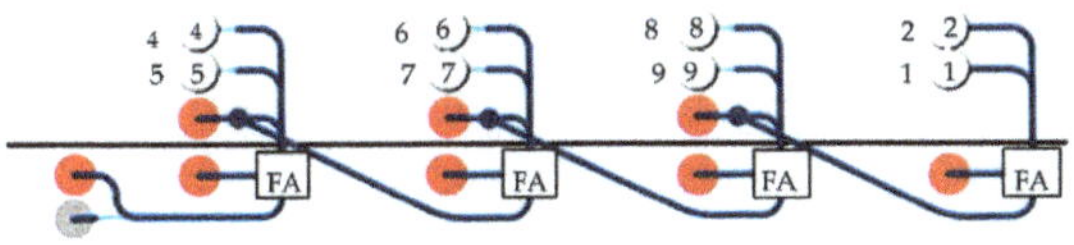

Decimal 4-digit ripple carry adder. FA = full adder, HA = half adder.

It is possible to create a logical circuit using multiple full adders to add N-bit numbers. Each full adder inputs a C_{in}, which is the C_{out} of the previous adder. This kind of adder is called a *ripple-carry adder*, since each carry bit "ripples" to the next full adder. Note that the first (and only the first) full adder may be replaced by a half adder (under the assumption that $C_{in} = 0$).

The layout of a ripple-carry adder is simple, which allows fast design time; however, the ripple-carry adder is relatively slow, since each full adder must wait for the carry bit to be calculated from the previous full adder. The gate delay can easily be calculated by inspection of the full adder circuit. Each full adder requires three levels of logic. In a 32-bit ripple-carry adder, there are 32 full adders, so the critical path (worst case) delay is 3 × 31 (from input to later adder) + 2 (in later adder) = 95 gate delays. The general equation for the worst-case delay for a n-bit carry-ripple adder is

$$T_{\mathrm{CRA}}(n) = (n-1)\cdot T_c + T_s = (n-1)\cdot 3D + 2D = (3n-1)D.$$

The delay from bit position 0 to the carry-out is a little different:

$$T_{\mathrm{CRA}_{[0:c_{\mathrm{out}}]}} = T_{\mathrm{CRA}_{[c_0:c_n]}}(n) = n\cdot T_c = n\cdot 3D = 3nD.$$

The carry-in must travel through n XOR-gates in adders and n carry-generator blocks to have an effect on the carry-out.

A design with alternating carry polarities and optimized AND-OR-Invert gates can be about twice as fast.

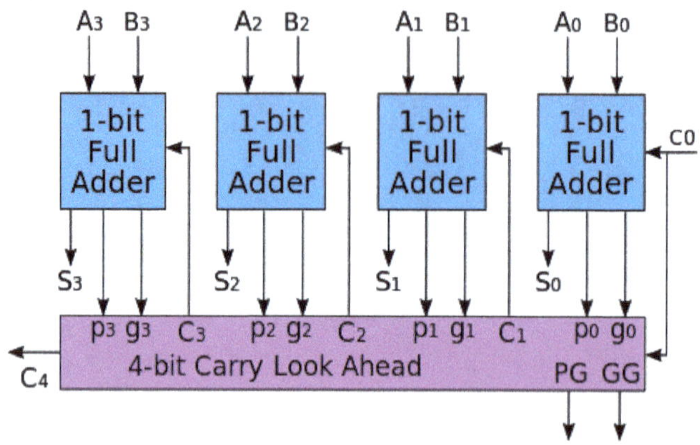

4-bit adder with carry lookahead.

Carry-lookahead Adder

To reduce the computation time, engineers devised faster ways to add two binary numbers by using carry-lookahead adders. They work by creating two signals (*P* and *G*) for each bit position, based on whether a carry is propagated through from a less significant bit position (at least one input is a 1), generated in that bit position (both inputs are 1), or killed in that bit position (both inputs are 0). In most cases, *P* is simply the sum output of a half adder and *G* is the carry output of the same adder. After *P* and *G* are generated the carries for every bit position are created. Some advanced carry-lookahead architectures are the Manchester carry chain, Brent–Kung adder, and the Kogge–Stone adder.

Some other multi-bit adder architectures break the adder into blocks. It is possible to vary the length of these blocks based on the propagation delay of the circuits to optimize computation time. These block based adders include the carry-skip (or carry-bypass) adder which will determine *P* and *G* values for each block rather than each bit, and the carry select adder which pre-generates the sum and carry values for either possible carry input (0 or 1) to the block, using multiplexers to select the appropriate result *when* the carry bit is known.

By combining multiple carry-lookahead adders, even larger adders can be created. This can be used at multiple levels to make even larger adders. For example, the following adder is a 64-bit adder that uses four 16-bit CLAs with two levels of LCUs.

Other adder designs include the carry-select adder, conditional sum adder, carry-skip adder, and carry-complete adder.

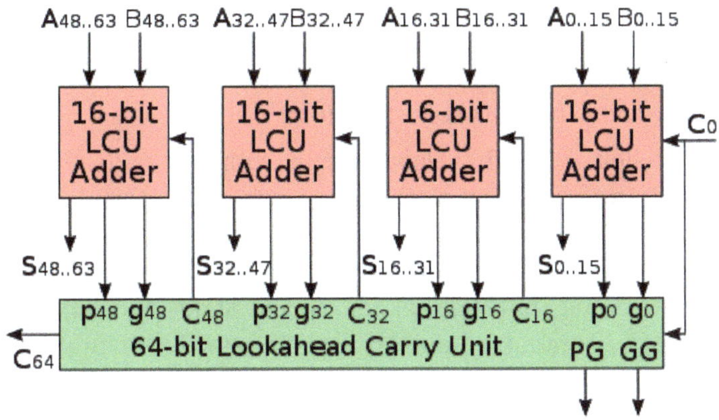

A 64-bit adder.

Carry-save Adders

If an adding circuit is to compute the sum of three or more numbers, it can be advantageous to not propagate the carry result. Instead, three-input adders are used, generating two results: a sum and a carry. The sum and the carry may be fed into two inputs of the subsequent 3-number adder without having to wait for propagation of a carry signal. After all stages of addition, however, a conventional adder (such as the ripple-carry or the lookahead) must be used to combine the final sum and carry results.

3:2 compressors

We can view a full adder as a *3:2 lossy compressor*: it sums three one-bit inputs and returns the result as a single two-bit number; that is, it maps 8 input values to 4 output values. Thus, for example, a binary input of 101 results in an output of 1 + 0 + 1 = 10 (decimal number 2). The carry-out represents bit one of the result, while the sum represents bit zero. Likewise, a half adder can be used as a *2:2 lossy compressor*, compressing four possible inputs into three possible outputs.

Such compressors can be used to speed up the summation of three or more addends. If the addends are exactly three, the layout is known as the carry-save adder. If the addends are four or more, more than one layer of compressors is necessary, and there are various possible design for the circuit: the most common are Dadda and Wallace trees. This kind of circuit is most notably used in multipliers, which is why these circuits are also known as Dadda and Wallace multipliers.

Flip-Flop (Electronics)

In electronics, a **flip-flop** or **latch** is a circuit that has two stable states and can be used to store state information. A flip-flop is a bistable multivibrator. The circuit can be made to change state by signals applied to one or more control inputs and will have one or two outputs. It is the basic storage element in sequential logic. Flip-flops and latches are fundamental building blocks of digital electronics systems used in computers, communications, and many other types of systems.

Flip-flops and latches are used as data storage elements. A flip-flop stores a single *bit* (binary digit) of data; one of its two states represents a "one" and the other represents a "zero". Such data storage can be used for storage of *state*, and such a circuit is described as sequential logic. When used in a finite-state machine, the output and next state depend not only on its current input, but also on its current state (and hence, previous inputs). It can also be used for counting of pulses, and for synchronizing variably-timed input signals to some reference timing signal.

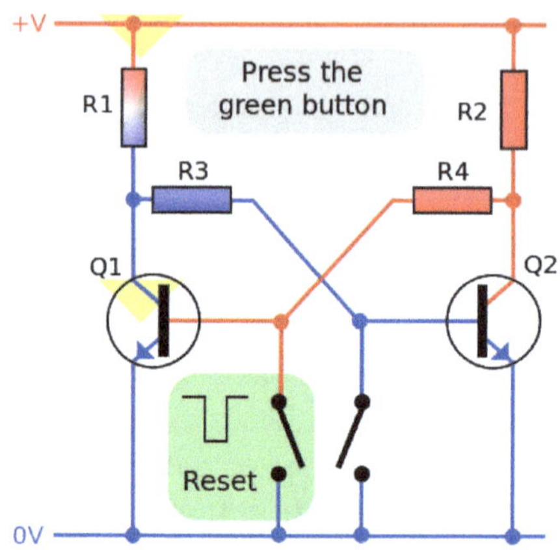

An animated interactive SR latch (*R1, R2* = 1 kΩ *R3, R4* = 10 kΩ).

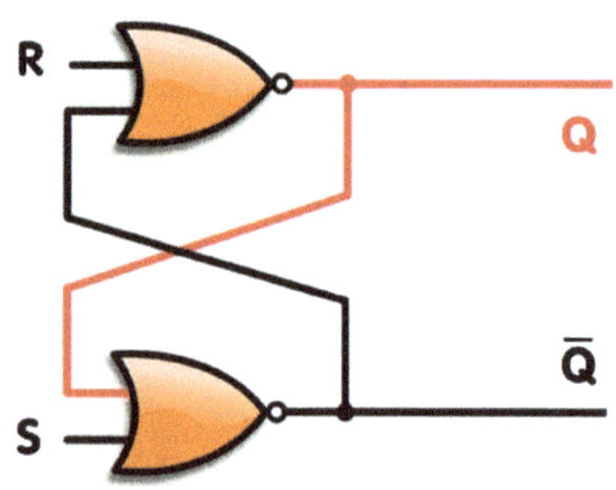

An SR latch, constructed from a pair of cross-coupled NOR gates.

Flip-flops can be either simple (transparent or opaque) or clocked (synchronous or edge-triggered). Although the term flip-flop has historically referred generically to both simple and clocked circuits, in modern usage it is common to reserve the term *flip-flop* exclusively for discussing clocked circuits; the simple ones are commonly called *latches*.

Using this terminology, a latch is level-sensitive, whereas a flip-flop is edge-sensitive. That is, when a latch is enabled it becomes transparent, while a flip flop's output only changes on a single type (positive going or negative going) of clock edge.

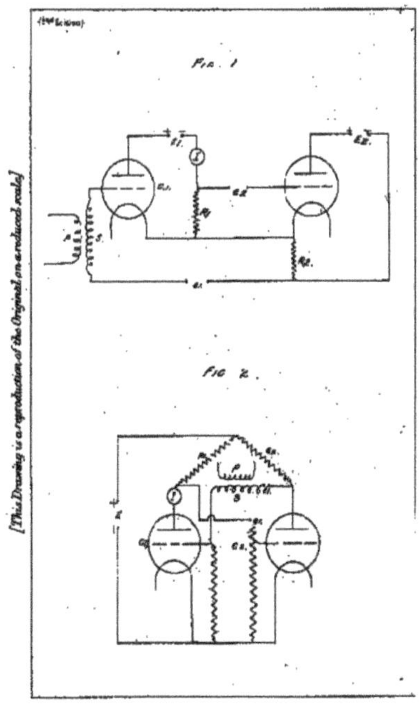

Flip-flop schematics from the Eccles and Jordan patent filed 1918, one drawn as a cascade of amplifiers with a positive feedback path, and the other as a symmetric cross-coupled pair.

The first electronic flip-flop was invented in 1918 by the British physicists William Eccles and F. W. Jordan. It was initially called the *Eccles–Jordan trigger circuit* and consisted of two active elements (vacuum tubes). The design was used in the 1943 British Colossus codebreaking computer and such circuits and their transistorized versions were common in computers even after the introduction of integrated circuits, though flip-flops made from logic gates are also common now. Early flip-flops were known variously as trigger circuits or multivibrators.

According to P. L. Lindley, an engineer at the US Jet Propulsion Laboratory, the flip-flop types detailed below (RS, D, T, JK) were first discussed in a 1954 UCLA course on computer design by Montgomery Phister, and then appeared in his book *Logical Design of Digital Computers.* Lindley was at the time working at Hughes Aircraft under Eldred Nelson, who had coined the term JK for a flip-flop which changed states when both inputs were on (a logical "one"). The other names were coined by Phister. They differ slightly from some of the definitions given below. Lindley explains that he heard the story of the JK flip-flop from Eldred Nelson, who is responsible for coining the term while working at Hughes Aircraft. Flip-flops in use at Hughes at the time were all of the type that came to be known as J-K. In designing a logical system, Nelson assigned letters to flip-flop inputs as follows: #1: A & B, #2: C & D, #3: E & F, #4: G & H, #5: J & K. Nelson used the notations "*j*-input" and "*k*-input" in a patent application filed in 1953.

Implementation

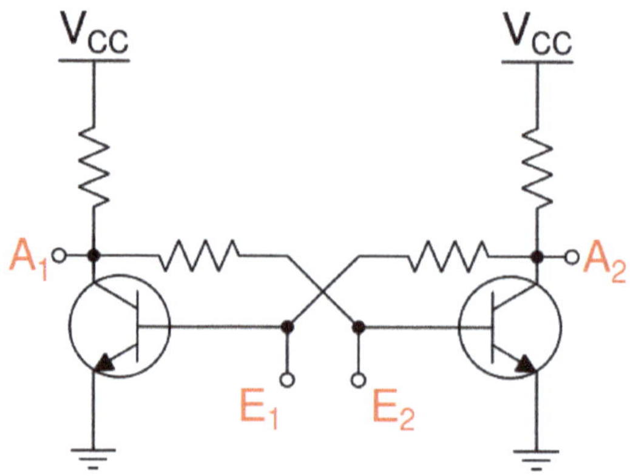

A traditional flip-flop circuit based on bipolar junction transistors.

Flip-flops can be either simple (transparent or asynchronous) or clocked (synchronous); the transparent ones are commonly called latches. The word *latch* is mainly used for storage elements, while clocked devices are described as *flip-flops*.

Simple flip-flops can be built around a pair of cross-coupled inverting elements: vacuum tubes, bipolar transistors, field effect transistors, inverters, and inverting logic gates have all been used in practical circuits. Clocked devices are specially designed for synchronous systems; such devices ignore their inputs except at the transition of a dedicated clock signal (known as clocking, pulsing, or strobing). Clocking causes the flip-flop either to change or to retain its output signal based upon the values of the input signals at the transition. Some flip-flops change output on the rising edge of the clock, others on the falling edge.

Since the elementary amplifying stages are inverting, two stages can be connected in succession (as a cascade) to form the needed non-inverting amplifier. In this configuration, each amplifier may be considered as an active inverting feedback network for the other inverting amplifier. Thus the two stages are connected in a non-inverting loop although the circuit diagram is usually drawn as a symmetric cross-coupled pair (both the drawings are initially introduced in the Eccles–Jordan patent).

Flip-flop Types

Flip-flops can be divided into common types: the **SR** ("set-reset"), **D** ("data" or "delay"), **T** ("toggle"), and **JK** types are the common ones. The behavior of a particular type can be described by what is termed the characteristic equation, which derives the "next" (i.e., after the next clock pulse) output, Q_{next}in terms of the input signal(s) and/or the current output, .Q.

Simple set-Reset Latches

SR NOR Latch

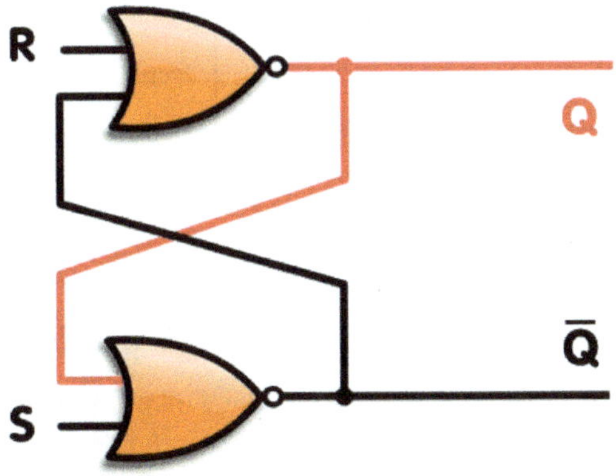

An animation of a SR latch, constructed from a pair of cross-coupled NOR gates. Red and black mean logical '1' and '0', respectively.

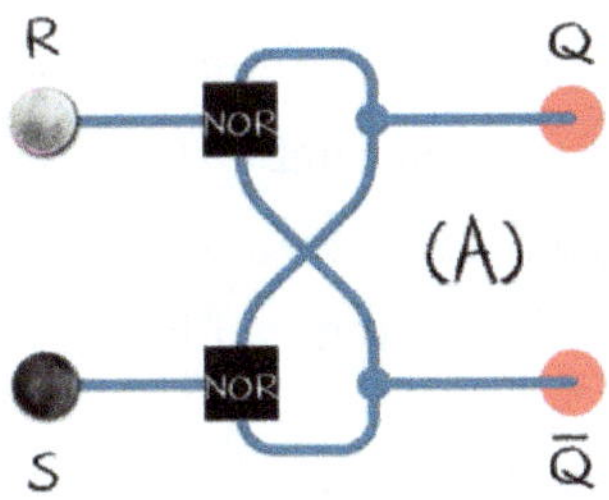

An animated SR latch. Black and white mean logical '1' and '0', respectively.

- S = 1, R = 0: set
- S = 0, R = 0: hold
- S = 0, R = 1: reset
- S = 1, R = 1: not allowed
- when S=1, Latched is in unstable state

When using static gates as building blocks, the most fundamental latch is the simple *SR latch*, where S and R stand for *set* and *reset*. It can be constructed from a pair of cross-coupled NOR logic gates. The stored bit is present on the output marked Q.

While the R and S inputs are both low, feedback maintains the Q and Q outputs in a constant state, with Q the complement of Q. If S (*Set*) is pulsed high while R (*Reset*) is held low, then the Q output is forced high, and stays high when S returns to low; similarly, if R is pulsed high while S is held low, then the Q output is forced low, and stays low when R returns to low.

SR latch operation								
Characteristic table				**Excitation table**				
S	**R**	**Q_{next}**	**Action**	**Q**	**Q_{next}**	**S**	**R**	
0	0	Q	hold state	0	0	0	X	
0	1	0	reset	0	1	1	0	
1	0	1	set	1	0	0	1	
1	1	X	not allowed	1	1	X	0	

Note: X means don't care, *that is, either 0 or 1 is a valid value.*

The R = S = 1 combination is called a **restricted combination** or a **forbidden state** because, as both NOR gates then output zeros, it breaks the logical equation Q = **not** Q. The combination is also inappropriate in circuits where *both* inputs may go low *simultaneously* (i.e. a transition from *restricted* to *keep*). The output would lock at either 1 or 0 depending on the propagation time relations between the gates (a race condition).

To overcome the restricted combination, one can add gates to the inputs that would convert (S,R) = (1,1) to one of the non-restricted combinations. That can be:

- Q = 1 (1,0) – referred to as an *S (dominated)-latch*
- Q = 0 (0,1) – referred to as an *R (dominated)-latch*

This is done in nearly every programmable logic controller.

- Keep state (0,0) – referred to as an *E-latch*

Alternatively, the restricted combination can be made to *toggle* the output. The result is the JK latch.

Characteristic: Q+ = R'Q + R'S or Q+ = R'Q + S.

SR NAND Latch

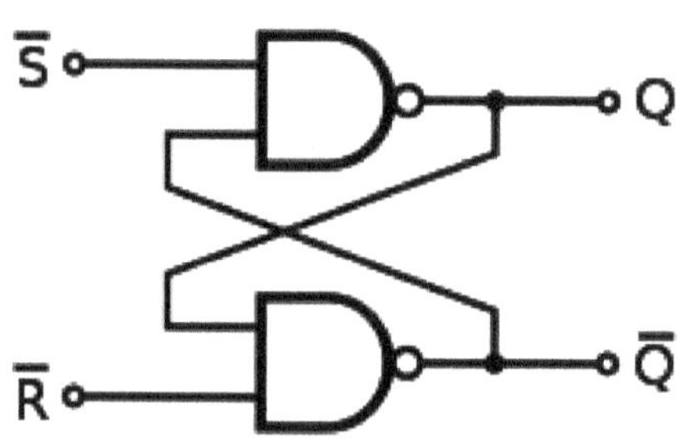

- This is an alternate model of the simple SR latch which is built with NAND logic gates. *Set* and *reset* now become active low signals, denoted S and R respectively. Otherwise, operation is identical to that of the SR latch. Historically,

SR-latches have been predominant despite the notational inconvenience of active-low inputs.

An SR latch.

SR latch operation						
S	R	Action				
0	0	Not allowed				
0	1	Q = 1				
1	0	Q = 0				
1	1	No change				

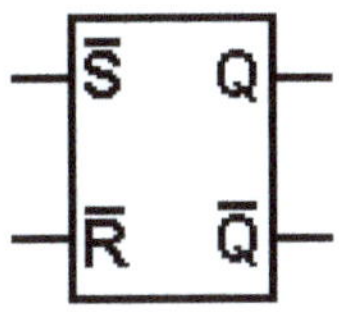

Symbol for an SR NAND latch

SR AND-OR latch

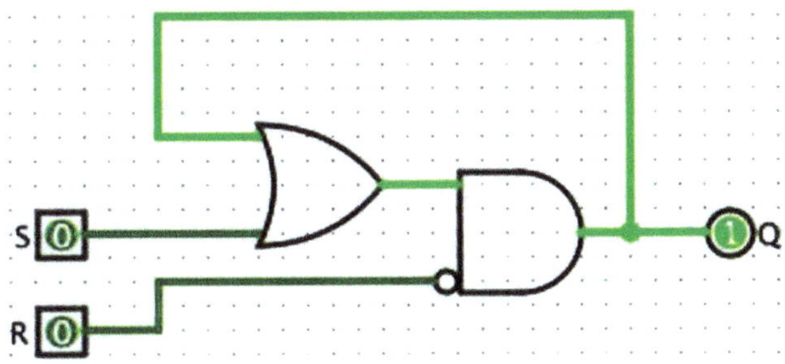

An SR AND-OR latch. Light green means logical '1' and dark green means logical '0'. The latch is currently in hold mode (no change).

From the teaching point of view, SR latches realised as a pair of cross-coupled components (transistors, gates, tubes, etc.) are rather hard to understand for beginners. A didactically easier to understand model uses a feedback loop instead of the cross-coupling. The following is an SR latch built with an AND gate with one inverted input and an OR gate.

SR AND-OR latch operation			
S	R	Action	
0	0	No change	
1	0	Q = 1	
X	1	Q = 0	

JK latch

The JK latch is much less frequently used than the JK flip-flop. The JK latch follows the following state table:

JK latch truth table			
J	K	Q_{next}	Comment
0	0	Q	No change
0	1	0	Reset
1	0	1	Set
1	1	Q	Toggle

Hence, the JK latch is an SR latch that is made to *toggle* its output (oscillate between 0 and 1) when passed the input combination of 11. Unlike the JK flip-flop, the 11 input combination for the JK latch is not very useful because there is no clock that directs toggling.

Gated latches and Conditional Transparency

Latches are designed to be *transparent.* That is, input signal changes cause immediate changes in output; when several *transparent* latches follow each other, using the same enable signal, signals can propagate through all of them at once. Alternatively, additional logic can be added to a simple transparent latch to make it *non-transparent* or *opaque* when another input (an "enable" input) is not asserted. By following a *transparent-high* latch with a *transparent-low* (or *opaque-high*) latch, a master–slave flip-flop is implemented.

Gated SR Latch

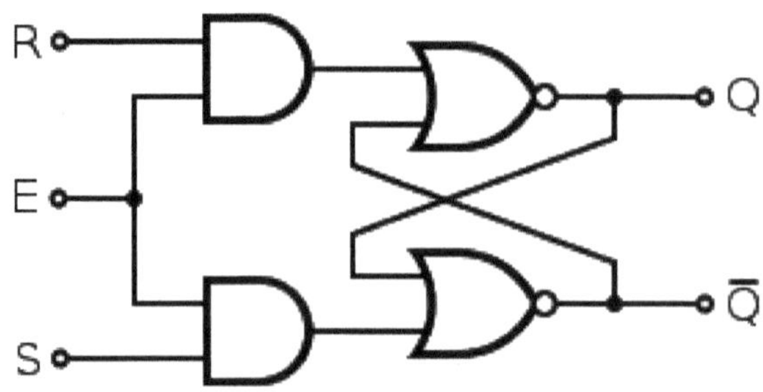

A gated SR latch circuit diagram constructed from NOR gates.

A *synchronous SR latch* (sometimes *clocked SR flip-flop*) can be made by adding a second level of NAND gates to the inverted SR latch (or a second level of AND gates to the direct SR latch). The extra NAND gates further invert the inputs so the simple SR latch becomes a gated SR latch (and a simple SR latch would transform into a gated SR latch with inverted enable).

With E high (*enable* true), the signals can pass through the input gates to the encapsulated latch; all signal combinations except for (0,0) = *hold* then immediately reproduce on the (Q,Q) output, i.e. the latch is *transparent*.

With E low (*enable* false) the latch is *closed (opaque)* and remains in the state it was left the last time E was high.

The *enable* input is sometimes a clock signal, but more often a read or write strobe.

Gated SR latch operation	
E/C	**Action**
0	No action (keep state)
1	The same as non-clocked SR latch

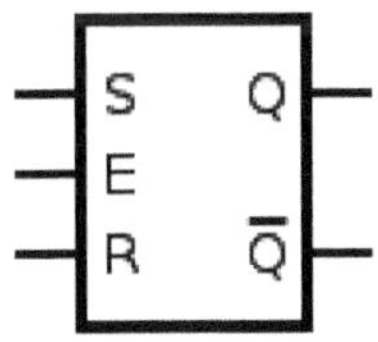

Symbol for a gated SR latch

Gated D latch

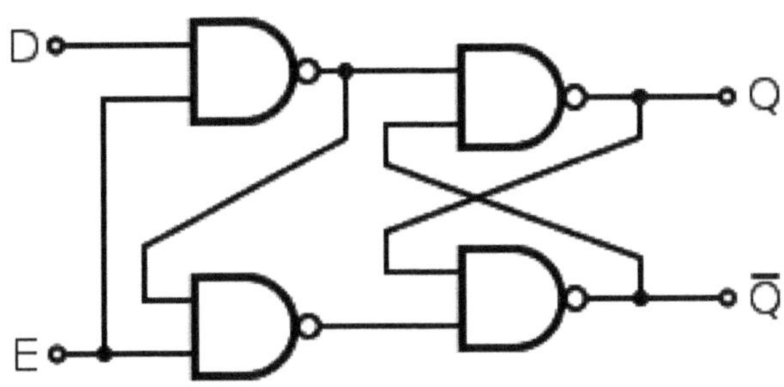

A gated D latch based on an SR NAND latch.

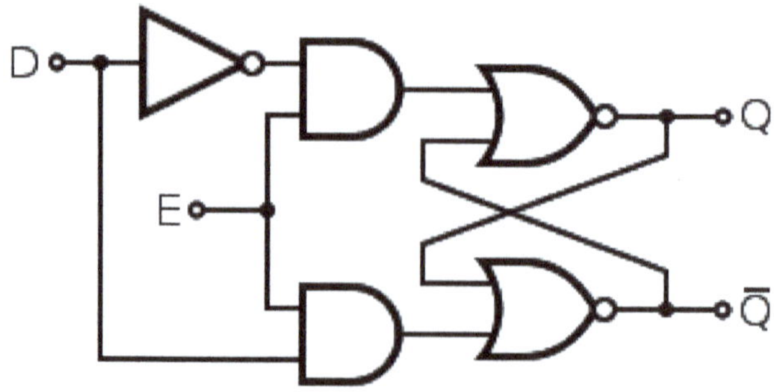

A gated D latch based on an SR NOR latch.

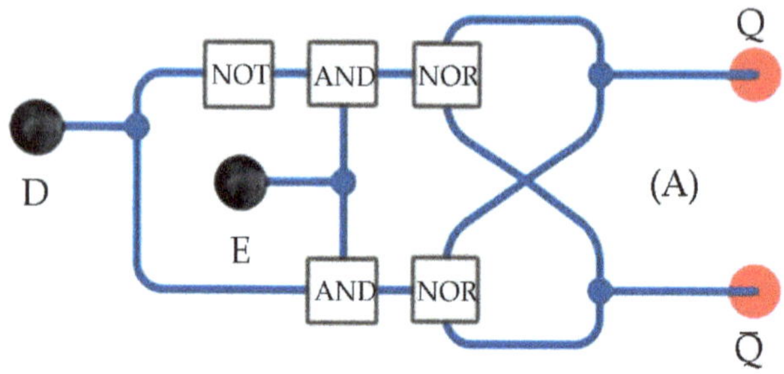

An animated gated D latch.

- D = 1, E = 1: set
- (B) D = 1, E = 0: hold
- (C) D = 0, E = 0: hold
- (D) D = 0, E = 1: reset

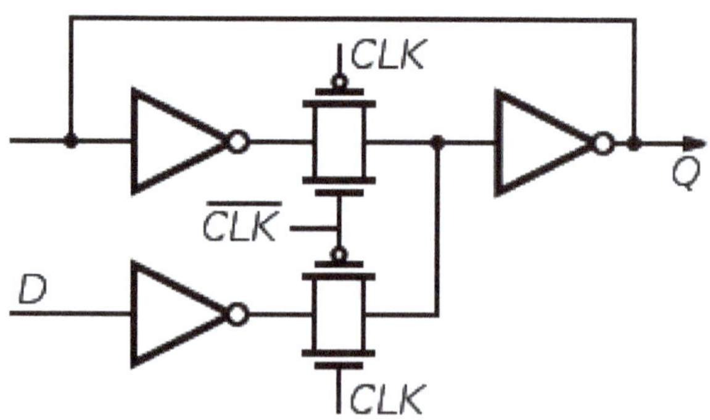

A gated D latch in pass transistor logic, similar to the ones in the CD4042 or the CD74HC75 integrated circuits.

This latch exploits the fact that, in the two active input combinations (01 and 10) of a gated SR latch, R is the complement of S. The input NAND stage converts the two D input states (0 and 1) to these two input combinations for the next SR latch by inverting the data input signal. The low state of the *enable* signal produces the inactive "11" combination. Thus a gated D-latch may be considered as a *one-input synchronous SR latch*. This configuration prevents application of the restricted input combination. It is also known as *transparent latch*, *data latch*, or simply *gated latch*. It has a *data* input and an *enable* signal (sometimes named *clock*, or *control*). The word *transparent* comes from the fact that, when the enable input is on, the signal propagates directly through the circuit, from the input D to the output Q.

Transparent latches are typically used as I/O ports or in asynchronous systems, or in synchronous two-phase systems (synchronous systems that use a two-phase clock), where two latches operating on different clock phases prevent data transparency as in a master–slave flip-flop.

Latches are available as integrated circuits, usually with multiple latches per chip. For example, 74HC75 is a quadruple transparent latch in the 7400 series.

Gated D latch truth table					
E/C	**D**		**Q**	**Q**	**Comment**
0	X		Q_{prev}	Q_{prev}	No change
1	0		0	1	Reset
1	1		1	0	Set

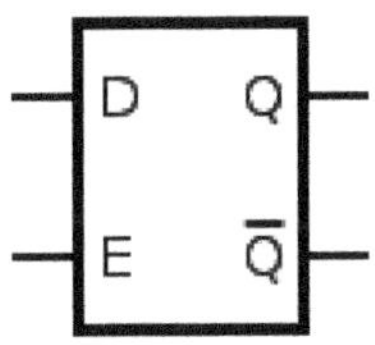

Symbol for a gated D latch

The truth table shows that when the *e*nable/*c*lock input is 0, the D input has no effect on the output. When E/C is high, the output equals D.

Earle latch

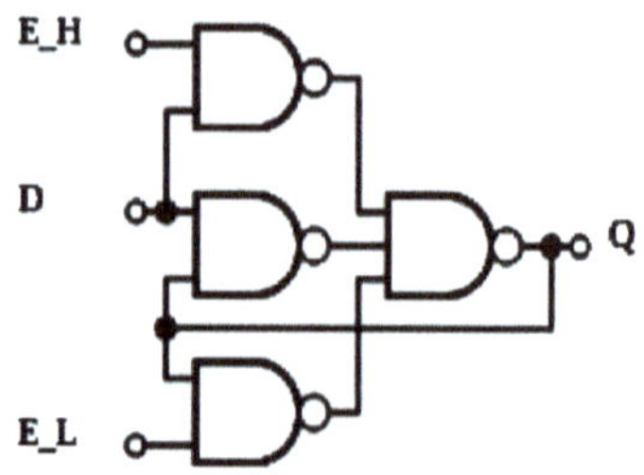

Earle latch uses complementary enable inputs: enable active low (E_L) and enable active high (E_H).

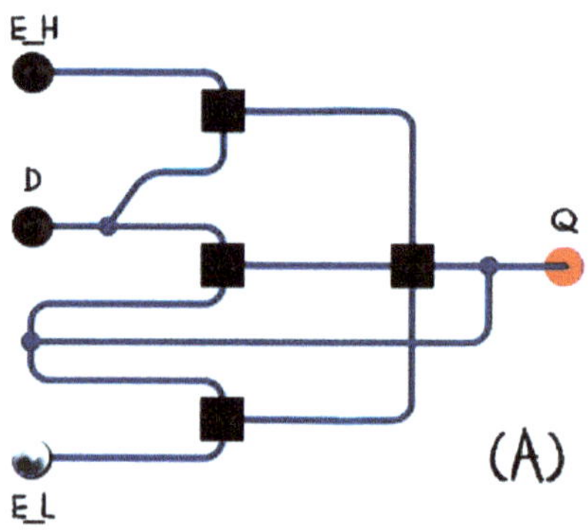

An animated Earle latch.

- D = 1, E_H = 1: set
- D = 0, E_H = 1: reset
- D = 1, E_H = 0: hold
- D = 1, E_H = 1: set

The classic gated latch designs have some undesirable characteristics. They require double-rail logic or an inverter. The input-to-output propagation may take up to three gate delays. The input-to-output propagation is not constant – some outputs take two gate delays while others take three.

Designers looked for alternatives. A successful alternative is the Earle latch. It requires only a single data input, and its output takes a constant two gate delays. In addition, the two gate levels of the Earle latch can, in some cases, be merged with the last two gate levels of the circuits driving the latch because many common computational circuits have an OR layer followed by an AND layer as their last two levels. Merging the latch function can implement the latch with no additional gate delays. The merge is commonly exploited in the design of pipelined computers, and, in fact, was originally developed by J. G. Earle to be used in the IBM System/360 Model 91 for that purpose.

The Earle latch is hazard free. If the middle NAND gate is omitted, then one gets the **polarity hold latch**, which is commonly used because it demands less logic. However, it is susceptible to logic hazard. Intentionally skewing the clock signal can avoid the hazard.

D flip-Flop

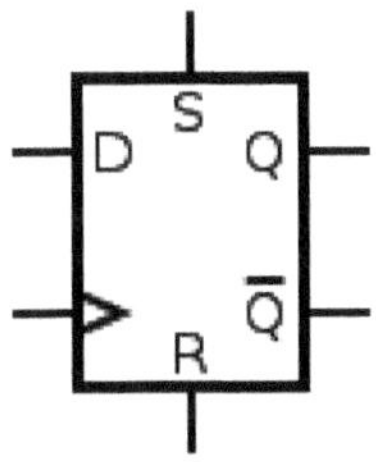

D flip-flop symbol

The D flip-flop is widely used. It is also known as a "data" or "delay" flip-flop.

The D flip-flop captures the value of the D-input at a definite portion of the clock cycle (such as the rising edge of the clock). That captured value becomes the Q output. At other times, the output Q does not change. The D flip-flop can be viewed as a memory cell, a zero-order hold, or a delay line.

Truth table:

Clock	**D**	**Q_{next}**
Rising edge	0	0
Rising edge	1	1
Non-Rising	X	Q

('X' denotes a *Don't care* condition, meaning the signal is irrelevant)

Most D-type flip-flops in ICs have the capability to be forced to the set or reset state (which ignores the D and clock inputs), much like an SR flip-flop. Usually, the illegal S = R = 1 condition is resolved in D-type flip-flops. By setting S = R = 0, the flip-flop can be used as described above. Here is the truth table for the others S and R possible configurations:

Inputs				Outputs	
S	R	D	>	Q	Q'
0	1	X	X	0	1
1	0	X	X	1	0
1	1	X	X	1	1

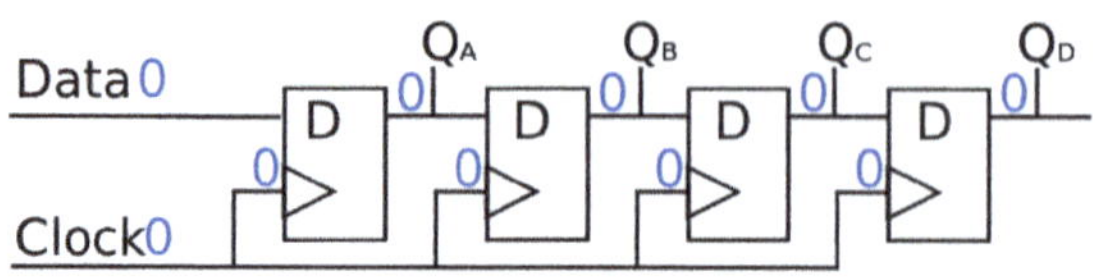

4-bit serial-in, parallel-out (SIPO) shift register.

These flip-flops are very useful, as they form the basis for shift registers, which are an essential part of many electronic devices. The advantage of the D flip-flop over the D-type "transparent latch" is that the signal on the D input pin is captured the moment the flip-flop is clocked, and subsequent changes on the D input will be ignored until the next clock event. An exception is that some flip-flops have a "reset" signal input, which will reset Q (to zero), and may be either asynchronous or synchronous with the clock.

The above circuit shifts the contents of the register to the right, one bit position on each active transition of the clock. The input X is shifted into the leftmost bit position.

Classical Positive-Edge-triggered D Flip-Flop

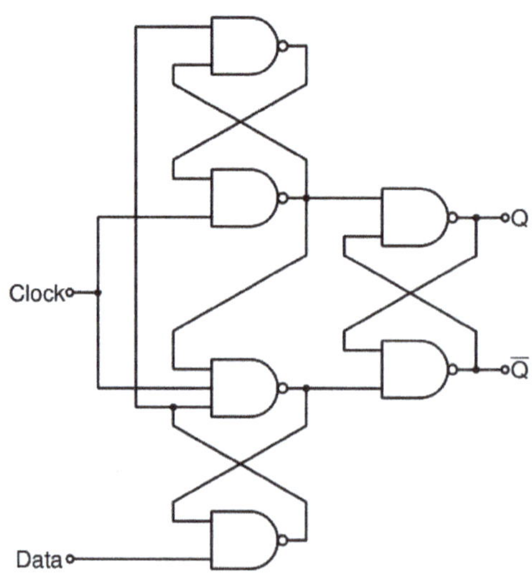

A positive-edge-triggered D flip-flop.

This circuit consists of two stages implemented by SR NAND latches. The input stage (the two latches on the left) processes the clock and data signals to ensure correct input signals for the output stage (the single latch on the right). If the clock is low, both the output signals of the input stage are high regardless of the data input; the output latch is unaffected and it stores the previous state. When the clock signal changes from low to high, only one of the output voltages (depending on the data signal) goes low and sets/resets the output latch: if D = 0, the lower output becomes low; if D = 1, the upper output becomes low. If the clock signal continues staying high, the outputs keep their states regardless of the data input and force the output latch to stay in the corresponding state as the input logical zero (of the output stage) remains active while the clock is high. Hence the role of the output latch is to store the data only while the clock is low.

The circuit is closely related to the gated D latch as both the circuits convert the two D input states (0 and 1) to two input combinations (01 and 10) for the output SR latch by inverting the data input signal (both the circuits split the single D signal in two complementary S and R signals). The difference is that in the gated D latch simple NAND logical gates are used while in the positive-edge-triggered D flip-flop SR NAND latches are used for this purpose. The role of these latches is to "lock" the active output producing low voltage (a logical zero); thus the positive-edge-triggered D flip-flop can also be thought of as a gated D latch with latched input gates.

Master–slave Edge-Triggered D Flip-Flop

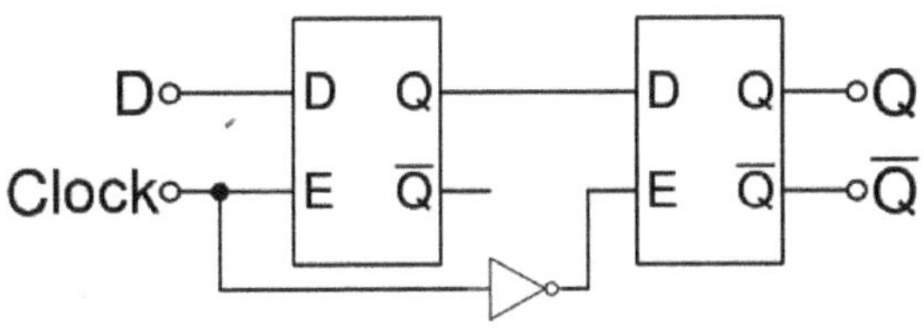

A master–slave D flip-flop. It responds on the falling edge of the *enable* input (usually a clock).

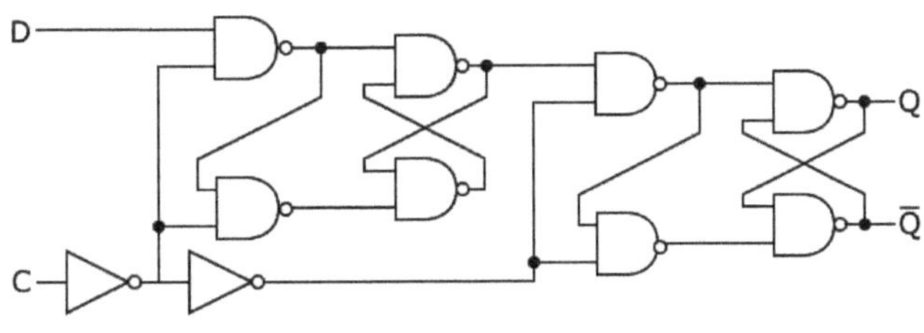

An implementation of a master–slave D flip-flop that is triggered on the rising edge of the clock.

A master–slave D flip-flop is created by connecting two gated D latches in series, and inverting the *enable* input to one of them. It is called master–slave because the second latch in the series only changes in response to a change in the first (master) latch.

For a positive-edge triggered master–slave D flip-flop, when the clock signal is low (logical 0) the "enable" seen by the first or "master" D latch (the inverted clock signal) is high (logical 1). This allows the "master" latch to store the input value when the clock signal transitions from low to high. As the clock signal goes high (0 to 1) the inverted "enable" of the first latch goes low (1 to 0) and the value seen at the input to the master latch is "locked". Nearly simultaneously, the twice inverted "enable" of the second or "slave" D latch transitions from low to high (0 to 1) with the clock signal. This allows the signal captured at the rising edge of the clock by the now "locked" master latch to pass through the "slave" latch. When the clock signal returns to low (1 to 0), the output of the "slave" latch is "locked", and the value seen at the last rising edge of the clock is held while the "master" latch begins to accept new values in preparation for the next rising clock edge.

By removing the leftmost inverter in the circuit at side, a D-type flip-flop that strobes on the *falling edge* of a clock signal can be obtained. This has a truth table like this:

D	Q	>	Q_{next}
0	X	Falling	0
1	X	Falling	1

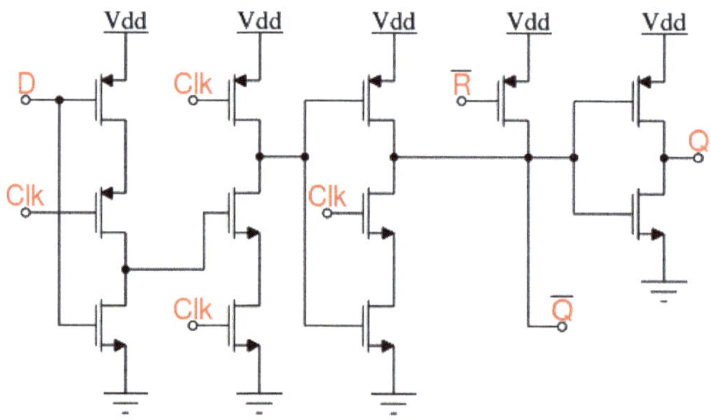

A CMOS IC implementation of a "true single-phase edge-triggered flip-flop with reset".

Edge-Triggered Dynamic D Storage Element

An efficient functional alternative to a D flip-flop can be made with dynamic circuits (where information is stored in a capacitance) as long as it is clocked often enough; while not a true flip-flop, it is still called a flip-flop for its functional role. While the master–slave D element is triggered on the edge of a clock, its components are each triggered by clock levels. The "edge-triggered D flip-flop", as it is called even though it is not a true flip-flop, does not have the master–slave properties.

Edge-triggered D flip-flops are often implemented in integrated high-speed operations using dynamic logic. This means that the digital output is stored on parasitic device capacitance while the device is not transitioning. This design of dynamic flip flops also enables simple resetting since the reset operation can be performed by simply discharging one or more internal nodes. A common dynamic flip-flop variety is the true single-phase clock (TSPC) type which performs the flip-flop operation with little power and at high speeds. However, dynamic flip-flops will typically not work at static or low clock speeds: given enough time, leakage paths may discharge the parasitic capacitance enough to cause the flip-flop to enter invalid states.

T flip-flop

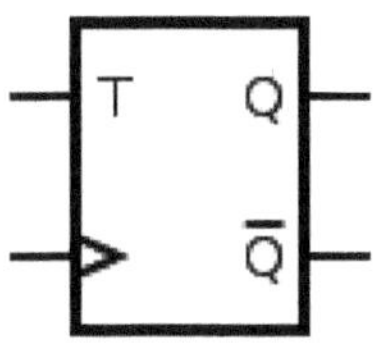

A circuit symbol for a T-type flip-flop.

If the T input is high, the T flip-flop changes state ("toggles") whenever the clock input is strobed. If the T input is low, the flip-flop holds the previous value. This behavior is described by the characteristic equation:

$Q_{\text{next}} = T \oplus Q = T\overline{Q} + \overline{T}Q$ (expanding the XOR operator)

and can be described in a truth table:

T flip-flop operation							
Characteristic table				**Excitation table**			
{\displaystyle T}	**{\displaystyle Q}**	**{\displaystyle Q_{\rm {next}}}**	**Comment**	**{\displaystyle Q}**	**{\displaystyle Q_{\rm {next}}}**	**{\displaystyle T}**	**Comment**
0	0	0	hold state (no clk)	0	0	0	No change
0	1	1	hold state (no clk)	1	1	0	No change
1	0	1	toggle	0	1	1	Complement
1	1	0	toggle	1	0	1	Complement

When T is held high, the toggle flip-flop divides the clock frequency by two; that is, if clock frequency is 4 MHz, the output frequency obtained from the flip-flop will be 2 MHz. This "divide by" feature has application in various types of digital counters. A T flip-flop can also be built using a JK flip-flop (J & K pins are connected together and act as T) or a D flip-flop (T input XOR Q_{previous} drives the D input).

JK flip-flop

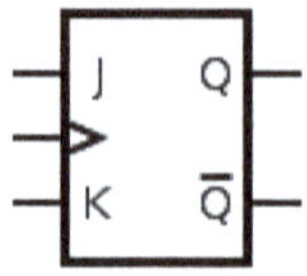

A circuit symbol for a positive-edge-triggered JK flip-flop

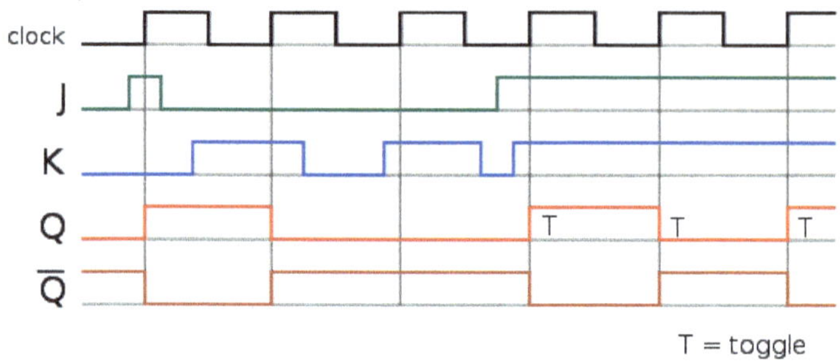

JK flip-flop timing diagram.

The JK flip-flop augments the behavior of the SR flip-flop (J=Set, K=Reset) by interpreting the J = K = 1 condition as a "flip" or toggle command. Specifically, the combination J = 1, K = 0 is a command to set the flip-flop; the combination J = 0, K = 1 is a command to reset the flip-flop; and the combination J = K = 1 is a command to toggle the flip-flop, i.e., change its output to the logical complement of its current value. Setting J = K = 0 maintains the current state. To synthesize a D flip-flop, simply set K equal to the complement of J. Similarly, to synthesize a T flip-flop, set K equal to J. The JK flip-flop is therefore a universal flip-flop, because it can be configured to work as an SR flip-flop, a D flip-flop, or a T flip-flop.

The characteristic equation of the JK flip-flop is:

$$Q_{\text{next}} = J\overline{Q} + \overline{K}Q$$

and the corresponding truth table is:

JK flip-flop operation								
Characteristic table				**Excitation table**				
J	**K**	**Comment**	**Q_{next}**	**Q**	**Q_{next}**	**Comment**	**J**	**K**
0	0	hold state	Q	0	0	No Change	0	X
0	1	reset	0	0	1	Set	1	X
1	0	set	1	1	0	Reset	X	1
1	1	toggle	$\overline{Q}$	1	1	No Change	X	0

Timing Considerations

Timing Parameters

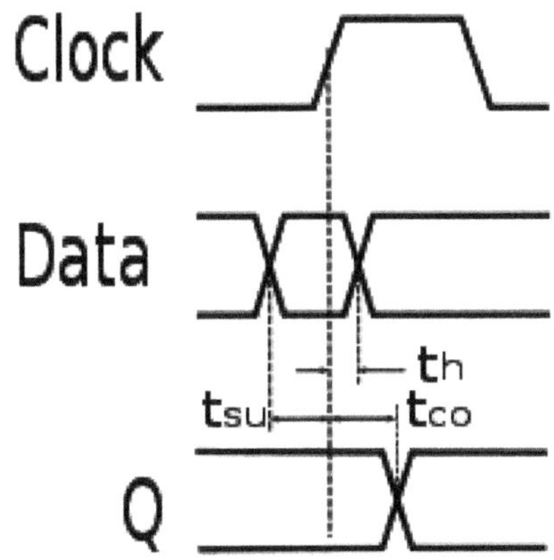

Flip-flop setup, hold and clock-to-output timing parameters.

The input must be held steady in a period around the rising edge of the clock known as the aperture. Imagine taking a picture of a frog on a lily-pad. Suppose the frog then jumps into the water. If you take a picture of the frog as it jumps into the water, you will get a blurry picture of the frog jumping into the water—it's not clear which state the frog was in. But if you take a picture while the frog sits steadily on the pad (or is steadily in the water), you will get a clear picture. In the same way, the input to a flip-flop must be held steady during the **aperture** of the flip-flop.

Setup time is the minimum amount of time the data input should be held steady **before** the clock event, so that the data is reliably sampled by the clock.

Hold time is the minimum amount of time the data input should be held steady **after** the clock event, so that the data is reliably sampled by the clock.

Aperture is the sum of setup and hold time. The data input should be held steady throughout this time period.

Recovery time is the minimum amount of time the asynchronous set or reset input should be inactive **before** the clock event, so that the data is reliably sampled by the clock. The recovery time for the asynchronous set or reset input is thereby similar to the setup time for the data input.

Removal time is the minimum amount of time the asynchronous set or reset input should be inactive **after** the clock event, so that the data is reliably sampled by the clock. The removal time for the asynchronous set or reset input is thereby similar to the hold time for the data input.

Short impulses applied to asynchronous inputs (set, reset) should not be applied completely within the recovery-removal period, or else it becomes entirely indeterminable whether the flip-flop will transition to the appropriate state. In another case, where an asynchronous signal simply makes one transition that happens to fall between the recovery/removal time, eventually the flip-flop will transition to the appropriate state, but a very short glitch may or may not appear on the output, dependent on the synchronous input signal. This second situation may or may not have significance to a circuit design.

Set and Reset (and other) signals may be either synchronous or asynchronous and therefore may be characterized with either Setup/Hold or Recovery/Removal times, and synchronicity is very dependent on the design of the flip-flop.

Differentiation between Setup/Hold and Recovery/Removal times is often necessary when verifying the timing of larger circuits because asynchronous signals may be found to be less critical than synchronous signals. The differentiation offers circuit designers the ability to define the verification conditions for these types of signals independently.

Metastability

Flip-flops are subject to a problem called metastability, which can happen when two inputs, such as data and clock or clock and reset, are changing at about the same time. When the order is not clear, within appropriate timing constraints, the result is that the output may behave unpredictably, taking many times longer than normal to settle to one state or the other, or even oscillating several times before settling. Theoretically, the time to settle down is not bounded. In a computer system, this metastability can cause corruption of data or a program crash if the state is not stable before another circuit uses its value; in particular, if two different logical paths use the output of a flip-flop, one path can interpret it as a 0 and the other as a 1 when it has not resolved to stable state, putting the machine into an inconsistent state.

The metastability in flip-flops can be avoided by ensuring that the data and control inputs are held valid and constant for specified periods before and after the clock pulse, called the **setup time** (t_{su}) and the **hold time** (t_h) respectively. These times are specified in the data sheet for the device, and are typically between a few nanoseconds and a few hundred picoseconds for modern devices. Depending upon the flip-flop's internal organization, it is possible to build a device with a zero (or even negative) setup or hold time requirement but not both simultaneously.

Unfortunately, it is not always possible to meet the setup and hold criteria, because the flip-flop may be connected to a real-time signal that could change at any time, outside the control of the designer. In this case, the best the designer can do is to reduce the probability of error to a certain level, depending on the required reliability of the circuit. One technique for suppressing metastability is to connect two or more flip-flops in a chain, so that the output of each one feeds the data input of the next, and all devices share a common clock. With this method, the probability of a metastable event can be reduced to a negligible value, but never to zero. The probability of metastability gets closer and closer to zero as the number of flip-flops connected in series is increased. The number of flip-flops being cascaded is referred to as the "ranking"; "dual-ranked" flip flops (two flip-flops in series) is a common situation.

So-called metastable-hardened flip-flops are available, which work by reducing the setup and hold times as much as possible, but even these cannot eliminate the problem entirely. This is because metastability is more than simply a matter of circuit design. When the transitions in the clock and the data are close together in time, the flip-flop is forced to decide which event happened first. However fast we make the device, there is always the possibility that the input events will be so close together that it cannot detect which one happened first. It is therefore logically impossible to build a perfectly metastable-proof flip-flop. Flip-flops are sometimes characterized for a maximum settling time (the maximum time they will remain metastable under specified conditions). In this case, dual-ranked flip-flops that are clocked slower than the maximum allowed metastability time will provide proper conditioning for asynchronous (e.g., external) signals.

Propagation Delay

Another important timing value for a flip-flop is the clock-to-output delay (common symbol in data sheets: t_{CO}) or propagation delay (t_P), which is the time a flip-flop takes to change its output after the clock edge. The time for a high-to-low transition (t_{PHL}) is sometimes different from the time for a low-to-high transition (t_{PLH}).

When cascading flip-flops which share the same clock (as in a shift register), it is important to ensure that the t_{CO} of a preceding flip-flop is longer than the hold time (t_h) of the following flip-flop, so data present at the input of the succeeding flip-flop is properly "shifted in" following the active edge of the clock. This relationship between t_{CO} and t_h is normally guaranteed if the flip-flops are physically identical. Furthermore, for correct operation, it is easy to verify that the clock period has to be greater than the sum $t_{su} + t_h$.

Generalizations

Flip-flops can be generalized in at least two ways: by making them 1-of-N instead of 1-of-2, and by adapting them to logic with more than two states. In the special cases of 1-of-3 encoding, or multi-valued ternary logic, these elements may be referred to as *flip-flap-flops*.

In a conventional flip-flop, exactly one of the two complementary outputs is high. This can be generalized to a memory element with N outputs, exactly one of which is high (alternatively, where exactly one of N is low). The output is therefore always a one-hot (respectively *one-cold*) representation. The construction is similar to a conventional cross-coupled flip-flop; each output, when high, inhibits all the other outputs.Alternatively, more or less conventional flip-flops can be used, one per output, with additional circuitry to make sure only one at a time can be true. Another generalization of the conventional flip-flop is a memory element for multi-valued logic. In this case the memory element retains exactly one of the logic states until the control inputs induce a change. In addition, a multiple-valued clock can also be used, leading to new possible clock transitions.

COUNTER (DIGITAL)

A counter circuit is usually constructed of a number of flip-flops connected in cascade. Counters are a very widely used component in digital circuits, and are manufactured as separate integrated circuits and also incorporated as parts of larger integrated circuits.

Electronic Counters

In electronics, counters can be implemented quite easily using register-type circuits such as the flip-flop, and a wide variety of classifications exist:

- Asynchronous (ripple) counter – changing state bits are used as clocks to subsequent state flip-flops
- Synchronous counter – all state bits change under control of a single clock
- Decade counter – counts through ten states per stage
- Up/down counter – counts both up and down, under command of a control input
- Ring counter – formed by a shift register with feedback connection in a ring
- Johnson counter – a *twisted* ring counter
- Cascaded counter
- Modulus counter.

Each is useful for different applications. Usually, counter circuits are digital in nature, and count in natural binary. Many types of counter circuits are available as digital building blocks, for example a number of chips in the 4000 series implement different counters.

Occasionally there are advantages to using a counting sequence other than the natural binary sequence—such as the binary coded decimal counter, a linear feedback shift register counter, or a Gray-code counter.

Counters are useful for digital clocks and timers, and in oven timers, VCR clocks, etc.

Asynchronous (ripple) Counter

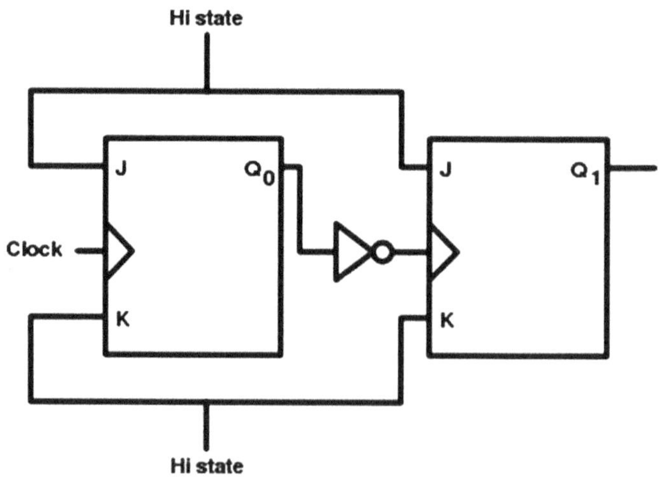

Asynchronous counter created from two JK flip-flops.

An asynchronous (ripple) counter is a single d-type flip-flop, with its J (data) input fed from its own inverted output. This circuit can store one bit, and hence can count from zero to one before it overflows (starts over from 0). This counter will increment once for every clock cycle and takes two clock cycles to overflow, so every cycle it will alternate between a transition from 0 to 1 and a transition from 1 to 0. Notice that this creates a new clock with a 50% duty cycle at exactly half the frequency of the input clock. If this output is then used as the clock signal for a similarly arranged D flip-flop (remembering to invert the output to the input), one will get another 1 bit counter that counts half as fast. Putting them together yields a two-bit counter:

Cycle	Q1	Q0	(Q1:Q0)dec
0	0	0	0
1	0	1	1
2	1	0	2
3	1	1	3
4	0	0	0

You can continue to add additional flip-flops, always inverting the output to its own input, and using the output from the previous flip-flop as the clock signal. The result is called a ripple counter, which can count to 2^n - 1 where n is the number of bits (flip-flop stages) in the counter. Ripple counters suffer from unstable outputs as the overflows "ripple" from stage to stage, but they do find frequent application as dividers for clock signals, where the instantaneous count is unimportant, but the division ratio overall is (to clarify this, a 1-bit counter is exactly equivalent to a divide by two circuit; the output frequency is exactly half that of the input when fed with a regular train of clock pulses).

The use of flip-flop outputs as clocks leads to timing skew between the count data bits, making this ripple technique incompatible with normal synchronous circuit design styles.

Synchronous Counter

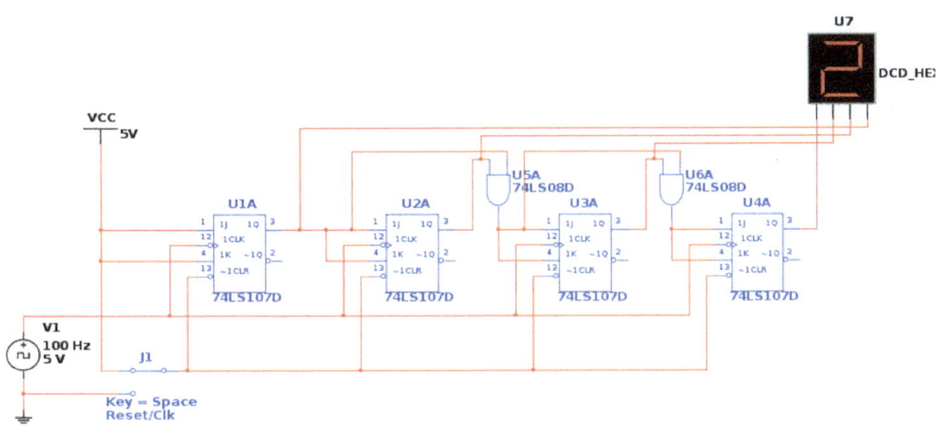

A 4-bit synchronous counter using JK flip-flops.

In synchronous counters, the clock inputs of all the flip-flops are connected together and are triggered by the input pulses. Thus, all the flip-flops change state simultaneously (in parallel). The circuit below is a 4-bit synchronous counter. The J and K inputs of FF0 are connected to HIGH. FF1 has its J and K inputs connected to the output of FF0, and the J and K inputs of FF2 are connected to the output of an AND gate that is fed by the outputs of FF0 and FF1. A simple way of implementing the logic for each bit of an ascending counter (which is what is depicted in the adjacent image) is for each bit to toggle when all of the less significant bits are at a logic high state. For example, bit 1 toggles when bit 0 is logic high; bit 2 toggles when both bit 1 and bit 0 are logic high; bit 3 toggles when bit 2, bit 1 and bit 0 are all high; and so on.

Synchronous counters can also be implemented with hardware finite-state machines, which are more complex but allow for smoother, more stable transitions.

Decade Counter

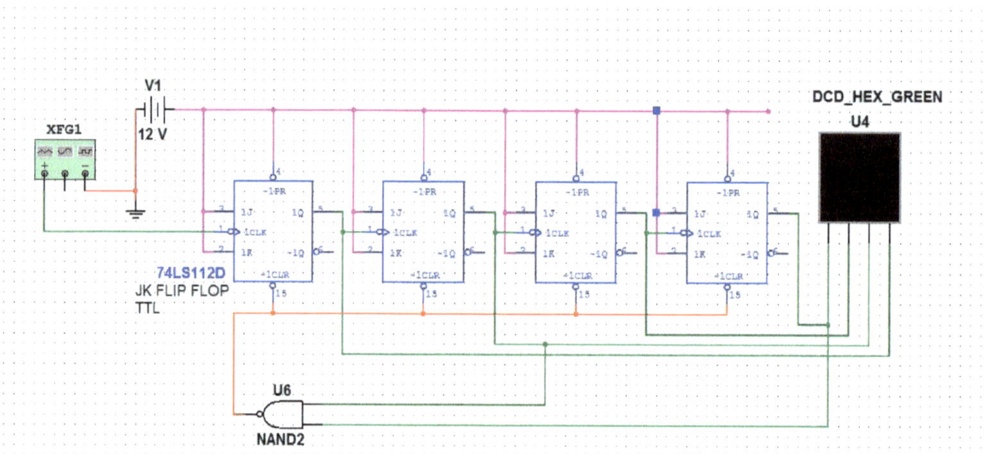

A circuit decade counter using JK Flip-flops (74LS112D).

A decade counter is one that counts in decimal digits, rather than binary. A decade counter may have each (that is, it may count in binary-coded decimal, as the 7490 integrated circuit did) or other binary encodings. "A decade counter is a binary counter that is designed to count to 1010b (decimal 10). An ordinary four-stage counter can be easily modified to a decade counter by adding a NAND gate as in the schematic to the right. Notice that FF2 and FF4 provide the inputs to the NAND gate. The NAND gate outputs are connected to the CLR input of each of the FFs." A decade counter is one that counts in decimal digits, rather than binary. It counts from 0 to 9 and then resets to zero. The counter output can be set to zero by pulsing the reset line low. The count then increments on each clock pulse until it reaches 1001 (decimal 9). When it increments to 1010 (decimal 10) both inputs of the NAND gate go high. The result is that the NAND output goes low, and resets the counter to zero. D going low can be a CARRY OUT signal, indicating that there has been a count of ten.

Ring Counter

A ring counter is a circular shift register which is initiated such that only one of its flip-flops is the state one while others are in their zero states.

A ring counter is a shift register (a cascade connection of flip-flops) with the output of the last one connected to the input of the first, that is, in a ring. Typically, a pattern consisting of a single bit is circulated so the state repeats every n clock cycles if n flip-flops are used.

Johnson Counter

A Johnson counter (or switch-tail ring counter, twisted ring counter, walking ring counter, or Möbius counter) is a modified ring counter, where the output from the last stage is inverted and fed back as input to the first stage. The register cycles through a sequence of bit-patterns, whose length is equal to twice the length of the shift register, continuing indefinitely. These counters find specialist applications, including those similar to the decade counter, digital-to-analog conversion, etc. They can be implemented easily using D- or JK-type flip-flops. It is also known as twisted ring counter.

Computer science counters

In computability theory, a **counter** is considered a type of memory. A counter stores a single natural number (initially zero) and can be arbitrarily long. A counter is usually considered in conjunction with a finite-state machine (FSM), which can perform the following operations on the counter:

- Check whether the counter is zero
- Increment the counter by one.
- Decrement the counter by one (if it's already zero, this leaves it unchanged).

The following machines are listed in order of power, with each one being strictly more powerful than the one below it:

- Deterministic or non-deterministic FSM plus two counters
- Non-deterministic FSM plus one stack
- Non-deterministic FSM plus one counter
- Deterministic FSM plus one counter
- Deterministic or non-deterministic FSM.

For the first and last, it doesn't matter whether the FSM is a deterministic finite automaton or a nondeterministic finite automaton. They have the same power. The first two and the last one are levels of the Chomsky hierarchy.

The first machine, an FSM plus two counters, is equivalent in power to a Turing machine. See the article on counter machines for a proof.

Web Counter

A **web counter** or **hit counter** is a computer software program that indicates the number of visitors, or hits, a particular webpage has received. Once set up, these counters will be incremented by one every time the web page is accessed in a web browser.

The number is usually displayed as an inline digital image or in plain text or on a physical counter such as a mechanical counter. Images may be presented in a variety of fonts, or styles; the classic example is the wheels of an odometer.

Web counter was popular in the 1980s and 1990s, later replaced by more detailed and complete web traffic measures.

Computer Based Counters

Many automation systems use PC and laptops to monitor different parameters of machines and production data. Counters may count parameters such as the number of pieces produced, the production batch number, and measurements of the amounts of material used.

Mechanical Counters

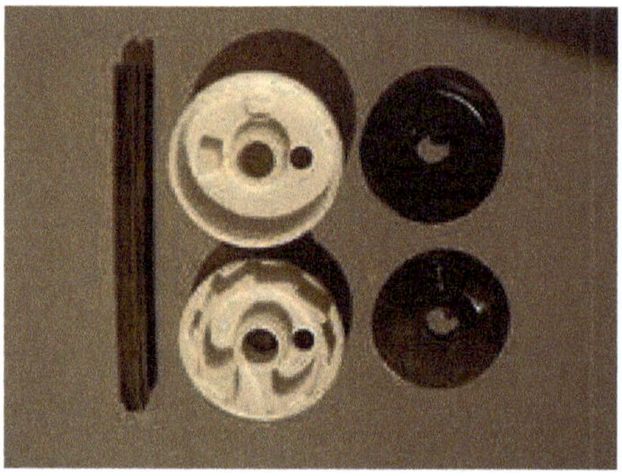

Mechanical counter wheels showing both sides. The bump on the wheel shown at the top engages the ratchet on the wheel below every turn.

Several mechanical counters

Early IBM tabulating machine using mechanical counters

Long before electronics became common, mechanical devices were used to count events. These are known as tally counters. They typically consist of a series of disks mounted on an axle, with the digits 0 through 9 marked on their edge. The right most disk moves one increment with each event. Each disk except the left-most has a protrusion that, after the completion of one revolution, moves the next disk to the left one increment. Such counters were used as odometers for bicycles and cars and in tape recorders, fuel dispensers, in production machinery as well as in other machinery. One of the largest manufacturers was the Veeder-Root company, and their name was often used for this type of counter.

Hand held tally counters are used mainly for stocktaking and for counting people attending events.

Electromechanical counters were used to accumulate totals in tabulating machines that pioneered the data processing industry.

PROCESSOR REGISTER

In computer architecture, a **processor register** is a quickly accessible location available to a digital processor's central processing unit (CPU). Registers usually consist of a small amount of fast storage, although some registers have specific hardware functions, and may be read-only or write-only. Registers are typically addressed by mechanisms other than main memory, but may in some cases be *memory mapped.*

Almost all computers, whether load/store architecture or not, load data from a larger memory into registers where it is used for arithmetic operations and is manipulated or tested by machine instructions. Manipulated data is then often stored back to main memory, either by the same instruction or by a subsequent one. Modern processors use either static or dynamic RAM as main memory, with the latter usually accessed via one or more cache levels.

Processor registers are normally at the top of the memory hierarchy, and provide the fastest way to access data. The term normally refers only to the group of registers that are directly encoded as part of an instruction, as defined by the instruction set. However, modern high-performance CPUs often have duplicates of these "architectural registers" in order to improve performance via register renaming, allowing parallel and speculative execution. Modern x86 design acquired these techniques around 1995 with the releases of Pentium Pro, Cyrix 6x86, Nx586, and AMD K5.

A common property of computer programs is locality of reference, which refers to accessing the same values repeatedly and holding frequently used values in registers to improve performance; this makes fast registers and caches meaningful. Allocating frequently used variables to registers can be critical to a program's performance; this register allocation is performed either by a compiler in the code generation phase, or manually by an assembly language programmer.

Categories of Registers

Registers are normally measured by the number of bits they can hold, for example, an "8-bit register" or a "32-bit register". A processor often contains several kinds of registers, that can be classified according to their content or instructions that operate on them:

- *User-accessible registers* can be read or written by machine instructions. The most common division of user-accessible registers is into data registers and address registers.
    - *Data registers* can hold numeric values such as integer and, in some architectures, floating-point values, as well as characters, small bit arrays and other data. In some older and low end CPUs, a special data register, known as the accumulator, is used implicitly for many operations.
    - *Address registers* hold addresses and are used by instructions that indirectly access primary memory.
        - Some processors contain registers that may only be used to hold an address or only to hold numeric values (in some cases used as an index register whose value is added as an offset from some address); others allow registers to hold either kind of quantity. A wide variety of possible addressing modes, used to specify the effective address of an operand, exist.
        - The stack pointer is used to manage the run-time stack. Rarely, other data stacks are addressed by dedicated address registers, see stack machine.
    - *General-purpose registers* (*GPR*s) can store both data and addresses, i.e., they are combined data/address registers and rarely the register file is *unified* to include floating point as well.

- *Status registers* hold truth values often used to determine whether some instruction should or should not be executed.
- *Floating-point registers* (*FPR*s) store floating point numbers in many architectures.
- *Constant registers* hold read-only values such as zero, one, or pi.
- *Vector registers* hold data for vector processing done by SIMD instructions (Single Instruction, Multiple Data).
- *Special-purpose registers* (*SPR*s) hold program state; they usually include the program counter, also called the instruction pointer, and the status register; the program counter and status register might be combined in a program status word (PSW) register. The aforementioned stack pointer is sometimes also included in this group. Embedded microprocessors can also have registers corresponding to specialized hardware elements.
- In some architectures, *model-specific registers* (also called *machine-specific registers*) store data and settings related to the processor itself. Because their meanings are attached to the design of a specific processor, they cannot be expected to remain standard between processor generations.
- *Memory Type Range Registers* (*MTRR*s)

■ *Internal registers* – registers not accessible by instructions, used internally for processor operations.

- *Instruction register*, holding the instruction currently being executed.
- Registers related to fetching information from RAM, a collection of storage registers located on separate chips from the CPU:
    - *Memory buffer register* (*MBR*)
    - *Memory data register* (*MDR*)
    - *Memory address register* (*MAR*)

Hardware registers are similar, but occur outside CPUs.

In some architectures, such as SPARC and MIPS, the first or last register in the integer register file is a pseudo-register in a way that it is hardwired to always return zero when read (mostly to simplify indexing modes), and it cannot be overwritten. In Alpha this is also done for the floating-point register file. As a result of this, register files are commonly quoted as having one register more than how many of them are actually usable; for example, 32 registers are quoted when only 31 of them fit within the above definition of a register.

Examples

The following table shows the number of registers in several mainstream architectures. Note that in x86-compatible processors the stack pointer (ESP) is counted as an integer register, even though there are a limited number of instructions that may be used to operate on its contents. Similar caveats apply to most architectures.

Although all of the above listed architectures are different, almost all are a basic arrangement known as the Von Neumann architecture, first proposed by mathematician John von Neumann. It is also noteworthy that the number of registers on GPUs is much higher than that on CPUs.

Architecture	GPRs/ data+address registers	FP registers	Notes
AT&T Hobbit	0	stack of 7	The AT&T Hobbit ATT92010 (around 1992) was a commercial version of the 32-bit CRISP(stack machine) processor with RISC style instruction, inspired by the Bell Labs C Machine project, aimed at a design optimized for the C language, Hobbit has no global registers. Addresses can be memory direct or indirect (for pointers) relative to the stack pointer without extra instructions or operand bits. The cache is not optimized for multiprocessors and FPU is consist with custom ASIC design in IEEE standard.
Cray-1	8 scalar data, 8 address	8 scalar, 8 vector (64 elements)	Scalar data registers can be integer or floating-point; also 64 scalar scratch-pad T registers and 64 address scratch-pad B registers
4004	1 accumulator, 16 others	0	Register A is for general purpose, while r0–r15 registers are for the address and segment.
8008	1 accumulator, 6 others	0	The A register is an accumulator to which all arithmetic is done; the H and L registers can be used in combination as an address register; all registers can be used as operands in load/store/move/increment/ decrement instructions and as the other operand in arithmetic instructions. There is no FP unit available.
8080	1 accumulator, 6 others	0	Plus a stack pointer. The A register is an accumulator to which all arithmetic is done; the register pairs B+C, D+E, and H+L, can be used as address registers in some instructions; all registers can be used as operands in load/store/move/ increment/decrement instructions and as the other operand in arithmetic instructions. Some instructions only use H+L; another instruction swaps H+L and D+E. There is no FP unit available.

Architecture	GPRs/ data+address registers	FP registers	Notes
iAPX432	0	stack of 6	The iAPX 432 was referred to as a micromainframe, designed to be programmed entirely in high-level languages. The instruction set architecture was also entirely new and a significant departure from Intel's previous 8008 and 8080 processors as the iAPX 432 programming model was a stack machine with no visible general-purpose registers. It supported object-oriented programming, garbage collection and multitasking as well as more conventional memory management directly in hardware and microcode. Direct support for various data structures was also intended to allow modern operating systems to be implemented using far less program code than for ordinary processors
16-bit x86	8	stack of 8 (if FP present)	8086/8088, 80186/80188, 80286, with 8087, 80187 or 80287 for floating-point, with an 80-bit wide, 8 deep register stack with some instructions able to use registers relative to the top of the stack as operands; without 8087/80187/80287, no floating-point registers
IA-32	8	stack of 8 (if FP present), 8 (if SSE/MMX present)	80386 required 80387 for floating-point, later processors had built-in floating point, with both having an 80-bit wide, 8 deep register stack with some instructions able to use registers relative to the top of the stack as operands. The Pentium III and later had the SSE with additional 128-bit XMM registers.
x86-64	16	16	FP registers are 128-bit XMM registers, later extended to 256-bit YMM registers with AVX.
Xeon Phi	16	32	Including 32 256/512-bit ZMM registers with AVX-512

Architecture	GPRs/ data+address registers	FP registers	Notes
Geode GX	1 data, 1 address	8	Geode GX/Media GX/4x86/5x86 is the emulation of 486/Pentium compatible processor made by Cyrix/National Semiconductor. Like Transmeta, the processor had a translation layer that translated x86 code to native code and executed it. It does not support 128-bit SSE registers, just the 80387 stack of eight 80-bit floating point registers, and partially support 3DNow! from AMD. The native processor only contains 1 data and 1 address register for all purpose and translated into 4 paths of 32-bit naming register r1 (base), r2 (data), r3 (back pointer), and r4 (stack pointer) within scratchpad sram for integer operation and uses the L1 cache for x86 code emulation(note that it's not compatible with some 286/386/486 instructions in real mode). Later the design was abandoned after AMD acquired the IP from National Semiconductor and branded it with Athlon core in embedded market.
SunPlus SPG	0	6	A 16-bit wide, 32-bit address space stack machine processor that made from Taiwanese semiconductor called "Sunplus", it can be found on Vtech's v'smile line for educational purpose and video game console like Mattel hyperscan, XaviXPORT. it does lack any general purpose register or internal register for naming/renaming but its Floating Point Unit has 80-bit 6 stage stack.
VM Labs Nuon	0	1	a 32-bit stack machine processor that developed by VM labs for specialized on multimedia purpose. It can be found on the company's own Nuon DVD player console line and Game Wave Family Entertainment System from ZaPit games. The design was heavy influence by Intel's MMX technology, it contained an 128 bytes unified stack cache for both vector and scalar instructions. the unified cache can be divided as 8 128-bit vector register or 32 32bit SIMD scalar register through bank renaming, no integer register found in this architecture.
Nios II	31	8	Nios II is based on MIPS IV instruction set and has 31 32-bit GPRs, with register 0 being hardwired to zero and 8 64-bit floating point registers
Motorola 6800	2 data, 1 index	0	Plus a stack pointer
Motorola 68k	8 data (d0-d7), 8 address (a0-a7)	8 (if FP present)	Address register 8 (a7) is the stack pointer. 68000, 68010, 68012, 68020, and 68030 require an FPU for floating point; 68040 had FPU built in. FP registers are 80-bit.
SH 16-bit	1	6	

Architecture	GPRs/ data+address registers	FP registers	Notes
Emotion Engine	4	32 SIMD + 32 Vector	The Emotion Engine's main core contains four entries 32-bit general-purpose registers for integer computation and 32 entries 128-bit SIMD registers for storing SIMD instruction, streaming data value and some integer calculation value. one accumulator register for connecting general floating-point computation to vector register file on co-processor. The coprocessor is built via 32 entries 128-bit vector register file(can only store vector value that pass from accumulator in cpu.) and no integer register is built in. Both vector co-processor(VPU 0/1) and emotion engine main processor are built based on modified MIPS instructions set. Accumulator in the this case is not general purpose but control status.
CUDA	1	8/16/32/64/128	Each CUDA core contains a single 32/64-bit integer data register while the floating point unit contains a much larger number of registers: • the Tesla 1.0-based G8x contains 8×128-bit HDR vector registers, • the Tesla 2.0-based GT200 increased the count to 16×128-bit • Fermi extended the register width to 256 bits and increased the register count to 32, (32×256-bit) • Kepler increased it to 64. (64×256-bit) • Maxwell contains a massive amount of 128 512-bit vector registers. (128×512-bit) • Pascal: ? Volta: ?
IBM/360	16	4 (if FP present)	This applies to S/360's successors, System/370 through System/390; FP was optional in System/360, and always present in S/370 and later. In processors with the Vector Facility, there are 16 vector registers containing a machine-dependent number of 32-bit elements.
z/Architecture	16	16	64-bit version of S/360 and successors
MMIX	256	256	An instruction set designed by Donald Knuth in the late 1990s for pedagogical purposes.

Architecture	GPRs/ data+address registers	FP registers	Notes
NS320xx	8	8 (if FP present)	
Xelerated X10	1	32	a 32/40 bit stack machine based network processor with modified MIPS instruction and 128 bit floating point unit.
Parallax Propeller	0	2	An eight core 8/16 bit sliced stack machine controller with simple logic circus inside, have eight cog counter(core) and each contain three 8/16 bit special control registers with 32 bit x 512 stack ram however it does not carrying any general register for integer purpose. unlike most of shadow register file in modern processor and multi core system, all these stack ram in cog can be accessed in instruction level which all these cog can act as one big single general purpose core if necessary. Floating point unit is external and it contain two 80 bit vector register.
Itanium	128	128	And 64 1-bit predicate registers and 8 branch registers. The FP registers are 82-bit.
SPARC	31	32	Global register 0 is hardwired to 0. Uses register windows.
IBM POWER	32	32	And 1 link and 1 count register.
Power Architecture	32	32	And 1 link and 1 count register. Processors supporting the Vector facility also have 32 128-bit vector registers,
Blackfin	8	16	containing two external uncore 40 bit accumulator, but non are general purpose. Support 64 bit RISC architecture ISA, vector register are 256 bit.
IBM Cell SPE	128		128 GPRs, which can hold integer, address, or floating-point values
Alpha	31	31	Registers R31 (integer) and F31 (floating-point) are hardwired to zero.
6502	1 data, 2 index	0	6502's content A (Accumulator) register for main purpose data store and memory address (8-bit data/16-bit address), X,Y are indirect and direct index registers (respectively) and SP register are specific index only.
W65C816S	1	0	65C816 is the 16-bit successor of the 6502. X,Y, D (Direct Page register) are condition register and SP register are specific index only. main accumulator extended to 16-bit (B) while keep 8-bit (A) for compatibility and main register can now address up to 24-bit (16-bit wide data instruction/24-bit memory address).

Architecture	GPRs/ data+address registers	FP registers	Notes
65k	1	0	Direct successor of 6502, 65002 only content A (Accumulator) register for main purpose data store and extend data wide to 32-bit and 64-bit instruction wide, support 48-bit virtual address in software mode, X,Y are still condition register and remain 8-bit and SP register are specific index but increase to 16-bit wide.
MeP	4	8	Media-embedded processor was a 32 bit processor developed by toshiba, a modded 8080 instruction set with only A, B, C, D register available through all mode(8/16/32 bit) and incompatible with x86, however it contain 80 bit floating point unit that is x87 compatible.
PIC microcontroller	1	0	
AVR microcontroller	32	0	
ARM 32-bit	14	Varies (up to 32)	r15 is the program counter, and not usable as a GPR; r13 is the stack pointer; r8-r13 can be switched out for others (banked) on a processor mode switch. Older versions had 26-bit addressing, and used upper bits of the program counter (r15) for status flags, making that register 32-bit.
ARM 64/32-bit	31	32	Register r31 is the stack pointer or hardwired to 0, depending on the context.
MIPS	31	32	Register 0 is hardwired to 0.
Epiphany	64 (per core)		Each instruction controls whether registers are interpreted as integers or single precision floating point. Architecture is scalable to 4096 cores with 16 and 64 core implementations currently available.

Register usage

The number of registers available on a processor and the operations that can be performed using those registers has a significant impact on the efficiency of code generated by optimizing compilers. The Strahler number of an expression tree gives the minimum number of registers required to evaluate that expression tree.

MULTIPLEXER

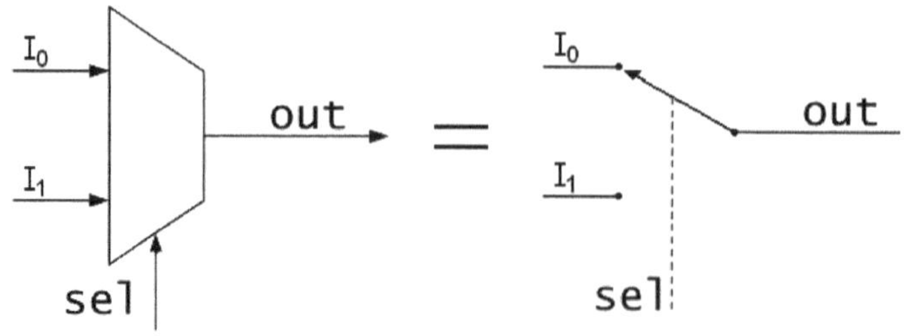

Schematic of a 2-to-1 Multiplexer. It can be equated to a controlled switch.

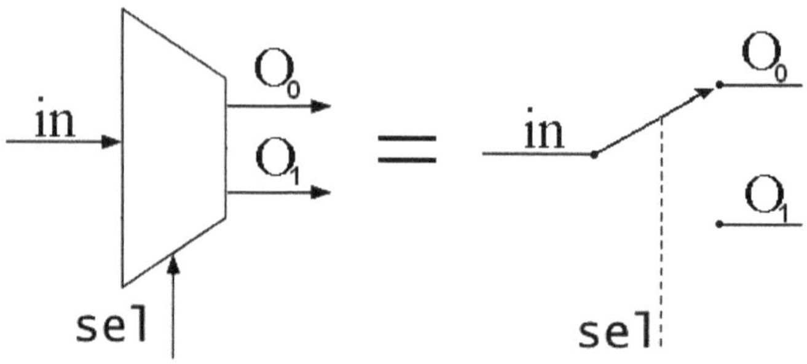

Schematic of a 1-to-2 Demultiplexer. Like a multiplexer, it can be equated to a controlled switch.

In electronics, a **multiplexer** (or **mux**) is a device that selects one of several analog or digital input signals and forwards the selected input into a single line. A multiplexer of 2^n inputs has n select lines, which are used to select which input line to send to the output. Multiplexers are mainly used to increase the amount of data that can be sent over the network within a certain amount of time and bandwidth. A multiplexer is also called a **data selector**. Multiplexers can also be used to implement Boolean functions of multiple variables.

An electronic multiplexer makes it possible for several signals to share one device or resource, for example one A/D converter or one communication line, instead of having one device per input signal.

Conversely, a **demultiplexer** (or **demux**) is a device taking a single input signal and selecting one of many data-output-lines, which is connected to the single input. A multiplexer is often used with a complementary demultiplexer on the receiving end.

An electronic multiplexer can be considered as a multiple-input, single-output switch, and a demultiplexer as a single-input, multiple-output switch. The schematic symbol for a multiplexer is an isosceles trapezoid with the longer parallel side containing the input pins and the short parallel side containing the output pin. The schematic on the right shows a 2-to-1 multiplexer on the left and an equivalent switch on the right. The ^{sel}wire connects the desired input to the output.

Cost Saving

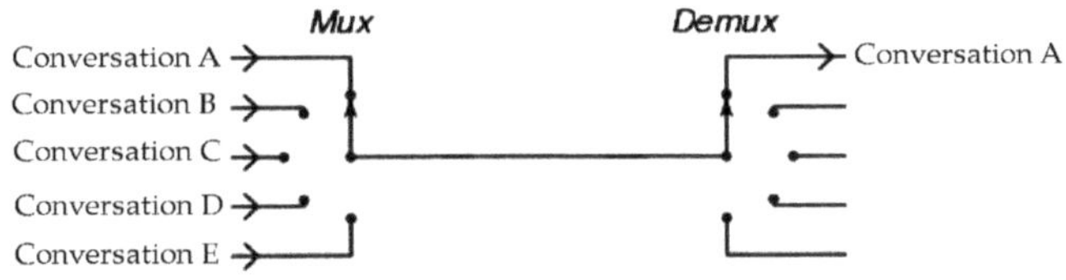

The basic function of a multiplexer: combining multiple inputs into a single data stream. On the receiving side, a demultiplexer splits the single data stream into the original multiple signals.

One use for multiplexers is economizing connections over a single channel, by connecting the multiplexer's single output to the demultiplexer's single input. The image to the right demonstrates this benefit. In this case, the cost of implementing separate channels for each data source is higher than the cost and inconvenience of providing the multiplexing/demultiplexing functions.

At the receiving end of the data link a complementary *demultiplexer* is usually required to break the single data stream back down into the original streams. In some cases, the far end system may have functionality greater than a simple demultiplexer; and while the demultiplexing still occurs technically, it may never be implemented discretely. This would be typical when: a multiplexer serves a number of IP network users; and then feeds directly into a router, which immediately reads the content of the entire link into its routing processor; and then does the demultiplexing in memory from where it will be converted directly into IP sections.

Often, a multiplexer and demultiplexer are combined together into a single piece of equipment, which is conveniently referred to as a "multiplexer". Both circuit elements are needed at both ends of a transmission link because most communications systems transmit in both directions.

In analog circuit design, a multiplexer is a special type of analog switch that connects one signal selected from several inputs to a single output.

Digital Multiplexers

In digital circuit design, the selector wires are of digital value. In the case of a 2-to-1 multiplexer, a logic value of 0 would connect I_0 to the output while a logic value of 1 would connect I_1 to the output. In larger multiplexers, the number of selector pins is equal to [$\log_2(n)$ where$_n$] is the number of inputs.

For example, 9 to 16 inputs would require no fewer than 4 selector pins and 17 to 32 inputs would require no fewer than 5 selector pins. The binary value expressed on these selector pins determines the selected input pin.

A 2-to-1 multiplexer has a boolean equation where A and B are the two inputs, is the selector input, and Z is the output:S

$Z = (A \cdot \overline{S}) + (B \cdot S)$

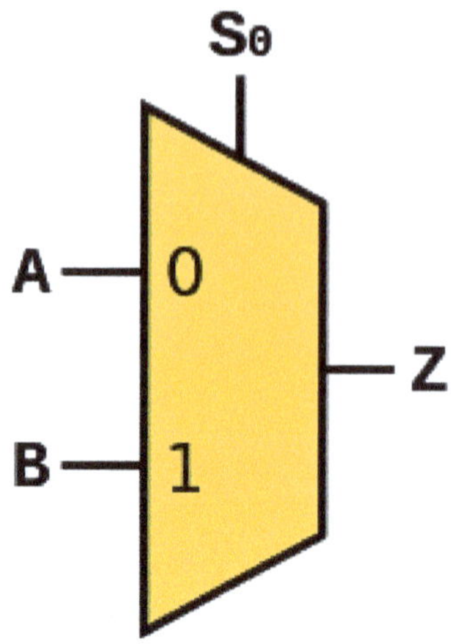

A 2-to-1 mux.

Which can be expressed as a truth table:

S	A	B	Z
0	0	0	0
0	0	1	0
0	1	0	1
0	1	1	1
1	0	0	0
1	0	1	1
1	1	0	0
1	1	1	1

Or, in simpler notation:

S	Z
0	A
1	B

These tables show that when s= 0 then Z = A but when S = 1 then Z = B. A straightforward realization of this 2-to-1 multiplexer would need 2 AND gates, an OR gate, and a NOT gate. While this is mathematically correct, it should be noted that a direct physical implementation would be prone to race conditions that require additional gates to suppress.

Larger multiplexers are also common and, as stated above, require [log2(n)] selector pins for n inputs. Other common sizes are 4-to-1, 8-to-1, and 16-to-1. Since digital logic uses binary values, powers of 2 are used (4, 8, 16) to maximally control a number of inputs for the given number of selector inputs.

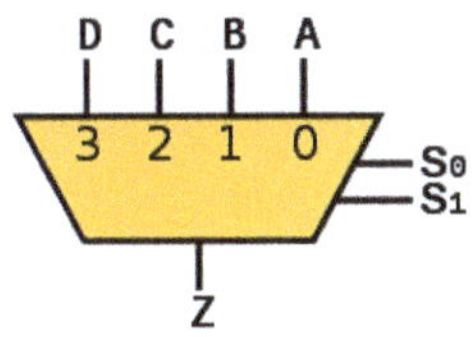

4-to-1 mux.

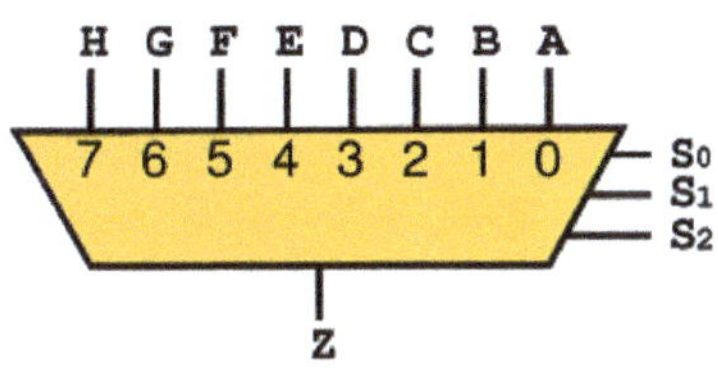

8-to-1 mux.

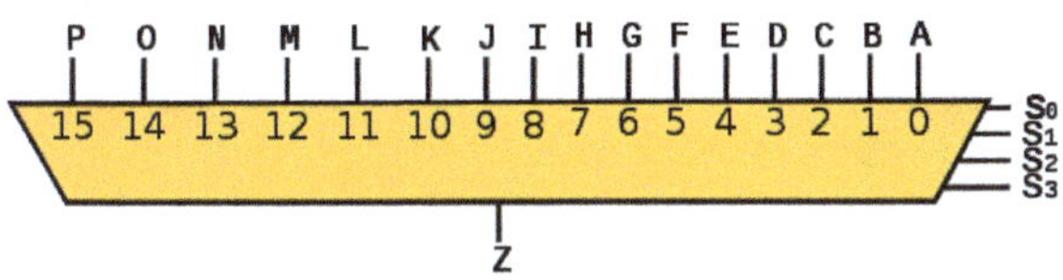

16-to-1 mux

The boolean equation for a 4-to-1 multiplexer is:

$$Z = (A \cdot \overline{S_0} \cdot \overline{S_1}) + (B \cdot S_0 \cdot \overline{S_1}) + (C \cdot \overline{S_0} \cdot S_1) + (D \cdot S_0 \cdot S_1)$$

The following 4-to-1 multiplexer is constructed from 3-state buffers and AND gates (the AND gates are acting as the decoder):

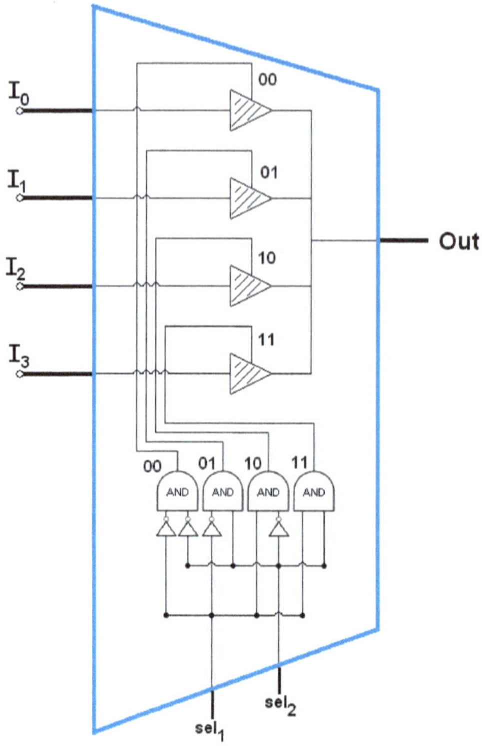

The subscripts on the I_n inputs indicate the decimal value of the binary control inputs at which that input is let through.

Chaining Multiplexers

Larger Multiplexers can be constructed by using smaller multiplexers by chaining them together. For example, an 8-to-1 multiplexer can be made with two 4-to-1 and one 2-to-1 multiplexers. The two 4-to-1 multiplexer outputs are fed into the 2-to-1 with the selector pins on the 4-to-1's put in parallel giving a total number of selector inputs to 3, which is equivalent to an 8-to-1.

List of ICs which Provide Multiplexing

The 7400 series has several ICs that contain multiplexer(s):

S.No.	IC No.	Function	Output State
1	74157	Quad 2:1 mux.	Output same as input given
2	74158	Quad 2:1 mux.	Output is inverted input
0	74153	Dual 4:1 mux.	Output same as input
5	74352	Dual 4:1 mux.	Output is inverted input

9	74151A	8:1 mux.	Both outputs available (i.e., complementary outputs)
6	74151	8:1 mux.	Output is inverted input
7	74150	16:1 mux.	Output is inverted input

Digital Demultiplexers

Demultiplexers take one data input and a number of selection inputs, and they have several outputs. They forward the data input to one of the outputs depending on the values of the selection inputs. Demultiplexers are sometimes convenient for designing general purpose logic, because if the demultiplexer's input is always true, the demultiplexer acts as a decoder. This means that any function of the selection bits can be constructed by logically OR-ing the correct set of outputs.

If X is the input and S is the selector, and A and B are the outputs:

$$A = (X \cdot \overline{S})$$
$$B = (X \cdot S)$$

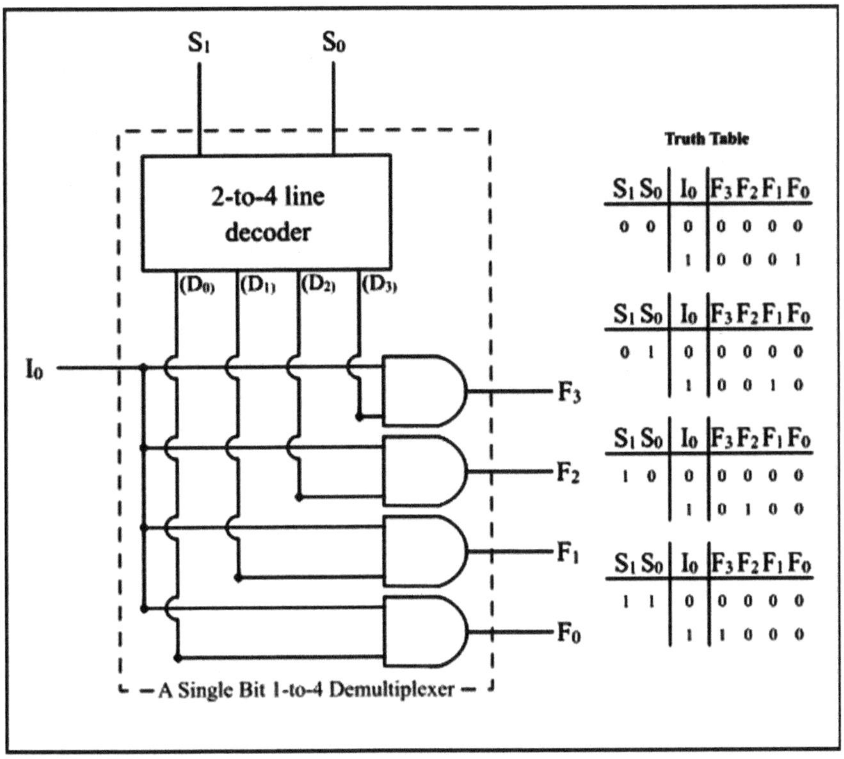

Example: A Single Bit 1-to-4 Line Demultiplexer.

List of ICs Which Provide Demultiplexing

The 7400 series has several ICs that contain demultiplexer(s):

S.No.	IC No. (7400)	IC No. (4000)	Function	Output State
1	74139		Dual 1:4 demux.	Output is inverted input
3	74156		Dual 1:4 demux.	Output is open collector
4	74138		1:8 demux.	Output is inverted input
5	74238		1:8 demux.	
6	74154		1:16 demux.	Output is inverted input
7	74159	CD4514/15	1:16 demux.	Output is open collector and same as input

Fairchild 74F138.

Multiplexers as PLDs

Multiplexers can also be used as programmable logic devices, specifically to implement Boolean functions. Any Boolean function of n variables and one result can be implemented with a multiplexer with n selector inputs. The variables are connected to the selector inputs, and the function result, 0 or 1, for each possible combination of selector inputs is connected to the corresponding data input. This is especially useful in situations when cost is a factor, for modularity, and for ease of modification. If one of the variables (for example, D) is also available inverted, a multiplexer with n-1 selector inputs is sufficient; the data inputs are connected to 0, 1, D, or $\sim D$, according to the desired output for each combination of the selector inputs.

HEAT DISSIPATION AND THERMAL MANAGEMENT

Thermal Management of Electronic Devices and Systems

Heat generated by electronic circuitry must be dissipated to prevent immediate failure and improve long term reliability. Heat dissipation is mostly achieved by passive conduction/ convection. Means to achieve greater dissipation include heat sinks and fans for air cooling, and other forms of computer cooling such as water cooling. These techniques use convection, conduction, and radiation of heat energy.

NOISE

Electronic noise is defined as unwanted disturbances superposed on a useful signal that tend to obscure its information content. Noise is not the same as signal distortion caused by a circuit. Noise is associated with all electronic circuits. Noise may be electromagnetically or thermally generated, which can be decreased by lowering the operating temperature of the circuit. Other types of noise, such as shot noise cannot be removed as they are due to limitations in physical properties.

In electronics, **noise** is a random fluctuation in an electrical signal, a characteristic of all electronic circuits. Noise generated by electronic devices varies greatly as it is produced by several different effects. Thermal noise is unavoidable at non-zero temperature (see fluctuation-dissipation theorem), while other types depend mostly on device type (such as shot noise, which needs a steep potential barrier) or manufacturing quality and semiconductor defects, such as conductance fluctuations, including 1/f noise.

In communication systems, noise is an error or undesired random disturbance of a useful information signal. The noise is a summation of unwanted or disturbing energy from natural and sometimes man-made sources. Noise is, however, typically distinguished from interference, for example in the signal-to-noise ratio (SNR), signal-to-interference ratio (SIR) and signal-to-noise plus interference ratio (SNIR) measures. Noise is also typically distinguished from distortion, which is an unwanted systematic alteration of the signal waveform by the communication equipment, for example in the signal-to-noise and distortion ratio (SINAD) and total harmonic distortion plus noise (THD+N).

While noise is generally unwanted, it can serve a useful purpose in some applications, such as random number generation or dither.

Noise Types

Thermal Noise

Johnson–Nyquist noise (sometimes thermal, Johnson or Nyquist noise) is unavoidable, and generated by the random thermal motion of charge carriers (usually electrons), inside an electrical conductor, which happens regardless of any applied voltage.

Thermal noise is approximately white, meaning that its power spectral density is nearly equal throughout the frequency spectrum. The amplitude of the signal has very nearly a Gaussian probability density function. A communication system affected by thermal noise is often modeled as an additive white Gaussian noise (AWGN) channel.

The root mean square (RMS) voltage due to thermal noise v_n, generated in a resistance R (ohms) over bandwidth Δf (hertz), is given by

$$v_n = \sqrt{4k_B T R \Delta f}$$

where k_B is Boltzmann's constant (joules per kelvin) and T is the resistor's absolute temperature (kelvin). As the amount of thermal noise generated depends upon the temperature of the circuit, very sensitive circuits such as preamplifiers in radio telescopes are sometimes cooled in liquid nitrogen to reduce the noise level.

Shot Noise

If electrons flow across a barrier, then they have discrete arrival times. Those discrete arrivals exhibit shot noise. The output of a shot noise generator is easily set by the current. Typically, the barrier in a diode is used.

Shot noise in electronic devices results from unavoidable random statistical fluctuations of the electric current when the charge carriers (such as electrons) traverse a gap. The current is a flow of discrete charges, and the fluctuation in the arrivals of those charges creates shot noise. Shot noise is similar to the noise created by rain falling on a tin roof. The flow of rain may be relatively constant, but the raindrops arrive discretely.

The root-mean-square value of the shot noise current i_n is given by the Schottky formula

$$i_n = \sqrt{2Iq\Delta B}$$

where I is the DC current, q is the charge of an electron, and B is the bandwidth in hertz.

The shot noise assumes independent arrivals. Vacuum tubes have shot noise because the electrons randomly leave the cathode and arrive at the anode (plate). A tube may not exhibit the full shot noise effect: the presence of a space charge tends to smooth out the arrival times (and thus reduce the randomness of the current).

Conductors and resistors typically do not exhibit shot noise because the electrons thermalize and move diffusively within the material; the electrons do not have discrete arrivial times. Shot noise has been demonstrated in mesoscopic resistors when the size of the resistive element becomes shorter than the electron-phonon scattering length.

Flicker Noise

Flicker noise, also known as $1/f$ noise, is a signal or process with a frequency spectrum that falls off steadily into the higher frequencies, with a pink spectrum. It occurs in almost all electronic devices, and results from a variety of effects, though always related to a direct current.

Burst Noise

Burst noise consists of sudden step-like transitions between two or more levels (non-Gaussian), as high as several hundred microvolts, at random and unpredictable times. Each shift in offset voltage or current lasts for several milliseconds, and the intervals between pulses tend to be in the audio range (less than 100 Hz), leading to the term *popcorn noise* for the popping or crackling sounds it produces in audio circuits.

Transit-time Noise

If the time taken by the electrons from traveling from emitter to collector becomes comparable to the period of the signal being amplified, that is, at frequencies above VHF and beyond, so-called transit-time effect takes place and noise input admittance of the transistor increases. From the frequency at which this effect becomes significant it goes on increasing with frequency and quickly dominates over other terms.

Coupled Noise

Energy external of the receiver can couple noise, also by energy conversion. Generally this is done by fundamental interaction, in electronics mainly by inductive coupling and/ or capacitive coupling.

Intermodulation noise

Intermodulation noise is caused when signals of different frequencies share the same non-linear medium.

Crosstalk

Phenomenon in which a signal transmitted in one circuit or channel of a transmission systems creates undesired interferance onto a signal in another channel.

Interference

Modification or disruption of a signal travelling along a medium

Atmospheric Noise (static noise)

This noise is also called static noise and it is the natural source of disturbance caused by lightning discharge in thunderstorm and the natural (electrical) disturbances occurring in nature.

Industrial Noise

Sources such as automobiles, aircraft, ignition electric motors and switching gear, High voltage wires and fluorescent lamps cause industrial noise. These noises are produced by the discharge present in all these operations.

Extraterrestrial Noise

Noise from outside the Earth includes:

Solar Noise

Noise that originates from the Sun is called solar noise. Under normal conditions there is constant radiation from the Sun due to its high temperature. Electrical disturbances such as corona discharges, as well as sunspots can produce additional noise.

Cosmic Noise

Distant stars generate noise called cosmic noise. While these stars are too far away to individually affect terrestrial communications systems, their large number leads to appreciable collective effects. Cosmic noise has been observed in a range from 8 MHz to 1.43 GHz.

Reduction of Electromagnetic Noise Coupling

In many cases noise found on a signal in a circuit is unwanted. When creating a circuit, one usually wants a true output of what the circuit has accomplished. There are many different noise reduction techniques that can change a noisy altered output signal to a more theoretical output signal.

- Faraday cage – A Faraday cage is a good way to reduce the overall noise in a complete circuit. The Faraday cage can be thought of as an enclosure that separates the complete circuit from outside power lines and any other signal that may alter the true signal. A Faraday cage will usually block out most electromagnetic and electrostatic noise.
- Capacitive coupling – A current through two resistors, or any other type of conductor, close to each other in a circuit can create unwanted capacitive coupling. If this happens an AC signal from one part of the circuit can be accidentally picked up in another part. The two resistors (conductors) act like a capacitor thus transferring AC signals. There may be other reasons for which capacitive coupling is wanted but then it would not be thought of as electronic noise.
- Ground loops – When grounding a circuit, it is important to avoid ground loops. Ground loops occur when there is a voltage drop between the two ground potentials. Since ground is thought of as 0V, the presence of a voltage is undesirable at any point of a ground bus. If this is the case, it would not be a true ground. A good way to fix this is to bring all the ground wires to the same potential in a ground bus.
- Shielding cables – In general, using shielded cables to protect the wires from unwanted noise frequencies in a sensitive circuit is good practice. A shielded wire can be thought of as a small Faraday cage for a specific wire as it uses a plastic or rubber enclosing the true wire. Just outside the rubber/plastic covering is a conductive metal that intercepts any noise signal. Because the conductive metal is grounded, the noise signal runs straight to ground before ever getting to the true wire. It is important to ground the shield at only one end to avoid a ground loop on the shield.
- Twisted pair wiring – Twisting wires very tightly together in a circuit will dramatically reduce electromagnetic noise. Twisting the wires decreases the loop size in which a magnetic field can run through to produce a current between the wires. Even if the wires are twisted very tightly, there may still be small loops somewhere between them, but because they are twisted the magnetic field going through the smaller loops induces a current flowing in opposite ways in each wire and thus cancelling them out.

- Notch filters – Notch filters or band-rejection filters are essential when eliminating a specific noise frequency. For example, in most cases the power lines within a building run at 60 Hz. Sometimes a sensitive circuit will pick up this 60 Hz noise through some unwanted antenna (could be as simple as a wire in the circuit). Running the output through a notch filter at 60 Hz will amplify the desired signal without amplifying the 60 Hz noise. So in a sense the noise will be lost at the output of the filter.

Quantification

The **noise level** in an electronic system is typically measured as an electrical power *N* in watts or dBm, a root mean square (RMS) voltage (identical to the noise standard deviation) in volts, dBμV or a mean squared error (MSE) in volts squared. Noise may also be characterized by its probability distribution and noise spectral density $N_0(f)$ in watts per hertz.

A noise signal is typically considered as a linear addition to a useful information signal. Typical signal quality measures involving noise are signal-to-noise ratio (SNR or *S*/*N*), signal-to-quantization noise ratio (SQNR) in analog-to-digital conversion and compression, peak signal-to-noise ratio (PSNR) in image and video coding, E_b/N_0 in digital transmission, carrier to noise ratio (CNR) before the detector in carrier-modulated systems, and noise figure in cascaded amplifiers.

Noise is a random process, characterized by stochastic properties such as its variance, distribution, and spectral density. The spectral distribution of noise can vary with frequency, so its power density is measured in watts per hertz (W/Hz). Since the power in a resistive element is proportional to the square of the voltage across it, noise voltage (density) can be described by taking the square root of the noise power density, resulting in volts per root hertz $(\mathrm{V}/\sqrt{\mathrm{Hz}})$. Integrated circuit devices, such as operational amplifiers commonly quote equivalent input noise level in these terms (at room temperature).

Noise power is measured in watts or decibels (dB) relative to a standard power, usually indicated by adding a suffix after dB. Examples of electrical noise-level measurement units are dBu, dBm0, dBrn, dBrnC, and dBrn($f_1 - f_2$), dBrn(144-line).

Noise levels are usually viewed in opposition to signal levels and so are often seen as part of a signal-to-noise ratio (SNR). Telecommunication systems strive to increase the ratio of signal level to noise level in order to effectively transmit data. In practice, if the transmitted signal falls below the level of the noise (often designated as the noise floor) in the system, data can no longer be decoded at the receiver. Noise in telecommunication systems is a product of both internal and external sources to the system.

In a carrier-modulated passband analog communication system, a certain carrier-to-noise ratio (CNR) at the radio receiver input would result in a certain signal-to-noise ratio in the detected message signal. In a digital communications system, a certain E_b/N_0 (normalized signal-to-noise ratio) would result in a certain bit error rate.

Dither

If the noise source is correlated with the signal, such as in the case of quantisation error, the intentional introduction of additional noise, called dither, can reduce overall noise in the bandwidth of interest. This technique allows retrieval of signals below the nominal detection threshold of an instrument. This is an example of stochastic resonance.

ELECTRONICS THEORY

Mathematical methods are integral to the study of electronics. To become proficient in electronics it is also necessary to become proficient in the mathematics of circuit analysis.

Circuit analysis is the study of methods of solving generally linear systems for unknown variables such as the voltage at a certain node or the current through a certain branch of a network. A common analytical tool for this is the SPICE circuit simulator.

Also important to electronics is the study and understanding of electromagnetic field theory.

ELECTRONICS LAB

Due to the complex nature of electronics theory, laboratory experimentation is an important part of the development of electronic devices. These experiments are used to test or verify the engineer's design and detect errors. Historically, electronics labs have consisted of electronics devices and equipment located in a physical space, although in more recent years the trend has been towards electronics lab simulation software, such as CircuitLogix, Multisim, and PSpice.

COMPUTER AIDED DESIGN (CAD)

Today's electronics engineers have the ability to design circuits using premanufactured building blocks such as power supplies, semiconductors (i.e. semiconductor devices, such as transistors), and integrated circuits. Electronic design automation software programs include schematic capture programs and printed circuit board design programs. Popular names in the EDA software world are NI Multisim, Cadence (ORCAD), EAGLE PCB and Schematic, Mentor (PADS PCB and LOGIC Schematic), Altium (Protel), LabCentre Electronics (Proteus), gEDA, KiCad and many others.

CONSTRUCTION METHODS

Many different methods of connecting components have been used over the years. For instance, early electronics often used point to point wiring with components attached to wooden breadboards to construct circuits. Cordwood construction and wire wrap were other methods used. Most modern day electronics now use printed circuit boards made of materials such as FR4, or the cheaper (and less hard-wearing) Synthetic Resin Bonded Paper (SRBP, also known as Paxoline/Paxolin (trade marks) and FR2) - characterised by its brown colour. Health and environmental concerns associated with electronics assembly have gained increased attention in recent years, especially for products destined to the European Union, with its Restriction of Hazardous Substances Directive (RoHS) and Waste Electrical and Electronic Equipment Directive (WEEE), which went into force in July 2006.

DEGRADATION

Rasberry crazy ants have been known to consume the insides of electrical wiring, and nest inside of electronics; they prefer DC to AC currents. This behavior is not well understood by scientists.

REFERENCES

1. "About the EDA Industry". Electronic Design Automation Consortium. Retrieved 29 July 2015.
2. *Lavagno, Martin, and Scheffer (2006). Electronic Design Automation For Integrated Circuits Handbook. Taylor and Francis. ISBN 0849330963.*
3. Company Comparison - Google Finance. Google.com. Retrieved on 2013-08-10.
4. Synopsys, Inc.: NASDAQ:SNPS quotes & news - Google Finance. Google.com (2013-05-22). Retrieved on 2013-08-10.
5. CDNS Key Statistics | Cadence Design Systems, Inc. Stock - Yahoo! Finance. Finance.yahoo.com. Retrieved on 2013-08-10.
6. *Dylan McGrath (30 Nov 2011). Synopsys to buy Magma for $507 million. EETimes.*
7. *Synopsys to Acquire Magma Design Automation".*
8. *Agilent EEsof EDA – Part I.*
9. *Kirti Sikri Desai (2006). EDA Innovation through Merger and Acquisitions. EDA Cafe. Retrieved March 23, 2010.*
10. List of EDA mergers and acquisitions at SemiWiki.com
11. Eda Consortium Reports Industry Revenue Down In Third Quarter 2009
12. Eda Consortium Reports Decrease For 2009 But Sequential Fourth Quarter Gains
13. Eda Consortium Reports Revenue Increase For Q1 2010
14. Eda Consortium Reports Revenue Increase For Q2 2010
15. Eda Consortium Reports Revenue Increase For Q3 2010
16. Eda Consortium Reports Revenue Increase For Q4 2010
17. Eda Consortium Reports Revenue Increase For Q1 2011
18. Eda Consortium Reports Revenue Increase For Q2 2011
19. Eda Consortium Reports Revenue Increase For Q3 2011
20. Eda Consortium Reports Revenue Increase Of Over 12 Percent For Q4 2011
21. Eda Consortium Reports Revenue Increase For Q1 2012

22. Eda Consortium Reports Revenue Increase For Q2 2012
23. Eda Consortium Reports Revenue Increase For Q3 2012
24. Eda Consortium Reports Revenue Increase For Q4 2012

CHAPTER 2

History of Electronic Engineering

Chapter Contents

- **Introduction**
- **Wireless Telegraphy and Radio**
- **Television**
- **Radar and Radio Location**
- **Computers**
- **Microprocessors**
- **Microprocessor Chronology**

An electronic engineering technician conceptualizes, assembles, and tests electronic systems and equipment. He or she helps an electronic engineer to refine product designs to make sure they work as efficiently as possible. Technicians also make careful calculations and keep detailed records during the testing phase and produce finalized reports that summarize results. Most technicians are employed by research and development (R&D) companies and electronic equipment manufacturing plants.

Electronic engineering technicians usually work with teams of other professionals to build and test products. After an engineer finalizes the schematics for a particular product or system, a team of technicians is asked to develop a working prototype. The team may work together on an entire prototype or work individually on different elements of construction. An experienced electronic engineering technician may be given additional responsibilities, such as consulting personally with an engineer to create original blueprints and computer models.

INTRODUCTION

The **history of electronic engineering** is a long one. *Chambers Twentieth Century Dictionary* (1972) defines electronics as "The science and technology of the conduction of electricity in a vacuum, a gas, or a semiconductor, and devices based thereon".

Electronic engineering as a profession sprang from technological improvements in the telegraph industry during the late 19th century and in the radio and telephone industries during the early 20th century. People gravitated to radio, attracted by the technical fascination it inspired, first in receiving and then in transmitting. Many who went into broadcasting in the 1920s had become "amateurs" in the period before World War I. The modern discipline of electronic engineering was to a large extent born out of telephone-, radio-, and television-equipment development and the large amount of electronic-systems development during World War II of radar, sonar, communication systems, and advanced munitions and weapon systems. In the interwar years, the subject was known as radio engineering. The word *electronics* began to be used in the 1940s In the late 1950s the term *electronic engineering* started to emerge.

The electronic laboratories (Bell Labs in the United States for instance) created and subsidized by large corporations in the industries of radio, television, and telephone equipment, began churning out a series of electronic advances. In 1948 came the transistor and in 1960 the integrated circuit, which would revolutionize the electronic industry. In the UK, the subject of electronic engineering became distinct from electrical engineering as a university-degree subject around 1960. (Before this time, students of electronics and related subjects like radio and telecommunications had to enroll in the electrical engineering department of the university as no university had departments of electronics. Electrical engineering was the nearest subject with which electronic engineering could be aligned, although the similarities in subjects covered (except mathematics and electromagnetism) lasted only for the first year of three-year courses.)

Electronic engineering (even before it acquired the name) facilitated the development of many technologies including wireless telegraphy, radio, television, radar, computers and microprocessors.

WIRELESS TELEGRAPHY AND RADIO

Some of the devices which would enable wireless telegraphy were invented before 1900. These include the spark-gap transmitter and the coherer with early demonstrations and published findings by David Edward Hughes (1880) and Heinrich Rudolf Hertz (1887 to 1890) and further additions to the field by Édouard Branly, Nikola Tesla, Oliver Lodge, Jagadish Chandra Bose, and Ferdinand Braun. In 1896, Guglielmo Marconi went on to develop a practical and widely used radio system.

In 1904, John Ambrose Fleming, the first professor of electrical Engineering at University College London, invented the first radio tube, the diode. Then, in 1906, Robert von Lieben and Lee De Forest independently developed the amplifier tube, called the triode. Electronics is often considered to have begun with the invention of the triode. Within 10 years, the device was used in radio transmitters and receivers as well as systems for long distance telephone calls.

The invention of the triode amplifier, generator, and detector made audio communication by radio practical. (Reginald Fessenden's 1906 transmissions used an electro-mechanical alternator.) In 1912, Edwin H. Armstrong invented the regenerative feedback amplifier and oscillator; he also invented the superheterodyne radio receiver and could be considered the father of modern radio.

The first known radio news program was broadcast 31 August 1920 by station 8MK, the unlicensed predecessor of WWJ (AM) in Detroit, Michigan. Regular wireless broadcasts for entertainment commenced in 1922 from the Marconi Research Centre at Writtle near Chelmsford, England. The station was known as 2MT and was followed by 2LO, broadcasting from Strand, London.

While some early radios used some type of amplification through electric current or battery, through the mid-1920s the most common type of receiver was the crystal set. In the 1920s, amplifying vacuum tubes revolutionized both radio receivers and transmitters.

Vacuum tubes remained the preferred amplifying device for 40 years, until researchers working for William Shockley at Bell Labs invented the transistor in 1947. In the following years, transistors made small portable radios, or transistor radios, possible as well as allowing more powerful mainframe computers to be built. Transistors were smaller and required lower voltages than vacuum tubes to work.

Before the invention of the integrated circuit in 1959, electronic circuits were constructed from discrete components that could be manipulated by hand. These non-integrated circuits consumed much space and power, were prone to failure and were limited in speed although they are still common in simple applications. By contrast, integrated circuits packed a large number — often millions — of tiny electrical components, mainly transistors, into a small chip around the size of a coin.

TELEVISION

In 1928 Philo Farnsworth made the first public demonstration of a purely electronic television. During the 1930s several countries began broadcasting, and after World War II it spread to millions of receivers, eventually worldwide. Ever since then, electronics have been fully present in television devices.

Modern televisions and video displays have evolved from bulky electron tube technology to use more compact devices, such as plasma and Liquid-crystal displays. The trend is for even lower power devices such as the organic light-emitting diode displays, and it is most likely to replace the LCD and plasma technologies.

RADAR AND RADIO LOCATION

History of radar

A Chain Home transmitter antenna, part of one of the first comprehensive radar systems.

The German *Freya* worked at higher frequencies, and was thus smaller than its Chain Home counterpart.

The anode block of the original cavity magnetron built by Randal and Boot, which provided a leap forward in radar design.

The **history of radar** starts with experiments by Heinrich Hertz in the late 19th century that showed that radio waves were reflected by metallic objects. This possibility was suggested in James Clerk Maxwell's seminal work on electromagnetism. However, it was not until the early 20th century that systems able to use these principles were becoming widely available, and it was German inventor Christian Hülsmeyer who first used them to build a simple ship detection device intended to help avoid collisions in fog (Reichspatent Nr. 165546). Numerous similar systems, which provided directional information to objects over short ranges, were developed over the next two decades.

The development of systems able to produce short pulses of radio energy was the key advance that allowed modern radar systems to come into existence. By timing the pulses on an oscilloscope, the range could be determined and the direction of the antenna revealed the angular location of the targets. The two, combined, produced a "fix", locating the target relative to the antenna. In the 1934–1939 period, eight nations developed independently, and in great secrecy, systems of this type: the United Kingdom, Germany, the United States, the USSR, Japan, the Netherlands, France, and Italy. In addition, Britain shared their information with the United States and four Commonwealth countries: Australia, Canada, New Zealand, and South Africa, and these countries also developed their own radar systems. During the war, Hungary was added to this list. The term *RADAR* was coined in 1939 by the United States Signal Corps as it worked on these systems for the Navy.

Progress during the war was rapid and of great importance, probably one of the decisive factors for the victory of the Allies. A key development was the magnetron in the UK, which allowed the creation of relatively small systems with sub-meter resolution. By the end of hostilities, Britain, Germany, the United States, the USSR, and Japan had a wide diversity of land- and sea-based radars as well as small airborne systems. After the war, radar use was widened to numerous fields including: civil aviation, marine navigation, radar guns for police, meteorology and even medicine. Key developments in the post-war period include the travelling wave tube as a way to produce large quantities of coherent microwaves, the development of signal delay systems that led to phased array radars, and ever-increasing frequencies that allow higher resolutions. Increases in signal processing capability due to the introduction of solid state computers has also had a large impact on radar use.

Significance

The place of radar in the larger story of science and technology is argued differently by different authors. On the one hand, radar contributed very little to theory, which was largely known since the days of Maxwell and Hertz. Therefore, radar did not advance science, but was simply a matter of technology and engineering. Maurice Ponte, one of the developers of radar in France, states:

The fundamental principle of the radar belongs to the common patrimony of the physicists; after all, what is left to the real credit of the technicians is measured by the effective realisation of operational materials.

But others point out the immense practical consequences of the development of radar. Far more than the atomic bomb, radar contributed to the Allied victory in World War II. Robert Buderi states that it was also the precursor of much modern technology. From a review of his book:

... radar has been the root of a wide range of achievements since the war, producing a veritable family tree of modern technologies. Because of radar, astronomers can map the contours of far-off planets, physicians can see images of internal organs, meteorologists can measure rain falling in distant places, air travel is hundreds of times safer than travel by road, long-distance telephone calls are cheaper than postage, computers have become ubiquitous and ordinary people can cook their daily dinners in the time between sitcoms, with what used to be called a *radar range.*

In later years radar was used in scientific instruments, such as weather radar and radar astronomy.

Early Contributors

Heinrich Hertz

In 1886-1888 the German physicist Heinrich Hertz conducted his series of experiments that proved the existence of electromagnetic waves (including radio waves), predicted in equations developed in 1862–4 by the Scottish physicist James Clerk Maxwell. In Hertz's

1887 experiment he found that these waves would transmit through different types of materials and also would reflect off metal surfaces in his lab as well as conductors and dielectrics. The nature of these waves being similar to visible light in their ability to be reflected, refracted, and polarized would be shown by Hertz and subsequent experiments by other physicists.

Guglielmo Marconi

Radio pioneer Guglielmo Marconi noticed radio waves were being reflected back to the transmitter by objects in radio beacon experiments he conducted on March 3, 1899 in Salisbury Plain. He would relate his findings 23 years later in a 1922 paper delivered before the Institution of Electrical Engineers in London:

I also described tests carried out in transmitting a beam of reflected waves across country ... and pointed out the possibility of the utility of such a system if applied to lighthouses and lightships, so as to enable vessels in foggy weather to locate dangerous points around the coasts ...

It [now] seems to me that it should be possible to design [an] apparatus by means of which a ship could radiate or project a divergent beam of these rays in any desired direction, which rays, if coming across a metallic object, such as another steamer or ship, would be reflected back to a receiver screened from the local transmitter on the sending ship, and thereby immediately reveal the presence and bearing of the other ship in fog or thick weather.

Christian Hülsmeyer

In 1904, Christian Hülsmeyer gave public demonstrations in Germany and the Netherlands of the use of radio echoes to detect ships so that collisions could be avoided. His device consisted of a simple spark gap used to generate a signal that was aimed using a dipole antenna with a cylindrical parabolic reflector. When a signal reflected from a ship was picked up by a similar antenna attached to the separate coherer receiver, a bell sounded. During bad weather or fog, the device would be periodically spun to check for nearby ships. The apparatus detected the presence of ships up to 3 kilometres (1.6 nmi), and Hülsmeyer planned to extend its capability to 10 kilometres (5.4 nmi). It did not provide range (distance) information, only warning of a nearby object. He patented the device, called the *telemobiloscope*, but due to lack of interest by the naval authorities the invention was not put into production.

Hülsmeyer also received a patent amendment for estimating the range to the ship. Using a vertical scan of the horizon with the *telemobiloscope* mounted on a tower, the operator would find the angle at which the return was the most intense and deduce, by simple triangulation, the approximate distance. This is in contrast to the later development of pulsed radar, which determines distance via two-way travel time.

United Kingdom

Robert Watson-Watt

In 1915, Robert Watson Watt joined the Meteorological Office as a meteorologist, working at an outstation at Aldershot in Hampshire. Over the next 20 years, he studied atmospheric phenomena and developed the use of radio signals generated by lightning strikes to map out the position of thunderstorms. The difficulty in pinpointing the direction of these fleeting signals using rotatable directional antennas led, in 1923, to the use of oscilloscopes in order to display the signals. The operation eventually moved to the outskirts of Slough in Berkshire, and in 1927 formed the Radio Research Station (RRS), Slough, an entity under the Department of Scientific and Industrial Research (DSIR). Watson Watt was appointed the RRS Superintendent.

As war clouds gathered over Britain, the likelihood of air raids and the threat of invasion by air and sea drove a major effort in applying science and technology to defence. In November 1934, the Air Ministry established the Committee for the Scientific Survey of Air Defence (CSSAD) with the official function of considering "how far recent advances in scientific and technical knowledge can be used to strengthen the present methods of defence against hostile aircraft". Commonly called the "Tizard Committee" after its Chairman, Sir Henry Tizard, this group had a profound influence on technical developments in Britain.

H. E. Wimperis, Director of Scientific Research at the Air Ministry and a member of the Tizard Committee, had read about a German newspaper article claiming that the Germans had built a death ray using radio signals, accompanied by an image of a very large radio antenna. Both concerned and potentially excited by this possibility, but highly skeptical at the same time, Wimperis looked for an expert in the field of radio propagation who might be able to pass judgement on the concept. Watt, Superintendent of the RRS, was now well established as an authority in the field of radio, and in January 1935, Wimperis contacted him asking if radio might be used for such a device. After discussing this with his scientific assistant, Arnold F. ‹Skip' Wilkins, Wilkins quickly produced a back-of-the-envelope calculation that showed the energy required would be enormous. Watt wrote back that this was unlikely, but added the following comment: "Attention is being turned to the still difficult, but less unpromising, problem of radio detection and numerical considerations on the method of detection by reflected radio waves will be submitted when required".

Over the following several weeks, Wilkins considered the radio detection problem. He outlined an approach and backed it with detailed calculations of necessary transmitter power, reflection characteristics of an aircraft, and needed receiver sensitivity. He proposed using a directional receiver based on Watt's lightning detection concept, listening for powerful signals from a separate transmitter. Timing, and thus distance measurements, would be accomplished by triggering the oscilloscope's trace with a muted signal from the transmitter, and then simply measuring the returns against a scale. Watson Watt sent this information to the Air Ministry on February 12, 1935, in a secret report titled "The Detection of Aircraft by Radio Methods".

Reflection of radio signals was critical to the proposed technique, and the Air Ministry asked if this could be proven. To test this, Wilkins set up receiving equipment in a field near Upper Stowe, Northamptonshire. On February 26, 1935, a Handley Page Heyford bomber flew along a path between the receiving station and the transmitting towers of a BBC shortwave station in nearby Daventry. The aircraft reflected the 6 MHz (49 m) BBC signal, and this was readily detected by Arnold "Skip" Wilkins using Doppler-beat interference at ranges up to 8 mi (13 km). This convincing test, known as the *Daventry Experiment*, was witnessed by a representative from the Air Ministry, and led to the immediate authorization to build a full demonstration system. This experiment was later reproduced by Wilkins for the 1977 BBC television series *The Secret War* episode "To See a Hundred Miles".

Based on pulsed transmission as used for probing the ionosphere, a preliminary system was designed and built at the RRS by the team. Their existing transmitter had a peak power of about 1 kW, and Wilkins had estimated that 100 kW would be needed. Edward George Bowen was added to the team to design and build such a transmitter. Bowens' transmitter operated at 6 MHz (50 m), had a pulse-repetition rate of 25 Hz, a pulse width of 25 μs, and approached the desired power.

Orfordness, a narrow 19-mile (31 km) peninsula in Suffolk along the coast of the North Sea, was selected as the test site. Here the equipment would be openly operated in the guise of an ionospheric monitoring station. In mid-May 1935, the equipment was moved to Orfordness. Six wooden towers were erected, two for stringing the transmitting antenna, and four for corners of crossed receiving antennas. In June, general testing of the equipment began.

On June 17, the first target was detected—a Supermarine Scapa flying boat at 17 mi (27 km) range. It is historically correct that, on June 17, 1935, radio-based detection and ranging was first demonstrated in Britain. Watson Watt, Wilkins, and Bowen are generally credited with initiating what would later be called radar in this nation.

In December 1935, the British Treasury appropriated £60,000 for a five-station system called Chain Home (CH), covering approaches to the Thames Estuary. The secretary of the Tizard Committee, Albert Percival Rowe, coined the acronym RDF as a cover for the work, meaning Range and Direction Finding but suggesting the already well-known Radio Direction Finding. Late in 1935, responding to Lindemann's recognition of the need for night detection and interception gear, and realizing existing transmitters were too heavy for aircraft, Bowen proposed fitting only receivers, what would later be called bistatic radar. Frederick Lindemann's proposals for infrared sensors and aerial mines would prove impractical. It would take Bowen's efforts, at the urging of Tizard, who became increasingly concerned about the need, to see Air to Surface Vessel (ASV), and through it Airborne Interception (AI), radar to fruition.

In 1937, Bowen's team set their crude ASV radar, the world's first airborne set, to detect the Home Fleet in dismal weather.Only in spring 1939, "as a matter of great urgency" after the failure of the searchlight system Silhouette, did attention turn to using ASV for air-to-air interception (AI). Demonstrated in June 1939, AI got a warm reception from Air Chief Marshal Hugh Dowding, and even more so from Churchill. This proved problematic. Its accuracy, dependent on the height of the aircraft, meant that CH, capable of only 4 sm (0.0068 km), was not accurate enough to place an aircraft within its detection range, and an additional system was required. Its wooden chassis had a disturbing tendency to catch fire (even with attention from expert technicians), so much so that Dowding, when told that Watson-Watt could provide hundreds of sets, demanded "ten that work". The Cossor and MetroVick sets were overweight for aircraft use and the RAF lacked night fighter pilots, observers, and suitable aircraft.

In 1940, John Randall and Harry Boot developed the cavity magnetron, which made ten-centimetre radar a reality. This device, the size of a small dinner plate, could be carried easily on aircraft and the short wavelength meant the antenna would also be small and hence suitable for mounting on aircraft. The short wavelength and high power made it very effective at spotting submarines from the air.

To aid Chain Home in making height calculations, at Dowding's request, the Electrical Calculator Type Q (commonly called the "Fruit Machine") was introduced in 1940.

The solution to night intercepts would be provided by Dr. W. B. "Ben" Lewis, who proposed a new, more accurate ground control display, the Plan Position Indicator (PPI), a new Ground-Controlled Interception (GCI) radar, and reliable AI radar.The AI sets would ultimately be built by EMI. GCI was unquestionably delayed by Watson-Watt's opposition to it and his belief that CH was sufficient, as well as by Bowen's preference for using ASV for navigation, despite Bomber Command disclaiming a need for it, and by Tizard's reliance on the faulty Silhouette system.

Air Ministry

In March 1936, the work at Orfordness was moved to Bawdsey Manor, nearby on the mainland. Until this time, the work had officially still been under the DSIR, but was now transferred to the Air Ministry. At the new Bawdsey Research Station, the Chain Home (CH) equipment was assembled as a prototype. There were equipment problems when the Royal Air Force (RAF) first exercised the prototype station in September 1936. These were cleared by the next April, and the Air Ministry started plans for a larger network of stations.

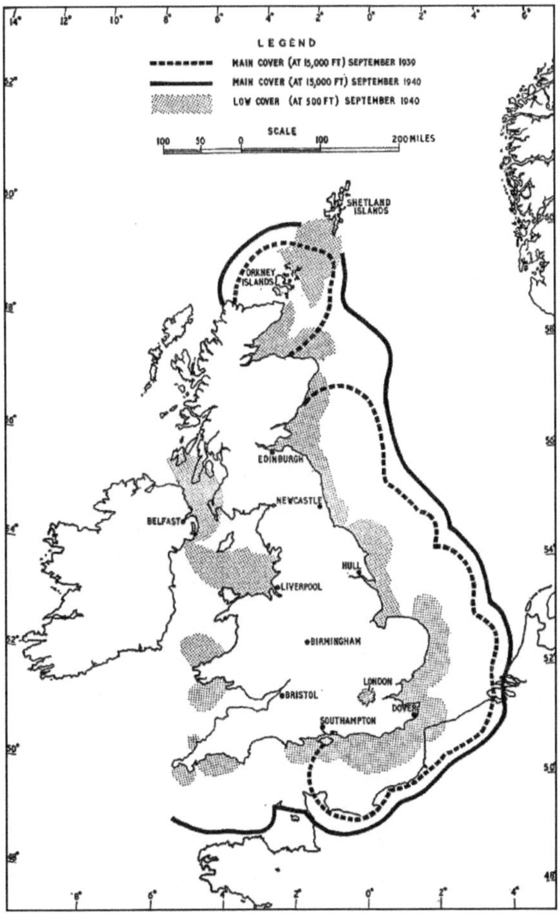

Chain Home Radar Coverage 1939–1940.

Initial hardware at CH stations was as follows: The transmitter operated on four pre-selected frequencies between 20 and 55 MHz, adjustable within 15 seconds, and delivered a peak power of 200 kW. The pulse duration was adjustable between 5 and 25 μs, with a repetition rate selectable as either 25 or 50 Hz. For synchronization of all CH transmitters, the pulse generator was locked to the 50 Hz of the British power grid. Four 360-foot (110 m) steel towers supported transmitting antennas, and four 240-foot (73 m) wooden towers supported cross-dipole arrays at three different levels. A goniometer was used to improve the directional accuracy from the multiple receiving antennas.

By the summer of 1937, 20 initial CH stations were in check-out operation. A major RAF exercise was performed before the end of the year, and was such a success that £10,000,000 was appropriated by the Treasury for an eventual full chain of coastal stations. At the start of 1938, the RAF took over control of all CH stations, and the network began regular operations.

In May 1938, Rowe replaced Watson Watt as Superintendent at Bawdsey. In addition to the work on CH and successor systems, there was now major work in airborne RDF equipment. This was led by E. G. Bowen and centered on 200-MHz (1.5 m) sets. The higher frequency allowed smaller antennas, appropriate for aircraft installation.

From the initiation of RDF work at Orfordness, the Air Ministry had kept the British Army and the Royal Navy generally informed; this led to both of these forces having their own RDF developments.

British Army

In 1931, at the Woolwich Research Station of the Army's Signals Experimental Establishment (SEE), W. A. S. Butement and P. E. Pollard had examined pulsed 600 MHz (50-cm) signals for detection of ships. Although they prepared a memorandum on this subject and performed preliminary experiments, for undefined reasons the War Office did not give it consideration.

As the Air Ministry's work on RDF progressed, Colonel Peter Worlledge of the Royal Engineer and Signals Board met with Watson Watt and was briefed on the RDF equipment and techniques being developed at Orfordness. His report, "The Proposed Method of Aeroplane Detection and Its Prospects", led the SEE to set up an "Army Cell" at Bawdsey in October 1936. This was under E. Talbot Paris and the staff included Butement and Pollard. The Cell's work emphasize two general types of RDF equipment: gun-laying (GL) systems for assisting anti-aircraft guns and searchlights, and coastal-defense (CD) systems for directing coastal artillery and defense of Army bases overseas.

Pollard led the first project, a gun-laying RDF code-named Mobile Radio Unit (MRU). This truck-mounted system was designed as a small version of a CH station. It operated at 23 MHz (13 m) with a power of 300 kW. A single 105-foot (32 m) tower supported a transmitting antenna, as well as two receiving antennas set orthogonally for estimating the signal bearing. In February 1937, a developmental unit detected an aircraft at a range of 60 miles (96 km). The Air Ministry also adopted this system as a mobile auxiliary to the CH system.

In early 1938, Butement started the development of a CD system based on Bowen's evolving 200-MHz (1.5-m) airborne sets. The transmitter had a 400 Hz pulse rate, a 2-μs pulse width, and 50 kW power (later increased to 150 kW). Although many of Bowen's transmitter and receiver components were used, the system would not be airborne so there were no limitations on antenna size.

Primary credit for introducing beamed RDF systems in Britain must be given to Butement. For the CD, he developed a large dipole array, 10 feet (3.0 m) high and 24 feet (7.3 m) wide, giving much narrower beams and higher gain. This could be rotated at a speed up to 1.5 revolutions per minute. For greater directional accuracy, lobe switching on the receiving antennas was adopted. As a part of this development, he formulated the first – at least in Britain – mathematical relationship that would later become well known as the "radar range equation".

By May 1939, the CD RDF could detect aircraft flying as low as 500 feet (150 m) and at a range of 25 mi (40 km). With an antenna 60 feet (18 m) above sea level, it could determine the range of a 2,000-ton ship at 24 mi (39 km) and with an angular accuracy of as little as a quarter of a degree.

Royal Navy

Although the Royal Navy maintained close contact with the Air Ministry work at Bawdsey, they chose to establish their own RDF development at the Experimental Department of His Majesty's Signal School (HMSS) in Portsmouth, Hampshire, on the south coast.

HMSS started RDF work in September 1935. Initial efforts, under R. F. Yeo, were in wavelengths ranging between 75 MHz (4 m) and 1.2 GHz (25 cm). All of the work was under the utmost secrecy; it could not even be discussed with other scientists and engineers at Portsmouth. A 75 MHz range-only set was eventually developed and designated Type 79X. Basic tests were done using a training ship, but the operation was unsatisfactory.

In August 1937, the RDF development at HMSS changed, with many of their best researchers brought into the activity. John D. S. Rawlinson was made responsible for improving the Type 79X. To increase the efficiency, he decreased the frequency to 43 MHz (7 m). Designated Type 79Y, it had separate, stationary transmitting and receiving antennas.

Prototypes of the Type 79Y air-warning system were successfully tested at sea in early 1938. The detection range on aircraft was between 30 and 50 mi (48 and 80 km), depending on height. The systems were then placed into service in August on the cruiser HMS *Sheffield* and in October on the battleship HMS *Rodney*. These were the first vessels in the Royal Navy with RDF systems.

Germany

A radio-based device for remotely indicating the presence of ships was built in Germany by Christian Hülsmeyer in 1904. Often referred to as the first radar system, this did not directly measure the range (distance) to the target, and thus did not meet the criteria to be given this name.

Over the following three decades in Germany, a number of radio-based detection systems were developed but none were true radars. This situation changed before World War II. Developments in three leading industries are described.

GEMA

In the early 1930s, physicist Rudolf Kühnhold, Scientific Director at the *Kriegsmarine* (German navy) *Nachrichtenmittel-Versuchsanstalt* (NVA—Experimental Institute of Communication Systems) in Kiel, was attempting to improve the acoustical methods of underwater detection of ships. He concluded that the desired accuracy in measuring distance to targets could be attained only by using pulsed electromagnetic waves.

During 1933, Kühnhold first attempted to test this concept with a transmitting and receiving set that operated in the microwave region at 13.5 cm (2.22 GHz). The transmitter used a Barkhausen-Kurz tube (the first microwave generator) that produced only 0.1 watt. Unsuccessful with this, he asked for assistance from Paul-Günther Erbslöh and Hans-Karl Freiherr von Willisen, amateur radio operators who were developing a VHF system for communications. They enthusiastically agreed, and in January 1934, formed a company,

Gesellschaft für Elektroakustische und Mechanische Apparate (GEMA), for the effort. From the start, the firm was always called simply GEMA.

Work on a *Funkmessgerät für Untersuchung* (radio measuring device for research) began in earnest at GEMA. Hans Hollmann and Theodor Schultes, both affiliated with the prestigious Heinrich Hertz Institute in Berlin, were added as consultants. The first apparatus used a split-anode magnetron purchased from Philips in the Netherlands. This provided about 70 W at 50 cm (600 MHz), but suffered from frequency instability. Hollmann built a regenerative receiver and Schultes developed Yagi antennas for transmitting and receiving. In June 1934, large vessels passing through the Kiel Harbor were detected by Doppler-beat interference at a distance of about 2 km (1.2 mi). In October, strong reflections were observed from an aircraft that happened to fly through the beam; this opened consideration of targets other than ships.

Kühnhold then shifted the GEMA work to a pulse-modulated system. A new 50 cm (600 MHz) Philips magnetron with better frequency stability was used. It was modulated with 2- μs pulses at a PRF of 2000 Hz. The transmitting antenna was an array of 10 pairs of dipoles with a reflecting mesh. The wide-band regenerative receiver used Acorn tubes from RCA, and the receiving antenna had three pairs of dipoles and incorporated lobe switching. A blocking device (a duplexer), shut the receiver input when the transmitter pulsed. A Braun tube (a CRT) was used for displaying the range.

The equipment was first tested at a NVA site at the Lübecker Bay near Pelzerhaken. During May 1935, it detected returns from woods across the bay at a range of 15 km (9.3 mi). It had limited success, however, in detecting a research ship, *Welle*, only a short distance away. The receiver was then rebuilt, becoming a super-regenerative set with two intermediate-frequency stages. With this improved receiver, the system readily tracked vessels at up to 8 km (5.0 mi) range.

In September 1935, a demonstration was given to the Commander-in-Chief of the *Kriegsmarine*. The system performance was excellent; the range was read off the Braun tube with a tolerance of 50 meters (less than 1 percent variance), and the lobe switching allowed a directional accuracy of 0.1 degree. Historically, this marked the first naval vessel equipped with radar. Although this apparatus was not put into production, GEMA was funded to develop similar systems operating around 50 cm (500 MHz). These became the Seetakt for the *Kriegsmarine* and the Freya for the *Luftwaffe* (German Air Force).

Kühnhold remained with the NVA, but also consulted with GEMA. He is considered by many in Germany as the Father of Radar. During 1933-6, Hollmann wrote the first comprehensive treatise on microwaves, *Physik und Technik der ultrakurzen Wellen* (Physics and Technique of Ultrashort Waves), Springer 1938.

Telefunken

In 1933, when Kühnhold at the NVA was first experimenting with microwaves, he had sought information from Telefunken on microwave tubes. (Telefunken was the largest supplier of radio products in Germany) There, Wilhelm Tolmé Runge had told him that no vacuum tubes were available for these frequencies. In fact, Runge was already experimenting with high-frequency transmitters and had Telefunken's tube department working on cm-wavelength devices.

In the summer of 1935, Runge, now Director of Telefunken's Radio Research Laboratory, initiated an internally funded project in radio-based detection. Using Barkhausen-Kurz tubes, a 50 cm (600 MHz) receiver and 0.5-W transmitter were built. With the antennas placed flat on the ground some distance apart, Runge arranged for an aircraft to fly overhead and found that the receiver gave a strong Doppler-beat interference signal.

Runge, now with Hans Hollmann as a consultant, continued in developing a 1.8 m (170 MHz) system using pulse-modulation. Wilhelm Stepp developed a transmit-receive device (a duplexer) for allowing a common antenna. Stepp also code-named the system *Darmstadt* after his home town, starting the practice in Telefunken of giving the systems names of cities. The system, with only a few watts transmitter power, was first tested in February 1936, detecting an aircraft at about 5 km (3.1 mi) distance. This led the *Luftwaffe* to fund the development of a 50 cm (600 MHz) gun-laying system, the *Würzburg*.

Lorenz

Since before the First World War, Standard Elektrik Lorenz had been the main supplier of communication equipment for the German military and was the main rival of Telefunken. In late 1935, when Lorenz found that Runge at Telefunken was doing research in radio-based detection equipment, they started a similar activity under Gottfried Müller. A pulse-modulated set called *Einheit für Abfragung* (DFA - Device for Detection) was built. It used a type DS-310 tube (similar to the Acorn) operating at 70 cm (430 MHz) and about 1 kW power, it had identical transmitting and receiving antennas made with rows of half-wavelength dipoles backed by a reflecting screen.

In early 1936, initial experiments gave reflections from large buildings at up to about 7 km (4.3 mi). The power was doubled by using two tubes, and in mid-1936, the equipment was set up on cliffs near Kiel, and good detections of ships at 7 km (4.3 mi) and aircraft at 4 km (2.5 mi) were attained. The success of this experimental set was reported to the *Kriegsmarine*, but they showed no interest; they were already fully engaged with GEMA for similar equipment. Also, because of extensive agreements between Lorenz and many foreign countries, the naval authorities had reservations concerning the company handling classified work. The DFA was then demonstrated to the *Heer* (German Army), and they contracted with Lorenz for developing *Kurfürst* (Elector), a system for supporting *Flugzeugabwehrkanone* (Flak, anti-aircraft guns).

United States

In the United States, both the Navy and Army needed means of remotely locating enemy ships and aircraft. In 1930, both services initiated the development of radio equipment that could meet this need. There was little coordination of these efforts; thus, they will be described separately.

United States Navy

In the autumn of 1922, Albert H. Taylor and Leo C. Young at the U.S. Naval Aircraft Radio Laboratory were conducting communication experiments when they noticed that a wooden ship in the Potomac River was interfering with their signals. They prepared a memorandum suggesting that this might be used for ship detection in a harbor defense, but their suggestion was not taken up. In 1930, Lawrence A. Hyland working with Taylor and Young, now at the U.S. Naval Research Laboratory (NRL) in Washington, D.C., used a similar arrangement of radio equipment to detect a passing aircraft. This led to a proposal and patent for using this technique for detecting ships and aircraft.

A simple wave-interference apparatus can detect the presence of an object, but it cannot determine its location or velocity. That had to await the invention of pulsed radar, and later, additional encoding techniques to extract this information from a CW signal. When Taylor's group at the NRL were unsuccessful in getting interference radio accepted as a detection means, Young suggested trying pulsing techniques. This would also allow the direct determination of range to the target. In 1924, Hyland and Young had built such a transmitter for Gregory Breit and Merle A. Tuve at the Carnegie Institution of Washington for successfully measuring the height of the ionosphere.

Robert Morris Page was assigned by Taylor to implement Young's suggestion. Page designed a transmitter operating at 60 MHz and pulsed 10 μs in duration and 90 μs between pulses. In December 1934, the apparatus was used to detect a plane at a distance of one mile (1.6 km) flying up and down the Potomac. Although the detection range was small and the indications on the oscilloscope monitor were almost indistinct, it demonstrated the basic concept of a pulsed radar system. Based on this, Page, Taylor, and Young are usually credited with building and demonstrating the world's first true radar.

An important subsequent development by Page was the duplexer, a device that allowed the transmitter and receiver to use the same antenna without overwhelming or destroying the sensitive receiver circuitry. This also solved the problem associated with synchronization of separate transmitter and receiver antennas which is critical to accurate position determination of long-range targets.

The experiments with pulsed radar were continued, primarily in improving the receiver for handling the short pulses. In June 1936, the NRL's first prototype radar system, now operating at 28.6 MHz, was demonstrated to government officials, successfully tracking an aircraft at distances up to 25 miles (40 km). Their radar was based on low frequency signals, at least by today's standards, and thus required large antennas, making it impractical for ship or aircraft mounting.

Antenna size is inversely proportional to the operating frequency; therefore, the operating frequency of the system was increased to 200 MHz, allowing much smaller antennas. The frequency of 200 MHz was the highest possible with existing transmitter tubes and other components. The new system was successfully tested at the NRL in April 1937, That same month, the first sea-borne testing was conducted. The equipment was temporarily installed on the USS *Leary*, with a Yagi antenna mounted on a gun barrel for sweeping the field of view.

Based on success of the sea trials, the NRL further improved the system. Page developed the ring oscillator, allowing multiple output tubes and increasing the pulse-power to 15 kW in 5-μs pulses. A 20-by-23 ft (6 x 7 m), stacked-dipole "bedspring" antenna was used. In laboratory test during 1938, the system, now designated XAF, detected planes at ranges up to 100 miles (160 km). It was installed on the battleship USS *New York* for sea trials starting in January 1939, and became the first operational radio detection and ranging set in the U.S. fleet.

In May 1939, a contract was awarded to RCA for production. Designated CXAM, deliveries started in May 1940. The acronym RADAR was coined from "Radio Detection And Ranging". One of the first CXAM systems was placed aboard the USS *California*, a battleship that was sunk in the Japanese attack on Pearl Harbor on December 7, 1941.

United States Army

As the Great Depression started, economic conditions led the U.S. Army Signal Corps to consolidate its widespread laboratory operations to Fort Monmouth, New Jersey. On June 30, 1930, these were designated the Signal Corps Laboratories (SCL) and Lt. Colonel (Dr.) William R. Blair was appointed the SCL Director.

Among other activities, the SCL was made responsible for research in the detection of aircraft by acoustical and infrared radiation means. Blair had performed his doctoral research in the interaction of electromagnet waves with solid materials, and naturally gave attention to this type of detection. Initially, attempts were made to detect infrared radiation, either from the heat of aircraft engines or as reflected from large searchlights with infrared filters, as well as from radio signals generated by the engine ignition.

Some success was made in the infrared detection, but little was accomplished using radio. In 1932, progress at the Naval Research Laboratory (NRL) on radio interference for aircraft detection was passed on to the Army. While it does not appear that any of this information was used by Blair, the SCL did undertake a systematic survey of what was then known throughout the world about the methods of generating, modulating, and detecting radio signals in the microwave region.

The SCL's first definitive efforts in radio-based target detection started in 1934 when the Chief of the Army Signal Corps, after seeing a microwave demonstration by RCA, suggested that radio-echo techniques be investigated. The SCL called this technique radio position-finding (RPF). Based on the previous investigations, the SCL first tried microwaves. During 1934 and 1935, tests of microwave RPF equipment resulted in Doppler-shifted signals being obtained, initially at only a few hundred feet distance and later greater than a mile. These tests involved a bi-static arrangement, with the transmitter at one end of the signal path and the receiver at the other, and the reflecting target passing through or near the path.

Blair was evidently not aware of the success of a pulsed system at the NRL in December 1934. In an internal 1935 note, Blair had commented:

Consideration is now being given to the scheme of projecting an interrupted sequence of trains of oscillations against the target and attempting to detect the echoes during the interstices between the projections.

In 1936, W. Delmar Hershberger, SCL's Chief Engineer at that time, started a modest project in pulsed microwave transmission. Lacking success with microwaves, Hershberger visited the NRL (where he had earlier worked) and saw a demonstration of their pulsed set. Back at the SCL, he and Robert H. Noyes built an experimental apparatus using a

75 watt, 110 MHz (2.73 m) transmitter with pulse modulation and a receiver patterned on the one at the NRL. A request for project funding was turned down by the War Department, but $75,000 for support was diverted from a previous appropriation for a communication project.

In October 1936, Paul E. Watson became the SCL Chief Engineer and led the project. A field setup near the coast was made with the transmitter and receiver separated by a mile. On December 14, 1936, the experimental set detected at up to 7 mi (11 km) range aircraft flying in and out of New York City.

Work then began on a prototype system. Ralph I. Cole headed receiver work and William S. Marks lead transmitter improvements. Separate receivers and antennas were used for azimuth and elevation detection. Both receiving and the transmitting antennas used large arrays of dipole wires on wooden frames. The system output was intended to aim a searchlight. The first demonstration of the full set was made on the night of May 26, 1937. A bomber was detected and then illuminated by the searchlight. The observers included the Secretary of War, Henry A. Woodring; he was so impressed that the next day orders were given for the full development of the system. Congress gave an appropriation of $250,000.

The frequency was increased to 200 MHz (1.5 m). The transmitter used 16 tubes in a ring oscillator circuit (developed at the NRL), producing about 75 kW peak power. Major James C. Moore was assigned to head the complex electrical and mechanical design of lobe switching antennas. Engineers from Western Electric and Westinghouse were brought in to assist in the overall development. Designated SCR-268, a prototype was successfully demonstrated in late 1938 at Fort Monroe, Virginia. The production of SCR-268 sets was started by Western Electric in 1939, and it entered service in early 1941.

Even before the SCR-268 entered service, it had been greatly improved. In a project led by Major (Dr.) Harold A. Zahl, two new configurations evolved – the SCR-270 (mobile) and the SCR-271 (fixed-site). Operation at 106 MHz (2.83 m) was selected, and a single water-cooled tube provided 8 kW (100 kW pulsed) output power. Westinghouse received a production contract, and started deliveries near the end of 1940.

The Army deployed five of the first SCR-270 sets around the island of Oahu in Hawaii. At 7:02 on the morning of December 7, 1941, one of these radars detected a flight of aircraft at a range of 136 miles (219 km) due north. The observation was passed on to an aircraft warning center where it was misidentified as a flight of U.S. bombers known to be approaching from the mainland. The alarm went unheeded, and at 7:48, the Japanese aircraft first struck at Pearl Harbor.

USSR

In 1895, Alexander Stepanovich Popov, a physics instructor at the Imperial Russian Navy school in Kronstadt, developed an apparatus using a coherer tube for detecting distant lightning strikes. The next year, he added a spark-gap transmitter and demonstrated the first radio communication set in Russia. During 1897, while testing this in communicating between two ships in the Baltic Sea, he took note of an interference beat caused by the passage of a third vessel. In his report, Popov wrote that this phenomenon might be used for detecting objects, but he did nothing more with this observation.

In a few years following the 1917 Russian Revolution and the establishment the Union of Soviet Socialist Republics (USSR or Soviet Union) in 1924, Germany's *Luftwaffe* had aircraft capable of penetrating deep into Soviet territory. Thus, the detection of aircraft at night or above clouds was of great interest to the Soviet Air Defense Forces (PVO).

The PVO depended on optical devices for locating targets, and had physicist Pavel K. Oshchepkov conducting research in possible improvement of these devices. In June 1933, Oshchepkov changed his research from optics to radio techniques and started the development of a *razvedyvlatl'naya elektromagnitnaya stantsiya* (reconnaissance electromagnetic station). In a short time, Oshchepkov was made responsible for a technical expertise sector of PVO devoted to *radiolokatory* (radio-location) techniques as well as heading a Special Design Bureau (SKB, spetsialnoe konstruktorskoe byuro) in Leningrad.

Radio-location Beginnings

The *Glavnoe Artilleriyskoe Upravlenie* (GAU, Main Artillery Administration) was considered the "brains" of the Red Army. It not only had competent engineers and physicists on its central staff, but also had a number of scientific research institutes. Thus, the GAU was also assigned the aircraft detection problem, and Lt. Gen. M. M. Lobanov was placed in charge.

After examining existing optical and acoustical equipment, Lobanov also turned to radio-location techniques. For this he approached the *Tsentral'naya Radiolaboratoriya* (TsRL, Central Radio Laboratory) in Leningrad. Here, Yu. K. Korovin was conducting research on VHF communications, and had built a 50 cm (600 MHz), 0.2 W transmitter using a Barkhausen-Kurz tube. For testing the concept, Korovin arranged the transmitting and receiving antennas along the flight path of an aircraft. On January 3, 1934, a Doppler signal was received by reflections from the aircraft at some 600 m range and 100–150 m altitude.

For further research in detection methods, a major conference on this subject was arranged for the PVO by the Russian Academy of Sciences (RAN). The conference was held in Leningrad in mid-January 1934, and chaired by Abram Fedorovich Ioffe, Director of the Leningrad Physical-Technical Institute (LPTI). Ioffe was generally considered the top Russian physicist of his time. All types of detection techniques were discussed, but radio-location received the greatest attention.

To distribute the conference findings to a wider audience, the proceedings were published the following month in a journal. This included all of the then-existing information on radio-location in the USSR, available (in Russian language) to researchers in this field throughout the world.

Recognizing the potential value of radio-location to the military, the GAU made a separate agreement with the Leningrad Electro-Physics Institute (LEPI), for a radio-location system. This technical effort was led by B. K. Shembel. The LEPI had built a transmitter and receiver to study the radio-reflection characteristics of various materials and targets. Shemlbel readily made this into an experimental bi-static radio-location system called *Bistro* (Rapid).

The *Bistro* transmitter, operating at 4.7 m (64 MHz), produced near 200 W and was frequency-modulated by a 1 kHz tone. A fixed transmitting antenna gave a broad coverage of what was called a *radioekran* (radio screen). A regenerative receiver, located some distance from the transmitter, had a dipole antenna mounted on a hand-driven reciprocating mechanism. An aircraft passing into the screened zone would reflect the radiation, and the receiver would detect the Doppler-interference beat between the transmitted and reflected signals.

Bistro was first tested during the summer of 1934. With the receiver up to 11 km away from the transmitter, the set could only detect an aircraft entering a screen at about 3 km (1.9 mi) range and under 1,000 m. With improvements, it was believed to have a potential range of 75 km, and five sets were ordered in October for field trials. *Bistro* is often cited as the USSR's first radar system; however, it was incapable of directly measuring range and thus could not be so classified.

LEPI and TsRL were both made a part of *Nauchno-issledovatelsky institut-9* (NII-9, Scientific Research Institute #9), a new GAU organization opened in Leningrad in 1935. Mikhail A. Bonch-Bruyevich, a renowned radio physicist previously with TsRL and the University of Leningrad, was named the NII-9 Scientific Director.

Research on magnetrons began at Kharkov University in Ukraine during the mid-1920s. Before the end of the decade this had resulted in publications with worldwide distribution, such as the German journal *Annalen der Physik* (*Annals of Physics*).Based on this work, Ioffe recommended that a portion of the LEPI be transferred to the city of Kharkov, resulting in the Ukrainian Institute of Physics and Technology (LIPT) being formed in 1930. Within the LIPT, the Laboratory of Electromagnetic Oscillations (LEMO), headed by Abram A. Slutskin, continued with magnetron development. Led by Aleksandr S. Usikov, a number of advanced segmented-anode magnetrons evolved. (It is noted that these and other early magnetrons developed in the USSR suffered from frequency instability, a problem in their use in Soviet radar systems.)

In 1936, one of Usikov's magnetrons producing about 7 W at 18 cm (1.7 GHz) was used by Shembel at the NII-9 as a transmitter in a *radioiskatel* (radio-seeker) called *Burya* (Storm). Operating similarly to *Bistro*, the range of detection was about 10 km, and provided azimuth and elevation coordinates estimated to within 4 degrees. No attempts were made to make this into a pulsed system, thus, it could not provide range and was not qualified to be classified as a radar. It was, however, the first microwave radio-detection system.

While work by Shembel and Bonch-Bruyevich on continuous-wave systems was taking place at NII-9, Oshehepkov at the SKB and V. V. Tsimbalin of Ioffe's LPTI were pursuing a pulsed system. In 1936, they built a radio-location set operating at 4 m (75 MHz) with a peak-power of about 500 W and a 10-μs pulse duration. Before the end of the year, tests using separated transmitting and receiving sites resulted in an aircraft being detected at 7 km. In April 1937, with the peak-pulse power increased to 1 kW and the antenna separation also increased, test showed a detection range of near 17 km at a height of 1.5 km. Although a pulsed system, it was not capable of directly providing range – the technique of using pulses for determining range had not yet been developed.

Pre-war Radio Location Systems

In June 1937, all of the work in Leningrad on radio-location suddenly stopped. The infamous Great Purge of dictator Joseph Stalin swept over the military high commands and its supporting scientific community. The PVO chief was executed. Oshchepkov, charged with "high crime", was sentenced to 10 years at a Gulag penal labor camp. NII-9 as an organization was saved, but Shenbel was dismissed and Bonch-Bruyevich was named the new director.

The *Nauchnoissledovatel'skii ispytalel'nyi institut svyazi RKKA* (NIIIS-KA, Scientific Research Institute of Signals of the Red Army), had initially opposed research in radio-location, favoring instead acoustical techniques. However, this portion of the Red Army gained power as a result of the Great Purge, and did an about face, pressing hard for speedy development of radio-location systems. They took over Oshchepkov's laboratory and were made responsible for all existing and future agreements for research and factory production. Writing later about the Purge and subsequent effects, General Lobanov commented that it led to the development being placed under a single organization, and the rapid reorganization of the work.

At Oshchepkov's former laboratory, work with the 4 m (75 MHz) pulsed-transmission system was continued by A. I. Shestako. Through pulsing, the transmitter produced a peak power of 1 kW, the highest level thus far generated. In July 1938, a fixed-position, bi-static experimental system detected an aircraft at about 30 km range at heights of 500 m, and at 95 km range, for high-flying targets at 7.5 km altitude. The system was still incapable of directly determining the range. The project was then taken up by Ioffe's LPTI, resulting in the development of a mobile system designated *Redut* (Redoubt). An arrangement of new transmitter tubes was used, giving near 50 kW peak-power with a 10 μs pulse-duration. Yagi antennas were adopted for both transmitting and receiving.

The *Redut* was first field tested in October 1939, at a site near Sevastopol, a port in Ukraine on the coast of the Black Sea. This testing was in part to show the NKKF (Soviet Navy) the value of early-warning radio-location for protecting strategic ports. With the equipment on a cliff about 160 meters above sea level, a flying boat was detected at ranges up to 150 km. The Yagi antennas were spaced about 1,000 meters; thus, close coordination was required to aim them in synchronization. An improved version of the Redut, the Redut-K, was developed by Aksel Berg in 1940 and placed aboard the light cruiser *Molotov* in April 1941. *Molotov* became the first Soviet warship equipped with radar.

At the NII-9 under Bonch-Bruyevich, scientists developed two types of very advanced microwave generators. In 1938, a linear-beam, velocity-modulated vacuum tube (a klystron) was developed by Nikolay Devyatkov, based on designs from Kharkpv. This device produced about 25 W at 15–18 cm (2.0–1.7 GHz) and was later used in experimental systems. Devyatkov followed this with a simpler, single-resonator device (a reflex klystron). At this same time, D. E. Malyarov and N. F. Alekseyev were building a series of magnetrons, also based on designs from Kharkov; the best of these produced 300 W at 9 cm (3 GHz).

Also at NII-9, D. S. Stogov was placed in charge of the improvements to the *Bistro* system. Redesignated as *Reven* (Rhubarb), it was tested in August 1938, but was only marginally better than the predecessor. With additional minor operational improvements, it was made into a mobile system called *Radio Ulavlivatel Samoletov* (RUS, Radio Catcher of Aircraft), soon designated as *RUS-1*. This continuous-wave, bi-static system had a truck-mounted transmitter operating at 4.7 m (64 MHz) and two truck-mounted receivers.

Although the *RUS-1* transmitter was in a cabin on the rear of a truck, the antenna had to be strung between external poles anchored to the ground. A second truck carrying the electrical generator and other equipment was backed against the transmitter truck. Two receivers were used, each in a truck-mounted cabin with a dipole antenna on a rotatable pole extended overhead. In use, the receiver trucks were placed about 40 km apart; thus, with two positions, it would be possible to make a rough estimate of the range by triangulation on a map.

The *RUS-1* system was tested and put into production in 1939, then entered service in 1940, becoming the first deployed radio-location system in the Red Army. About 45 *RUS-1* systems were built at the Svetlana Factory in Leningrad before the end of 1941, and deployed along the western USSR borders and in the Far East. Without direct ranging capability, however, the military found the *RUS-1* to be of little value.

Even before the demise of efforts in Leningrad, the NIIIS-KA had contracted with the UIPT in Kharkov to investigate a pulsed radio-location system for anti-aircraft applications. This led the LEMO, in March 1937, to start an internally funded project with the code name *Zenit* (a popular football team at the time). The transmitter development was led by Usikov, supplier of the magnetron used earlier in the *Burya*. For the *Zenit*, Usikov used a 60 cm (500 MHz) magnetron pulsed at 10–20 μs duration and providing 3 kW pulsed power, later increased to near 10 kW. Semion Braude led the development of a superheterodyne receiver using a tunable magnetron as the local oscillator. The system had separate transmitting and receiving antennas set about 65 m apart, built with dipoles backed by 3-meter parabolic reflectors.

Zenit was first tested in October 1938. In this, a medium-sized bomber was detected at a range of 3 km. The testing was observed by the NIIIS-KA and found to be sufficient for starting a contracted effort. An agreement was made in May 1939, specifying the required performance and calling for the system to be ready for production by 1941. The transmitter was increased in power, the antennas had selsens added to allow them to track, and the receiver sensitivity was improved by using an RCA 955 acorn triode as the local oscillator.

A demonstration of the improved *Zenit* was given in September 1940. In this, it was shown that the range, altitude, and azimuth of an aircraft flying at heights between 4,000 and 7,000 meters could be determined at up to 25 km distance. The time required for these measurements, however, was about 38 seconds, far too long for use by anti-aircraft batteries. Also, with the antennas aimed at a low angle, there was a dead zone of some distance caused by interference from ground-level reflections. While this performance was not satisfactory for immediate gun-laying applications, it was the first full three-coordinate radio-location system in the Soviet Union and showed the way for future systems.

Work at the LEMO continued on *Zenit*, particularly in converting it into a single-antenna system designated *Rubin*. This effort, however, was disrupted by the invasion of the USSR by Germany in June 1941. In a short while, the development activities at

Kharkov were ordered to be evacuated to the Far East. The research efforts in Leningrad were similarly dispersed.

After eight years of effort by highly qualified physicists and engineers, the USSR entered World War II without a fully developed and fielded radar system.

Japan

As a seafaring nation, Japan had an early interest in wireless (radio) communications. The first known use of wireless telegraphy in warfare at sea was by the Imperial Japanese Navy, in defeating the Russian Imperial Fleet in 1904 at the Battle of Port Arthur. There was an early interest in equipment for radio direction-finding, for use in both navigation and military surveillance. The Imperial Navy developed an excellent receiver for this purpose in 1921, and soon most of the Japanese warships had this equipment.

In the two decades between the two World Wars, radio technology in Japan made advancements on a par with that in the western nations. There were often impediments, however, in transferring these advancements into the military. For a long time, the Japanese had believed that they had the best fighting capability of any military force in the world. The military leaders, who were then also in control of the government, sincerely felt that the weapons, aircraft, and ships that they had built were fully sufficient and, with these as they were, the Japanese Army and Navy were invincible. In 1936, Japan joined Nazi Germany and Fascist Italy in a Tripartite Pact.

Technology Background

Radio engineering was strong in Japan's higher education institutions, especially the Imperial (government-financed) universities. This included undergraduate and graduate study, as well as academic research in this field. Special relationships were established with foreign universities and institutes, particularly in Germany, with Japanese teachers and researchers often going overseas for advanced study.

The academic research tended toward the improvement of basic technologies, rather than their specific applications. There was considerable research in high-frequency and high-power oscillators, such as the magnetron, but the application of these devices was generally left to industrial and military researchers.

One of Japan's best-known radio researchers in the 1920s-1930s era was Professor Hidetsugu Yagi. After graduate study in Germany, England, and America, Yagi joined Tohoku University, where his research centered on antennas and oscillators for high-frequency communications. A summary of the radio research work at Tohoku University was contained in a 1928 seminal paper by Yagi.

Jointly with Shintaro Uda, one of Yagi's first doctoral students, a radically new antenna emerged. It had a number of parasitic elements (directors and reflectors) and would come to be known as the Yagi-Uda or Yagi antenna. A U.S. patent, issued in May 1932, was assigned to RCA. To this day, this is the most widely used directional antenna worldwide.

The cavity magnetron was also of interest to Yagi. This HF (~10-MHz) device had been invented in 1921 by Albert W. Hull at General Electric, and Yagi was convinced that it could function in the VHF or even the UHF region. In 1927, Kinjiro Okabe, another of Yagi's early doctoral students, developed a split-anode device that ultimately generated oscillations at wavelengths down to about 12 cm (2.5 GHz).

Researchers at other Japanese universities and institutions also started projects in magnetron development, leading to improvements in the split-anode device. These included Kiyoshi Morita at the Tokyo Institute of Technology, and Tsuneo Ito at Tokoku University.

Shigeru Nakajima at Japan Radio Company (JRC) saw a commercial potential of these devices and began the further development and subsequent very profitable production of magnetrons for the medical dielectric heating (diathermy) market. The only military interest in magnetrons was shown by Yoji Ito at the Naval Technical Research Institute (NTRI).

The NTRI was formed in 1922, and became fully operational in 1930. Located at Meguro, Tokyo, near the Tokyo Institute of Technology, first-rate scientists, engineers, and technicians were engaged in activities ranging from designing giant submarines to building new radio tubes. Included were all of the precursors of radar, but this did not mean that the heads of the Imperial Navy accepted these accomplishments.

In 1936, Tsuneo Ito (no relationship to Yoji Ito) developed an 8-split-anode magnetron that produced about 10 W at 10 cm (3 GHz). Based on its appearance, it was named *Tachibana* (or Mandarin, an orange citrus fruit). Tsuneo Ito also joined the NTRI and continued his research on magnetrons in association with Yoji Ito. In 1937, they developed the technique of coupling adjacent segments (called push-pull), resulting in frequency stability, an extremely important magnetron breakthrough.

By early 1939, NTRI/JRC had jointly developed a 10-cm (3-GHz), stable-frequency Mandarin-type magnetron (No. M3) that, with water cooling, could produce 500-W power. In the same time period, magnetrons were built with 10 and 12 cavities operating as low as 0.7 cm (40 GHz). The configuration of the M3 magnetron was essentially the same as that used later in the magnetron developed by Boot and Randall at Birmingham University in early 1940, including the improvement of strapped cavities. Unlike the high-power magnetron in Britain, however, the initial device from the NTRI generated only a few hundred watts.

In general, there was no lack of scientific and engineering capabilities in Japan; their warships and aircraft clearly showed high levels of technical competency. They were ahead of Britain in the development of magnetrons, and their Yagi antenna was the world standard for VHF systems. It was simply that the top military leaders failed to recognize how the application of radio in detection and ranging – what was often called the Radio Range Finder (RRF) – could be of value, particularly in any defensive role; offense not defense, totally dominated their thinking.

Imperial Army

In 1938, engineers from the Research Office of Nippon Electric Company (NEC) were making coverage tests on high-frequency transmitters when rapid fading of the signal was observed. This occurred whenever an aircraft passed over the line between the transmitter and receiving meter. Masatsugu Kobayashi, the Manager of NEC's Tube Department, recognized that this was due to the beat-frequency interference of the direct signal and the Doppler-shifted signal reflected from the aircraft.

Kobayashi suggested to the Army Science Research Institute that this phenomenon might be used as an aircraft warning method. Although the Army had rejected earlier proposals for using radio-detection techniques, this one had appeal because it was based on an easily understandable method and would require little developmental cost and risk to prove its military value. NEC assigned Kinji Satake of their Research Institute to develop a system called the Bi-static Doppler Interference Detector (BDID).

For testing the prototype system, it was set up on an area recently occupied by Japan along the coast of China. The system operated between 4.0-7.5 MHz (75–40 m) and involved a number of widely spaced stations; this formed a radio screen that could detect the presence (but nothing more) of an aircraft at distances up to 500 km (310 mi). The BDID was the Imperial Army's first deployed radio-based detection system, placed into operation in early 1941.

A similar system was developed by Satake for the Japanese homeland. Information centers received oral warnings from the operators at BDID stations, usually spaced between 65 and 240 km (40 and 150 mi). To reduce homing vulnerability – a great fear of the military – the transmitters operated with only a few watts power. Although originally intended to be temporary until better systems were available, they remained in operation throughout the war. It was not until after the start of war that the Imperial Army had equipment that could be called radar.

Imperial Navy

In the mid-1930s, some of the technical specialists in the Imperial Navy became interested in the possibility of using radio to detect aircraft. For consultation, they turned to Professor Yagi who was the Director of the Radio Research Laboratory at Osaka Imperial University. Yagi suggested that this might be done by examining the Doppler frequency-shift in a reflected signal.

Funding was provided to the Osaka Laboratory for experimental investigation of this technique. Kinjiro Okabe, the inventor of the split-anode magnetron and who had followed Yagi to Osaka, led the effort. Theoretical analyses indicated that the reflections would be greater if the wavelength was approximately the same as the size of aircraft structures. Thus, a VHF transmitter and receiver with Yagi antennas separated some distance were used for the experiment.

In 1936, Okabe successfully detected a passing aircraft by the Doppler-interference method; this was the first recorded demonstration in Japan of aircraft detection by radio. With this success, Okabe's research interest switched from magnetrons to VHF equipment for target detection. This, however, did not lead to any significant funding. The top levels of the Imperial Navy believed that any advantage of using radio for this purpose were greatly outweighed by enemy intercept and disclosure of the sender's presence.

Historically, warships in formation used lights and horns to avoid collision at night or when in fog. Newer techniques of VHF radio communications and direction-finding might also be used, but all of these methods were highly vulnerable to enemy interception. At the NTRI, Yoji Ito proposed that the UHF signal from a magnetron might be used to generate a very narrow beam that would have a greatly reduced chance of enemy detection.

Development of microwave system for collision avoidance started in 1939, when funding was provided by the Imperial Navy to JRC for preliminary experiments. In a cooperative effort involving Yoji Ito of the NTRI and Shigeru Nakajima of JRC, an apparatus using a 3-cm (10-GHz) magnetron with frequency modulation was designed and built. The equipment was used in an attempt to detect reflections from tall structures a few kilometers away. This experiment gave poor results, attributed to the very low power from the magnetron.

The initial magnetron was replaced by one operating at 16 cm (1.9 GHz) and with considerably higher power. The results were then much better, and in October 1940, the equipment obtained clear echoes from a ship in Tokyo Bay at a distance of about 10 km (6.2 mi). There was still no commitment by top Japanese naval officials for using this technology aboard warships. Nothing more was done at this time, but late in 1941, the system was adopted for limited use.

In late 1940, Japan arranged for two technical missions to visit Germany and exchange information about their developments in military technology. Commander Yoji Ito represented the Navy's interest in radio applications, and Lieutenant Colonel Kinji Satake did the same for the Army. During a visit of several months, they exchanged significant general information, as well as limited secret materials in some technologies, but little directly concerning radio-detection techniques. Neither side even mentioned magnetrons, but the Germans did apparently disclose their use of pulsed techniques.

After receiving the reports from the technical exchange in Germany, as well as intelligence reports concerning the success of Britain with firing using RDF, the Naval General Staff reversed itself and tentatively accepted pulse-transmission technology. On August 2, 1941, even before Yoji Ito returned to Japan, funds were allocated for the initial development of pulse-modulated radars. Commander Chuji Hashimoto of the NTRI was responsible for initiating this activity.

A prototype set operating at 4.2 m (71 MHz) and producing about 5 kW was completed on a crash basis. With the NTRI in the lead, the firm NEC and the Research Laboratory of Japan Broadcasting Corporation (NHK) made major contributions to the effort. Kenjiro Takayanagi, Chief Engineer of NHK's experimental television station and called "the father of Japanese television", was especially helpful in rapidly developing the pulse-forming and timing circuits, as well as the receiver display. In early September 1941, the prototype set was first tested; it detected a single bomber at 97 km (60 mi) and a flight of aircraft at 145 km (90 mi).

The system, Japan's first full Radio Range Finder (RRF – radar), was designated Mark 1 Model 1. Contracts were given to three firms for serial production; NEC built the transmitters and pulse modulators, Japan Victor the receivers and associated displays, and Fuji Electrical the antennas and their servo drives. The system operated at 3.0 m (100 MHz) with a peak-power of 40 kW. Dipole arrays with matte+-type reflectors were used in separate antennas for transmitting and receiving.

In November 1941, the first manufactured RRF was placed into service as a land-based early-warning system at Katsuura, Chiba, a town on the Pacific coast about 100 km (62 mi) from Tokyo. A large system, it weighed close to 8,700 kg (19,000 lb). The detection range was about 130 km (81 mi) for single aircraft and 250 km (160 mi) for groups.

Netherlands

Early radio-based detection in the Netherlands was along two independent lines: one a microwave system at the firm Philips and the other a VHF system at a laboratory of the Armed Forces.

The Philips Company in Eindhoven, Netherlands, operated *Natuurkundig Laboratorium* (NatLab) for fundamental research related to its products. NatLab researcher Klaas Posthumus developed a magnetron split into four elements. In developing a communication system using this magnetron, C.H.J.A. Staal was testing the transmission by using parabolic transmitting and receiving antennas set side-by-side, both aimed at a large plate some distance away. To overcome frequency instability of the magnetron, pulse modulation was used. It was found that the plate reflected a strong signal.

Recognizing the potential importance of this as a detection device, NatLab arranged a demonstration for the *Koninklijke Marine* (Royal Netherlands Navy). This was conducted in 1937 across the entrance to the main naval port at Marsdiep. Reflections from sea waves obscured the return from the target ship, but the Navy was sufficiently impressed to initiate sponsorship of the research. In 1939, an improved set was demonstrated at Wijk aan Zee, detecting a vessel at a distance of 3.2 km (2.0 mi).

A prototype system was built by Philips, and plans were started by the firm Nederlandse Seintoestellen Fabriek (a Philips subsidiary) for building a chain of warning stations to protect the primary ports. Some field testing of the prototype was conducted, but the project was discontinued when Germany invaded the Netherlands on May 10, 1940. Within the NatLab, however, the work was continued in great secrecy until 1942.

During the early 1930s, there were widespread rumours of a "death ray" being developed. The Dutch Parliament set up a Committee for the Applications of Physics in Weaponry under G.J. Elias to examine this potential, but the Committee quickly discounted death rays. The Committee did, however, establish the *Laboratorium voor Fysieke Ontwikkeling* (LFO, Laboratory for Physical Development), dedicated to supporting the Netherlands Armed Forces.

Operating in great secrecy, the LFO opened a facility called the *Meetgebouw* (Measurements Building) located on the Plain of Waalsdorp. In 1934, J.L.W.C. von Weiler joined the LFO and, with S.G. Gratama, began research on a 1.25-m (240-MHz) communication system to be used in artillery spotting.

In 1937, while tests were being conducted on this system, a passing flock of birds disturbed the signal. Realizing that this might be a potential method for detecting aircraft, the Minister of War ordered continuation of the experiments. Weiler and Gratama set about developing a system for directing searchlights and aiming anti-aircraft guns.

The experimental "electrical listening device" operated at 70 cm (430 MHz) and used pulsed transmission at an RPF of 10 kHz. A transmit-receive blocking circuit was developed to allow a common antenna. The received signal was displayed on a CR tube with a circular time base. This set was demonstrated to the Army in April 1938 and detected an aircraft at a range of 18 km (11 mi). The set was rejected, however, because it could not withstand the harsh environment of Army combat conditions.

The Navy was more receptive. Funding was provided for final development, and Max Staal was added to the team. To maintain secrecy, they divided the development into parts. The transmitter was built at the Delft Technical College and the receiver at the University of Leiden. Ten sets would be assembled under the personal supervision of J.J.A. Schagen van Leeuwen, head of the firm Hazemeijer Fabriek van Signaalapparaten.

The prototype had a peak-power of 1 kW, and used a pulse length of 2 to 3 μs with a 10- to 20 kHz PRF. The receiver was a super-heterodyne type using Acorn tubes and a 6 MHz IF stage. The antenna consisted of 4 rows of 16 half-wave dipoles backed by a 3- by 3-meter mesh screen. The operator used a bicycle-type drive to rotate the antenna, and the elevation could be changed using a hand crank.

Several sets were completed, and one was put into operation on the Malieveld in The Hague just before the Netherlands fell to Germany in May 1940. The set worked well, spotting enemy aircraft during the first days of fighting. To prevent capture, operating units and plans for the system were destroyed. Von Weiler and Max Staal fled to England aboard one of the last ships able to leave, carrying two disassembled sets with them. Later, Gratama and van Leeuwen also escaped to England.

France

In 1927, French physicists Camille Gutton and Emile Pierret experimented with magnetrons and other devices generating wavelengths going down to 16 cm. Camille's son, Henri Gutton, was with the *Compagnie Générale de Télégraphie Sans Fil* (CSF) where he and Robert Warneck improved his father's magnetrons.

In 1934, following systematic studies on the magnetron, the research branch of the CSF, headed by Maurice Ponte, submitted a patent application for a device designed to detect obstacles using continuous radiation of ultra-short wavelengths produced by a magnetron. These were still CW systems and depended on Doppler interference for detection. However, as most modern radars, antennas were collocated. The device was measuring distance and azimuth but not directly as in the later "radar" on a screen (1939). Still, this was the first patent of an operational radio-detection apparatus using centimetric wavelengths.

The system was tested in late 1934 aboard the cargo ship *Oregon*, with two transmitters working at 80 cm and 16 cm wavelengths. Coastlines and boats were detected from a range of 10-12 nautical miles. The shortest wavelength was chosen for the final design, which equipped the liner *SS Normandie* as early as mid-1935 for operational use.

In late 1937, Maurice Elie at SFR developed a means of pulse-modulating transmitter tubes. This led to a new 16-cm system with a peak power near 500 W and a pulse width of 6 μs. French and U.S. patents were filed in December 1939. The system was planned to be sea-tested aboard the *Normandie*, but this was cancelled at the outbreak of war.

At the same time, Pierre David at the *Laboratoire National de Radioélectricité* (National Laboratory of Radioelectricity, LNR) experimented with reflected radio signals at about a meter wavelength. Starting in 1931, he observed that aircraft caused interference to the signals. The LNR then initiated research on a detection technique called *barrage électromagnétique* (electromagnetic curtain). While this could indicate the general location of penetration, precise determination of direction and speed was not possible.

In 1936, the *Défense Aérienne du Territoire* (Defence of Air Territory), ran tests on David's electromagnetic curtain. In the tests, the system detected most of the entering aircraft, but too many were missed. As the war grew closer, the need for an aircraft detection was critical. David realized the advantages of a pulsed system, and in October 1938 he designed a 50 MHz, pulse-modulated system with a peak-pulse power of 12 kW. This was built by the firm SADIR.

France declared war on Germany on September 1, 1939, and there was a great need for an early-warning detection system. The SADIR system was taken to near Toulon, and detected and measured the range of invading aircraft as far as 55 km (34 mi). The SFR pulsed system was set up near Paris where it detected aircraft at ranges up to 130 km (81 mi). However, the German advance was overwhelming and emergency measures had to be taken; it was too late for France to develop radars alone and it was decided that her breakthroughs would be shared with her allies.

In mid-1940, Maurice Ponte, from the laboratories of CSF in Paris, presented a cavity magnetron designed by Henri Gutton at SFR (see above) to the GEC laboratories at Wembley, Britain. This magnetron was designed for pulsed operation at a wavelength of 16 cm. Unlike other magnetron designs to that day, such as the Boots and Randall magnetron (see British contributions above), this tube used an oxide-coated cathode with a peak power output of 1 kW, demonstrating that oxide cathodes were the solution for producing high-power pulses at short wavelengths, a problem which had eluded British and American researchers for years. The significance this event was underlined by Eric Megaw, in a 1946 review of early radar developments: "This was the starting point of the use of the oxide cathode in practically all our subsequent pulsed transmitting waves and as such was a significant contribution to British radar. The date was the 8th May 1940". A tweaked version of this magnetron reached a peak output of 10 kW by August 1940. It was that model which, in turn, was handed to the Americans as a token of good faith during the negotiations made by the Tizard delegation in 1940 to obtain from the U.S. the resources necessary for Britain to exploit the full military potential of her research and development work.

Italy

Guglielmo Marconi initiated the research in Italy on radio-based detection technology. In 1933, while participating with his Italian firm in experiments with a 600 MHz communications link across Rome, he noted transmission disturbances caused by moving objects adjacent to its path. This led to the development at his laboratory at Cornegliano of a 330-MHz (0.91-m) CW Doppler detection system that he called *radioecometro*. Barkhausen-Kurz tubes were used in both the transmitter and receiver.

In May 1935, Marconi demonstrated his system to the Fascist dictator Benito Mussolini and members of the military General Staff; however the output power was insufficient for military use. While Marconi's demonstration raised considerable interest, little more was done with his apparatus.

Mussolini directed that radio-based detection technology be further developed, and it was assigned to the *Regio Instituto Electrotecnico e delle Comunicazioni* (RIEC, Royal Institute for Electro-technics and Communications). The RIEC had been established in 1916 on the campus of the Italian Naval Academy in Livorno. Lieutenant Ugo Tiberio, a physics and radio-technology instructor at the Academy, was assigned to head the project on a part-time basis.

Tiberio prepared a report on developing an experimental apparatus that he called *telemetro radiofonico del rivelatore* (RDT, Radio-Detector Telemetry). The report, submitted in mid-1936, included what was later known as the radar range equation. When the work got underway, Nello Carrara, a civilian physics instructor who had been doing research at the RIEC in microwaves, was added to be responsible for developing the RDT transmitter.

Before the end of 1936, Tiberio and Carrara had demonstrated the EC-1, the first Italian RDT system. This had an FM transmitter operating at 200 MHz (1.5 m) with a single parabolic cylinder antenna. It detected by mixing the transmitted and the Doppler-shifted reflected signals, resulting in an audible tone.

The EC-1 did not provide a range measurement; to add this capability, development of a pulsed system was initiated in 1937. Captain Alfeo Brandimarte joined the group and primarily designed the first pulsed system, the EC-2. This operated at 175 MHz (1.7 m) and used a single antenna made with a number of equi-phased dipoles. The detected signal was intended to be displayed on an oscilloscope. There were many problems, and the system never reached the testing stage.

Work then turned to developing higher power and operating frequencies. Carrara, in cooperation with the firm FIVRE, developed a magnetron-like device. This was composed of a pair of triodes connected to a resonate cavity and produced 10 kW at 425 MHz (70 cm). It was used in designing two versions of the EC-3, one for shipboard and the other for coastal defense.

Italy, joining Germany, entered WWII in June 1940 without an operational RDT. A breadboard of the EC-3 was built and tested from atop a building at the Academy, but most RDT work was stopped as direct support of the war took priority.

Others

In early 1939, the British Government invited representatives from the most technically advanced Commonwealth Nations to visit England for briefings and demonstrations on the highly secret RDF (radar) technology. Based on this, RDF developments were started in Australia, Canada, New Zealand, and South Africa by September 1939. In addition, this technology was independently developed in Hungary early in the war period.

In Australia, the Radiophysics Laboratory was established at Sydney University under the Council for Scientific and Industrial Research; John H. Piddington was responsible for RDF development. The first project was a 200-MHz (1.5-m) shore-defense system for the Australian Army. Designated ShD, this was first tested in September 1941, and eventually installed at 17 ports. Following the Japanese attack on Pearl Harbor, the Royal Australian Air Force urgently needed an air-warning system, and Piddington's team, using the ShD as a basis, put the AW Mark I together in five days. It was being installed in

Darwin, Northern Territory, when Australia received the first Japanese attack on February 19, 1942. A short time later, it was converted to a light-weight transportable version, the LW-AW Mark II; this was used by the Australian forces, as well as the U.S. Army, in early island landings in the South Pacific.

The early RDF developments in Canada were at the Radio Section of the National Research Council of Canada. Using commercial components and with essentially no further assistance from Britain, John Tasker Henderson led a team in developing the Night Watchman, a surface-warning system for the Royal Canadian Navy to protect the entrance to the Halifax Harbour. Successfully tested in July 1940, this set operated at 200 MHz (1.5 m), had a 1 kW output with a pulse length of 0.5 μs, and used a relatively small, fixed antenna. This was followed by a ship-borne set designated Surface Warning 1st Canadian (SW1C) with the antenna hand-rotated through the use of a Chevrolet steering wheel in the operator's compartment. The SW1C was first tested at sea in mid-May 1941, but the performance was so poor compared to the Royal Navy's Model 271 ship-borne radar that the Royal Canadian Navy eventually adopted the British 271 in place of the SW1C.

For coastal defense by the Canadian Army, a 200 MHz set with a transmitter similar to the Night Watchman was developed. Designated CD, it used a large, rotating antenna atop a 70-foot (21 m) wooden tower. The CD was put into operation in January 1942.

Ernest Marsden represented New Zealand at the briefings in England, and then established two facilities for RDF development – one in Wellington at the Radio Section of the Central NZ Post Office, and another at Canterbury University College in Christchurch. Charles N. Watson-Munro led the development of land-based and airborne sets at Wellington, while Frederick W. G. White led the development of shipboard sets at Christchurch.

Before the end of 1939, the Wellington group had converted an existing 180-MHz (1.6-m), 1 kW transmitter to produce 2-μs pulses and tested it to detect large vessels at up to 30 km; this was designated CW (Coastal Watching). A similar set, designated CD (Coast Defense) used a CRT for display and had lobe-switching on the receiving antenna; this was deployed in Wellington in late 1940. A partially completed ASV 200 MHz set was brought from Britain by Marsden, and another group at Wellington built this into an aircraft set for the Royal New Zealand Air Force; this was first flown in early 1940. At Christchurch, there was a smaller staff and work went slower, but by July 1940, a 430-MHz (70-cm), 5 kW set was tested. Two types, designated SW (Ship Warning) and SWG (Ship Warning, Gunnery), were placed into service by the Royal New Zealand Navy starting in August 1941. In all some 44 types were developed in New Zealand during WW1.

South Africa did not have a representative at the 1939 meetings in England, but in mid-September, as Ernest Marsden was returning by ship to New Zealand, Basil F. J. Schonland came aboard and received three days of briefings. Schonland, a world authority on lightning and Director of the Bernard Price Institute of Geophysics at Witwatersrand University, immediately started an RDF development using amateur radio components and Institute's lightning-monitoring equipment. Designated JB (for Johannesburg), the 90-MHz (3.3-m), 500-W mobile system was tested in November 1939, just two months after its start. The prototype was operated in Durban before the end of 1939, detecting ships and aircraft at distances up to 80 km, and by the next March a system was fielded by anti-aircraft brigades of the South African Defence Force.

In Hungary, Zoltán Lajos Bay was a Professor of Physics at the Technical University of Budapest as well as the Research Director of Egyesült Izzolampa (IZZO), a radio and electrical manufacturing firm. In late 1942, IZZO was directed by the Minister of Defense to develop a radio-location (*rádiólokáció*, radar) system. Using journal papers on ionospheric measurements for information on pulsed transmission, Bay developed a system called *Sas* (Eagle) around existing communications hardware.

The *Sas* operated at 120 MHz (2.5 m) and was in a cabin with separate transmitting and receiving dipole arrays attached; the assembly was all on a rotatable platform. According to published records, the system was tested in 1944 atop Mount János and had a range of "better than 500 km". A second *Sas* was installed at another location. There is no indication that either *Sas* installation was ever in regular service. After the war, Bay used a modified *Sas* to successfully bounce a signal off the moon.

World War II radar

At the start of World War II in September 1939, both the United Kingdom and Germany knew of each other's ongoing efforts in radio navigation and its countermeasures – the "Battle of the beams". Also, both nations were generally aware of, and intensely interested in, the other's developments in radio-based detection and tracking, and engaged in an active campaign of espionage and false leaks about their respective equipment. By the time of the Battle of Britain, both sides were deploying range and direction-finding units (radars) and control stations as part of integrated air defense capability. However, the German *Funkmessgerät* (radio measuring device) systems could not assist in an offensive role and was thus not supported by Adolf Hitler. Also, the *Luftwaffe* did not sufficiently appreciate the importance of British Range and Direction Finding (RDF) stations as part of RAF's air defense capability, contributing to their failure.

While the United Kingdom and Germany led in pre-war advances in the use of radio for detection and tracking of aircraft, there were also developments in the United States, the Soviet Union, and Japan. Wartime systems in all of these nations will be summarized. The acronym RADAR (for RAdio Detection And Ranging) was coined by the U.S. Navy in 1940, and the subsequent name "radar" was soon widely used.

When France had just fallen to the Nazis and Britain had no money to develop the magnetron on a massive scale, Churchill agreed that Sir Henry Tizard should offer the magnetron to the Americans in exchange for their financial and industrial help (the Tizard Mission). An early 6 kW version, built in England by the General Electric Company Research Laboratories, Wembley, London (not to be confused with the similarly named American company General Electric), was given to the US government in September 1940. The British magnetron was a thousand times more powerful than the best American transmitter at the time and produced accurate pulses. At the time the most powerful equivalent microwave producer available in the US (a klystron) had a power of only ten watts. The cavity magnetron was widely used during World War II in microwave radar equipment and is often credited with giving Allied radar a considerable performance advantage over German and Japanese radars, thus directly influencing the outcome of the war. It was later described by noted Historian James Phinney Baxter III as "The most valuable cargo ever brought to our shores".

The Bell Telephone Laboratories made a producible version from the magnetron delivered to America by the Tizard Mission, and before the end of 1940, the Radiation Laboratory had been set up on the campus of the Massachusetts Institute of Technology to develop various types of radar using the magnetron. By early 1941, portable centimetric airborne radars were being tested in American and British aircraft. In late 1941, the Telecommunications Research Establishment in Great Britain used the magnetron to develop a revolutionary airborne, ground-mapping radar codenamed H2S. The H2S radar was in part developed by Alan Blumlein and Bernard Lovell. The magnetron radars used by the US and Britain could spot the periscope of a U boat

Post-war Radar

World War II, which gave impetus to the great surge in radar development, ended between the Allies and Germany in May 1945, followed by Japan in August. With this, radar activities in Germany and Japan ceased for a number of years. In other countries, particularly the United States, Britain, and the USSR, the politically unstable post-war years saw continued radar improvements for military applications. In fact, these three nations all made significant efforts in bringing scientists and engineers from Germany to work in their weapon programs; in the U.S., this was under Operation Paperclip.

Even before the end of the war, various project directed toward non-military applications of radar and closely related technologies were initiated. The US Army Air Forces and the British RAF had made wartime advances in using radar for handling aircraft landing, and this was rapidly expanded into the civil sector. The field of radio astronomy was one of the related technologies; although discovered before the war, it immediately flourished in the late 1940s with many scientists around the world establishing new careers based on their radar experience.

Four techniques, highly important in post-war radars, were matured in the late 1940s-early 1950s: pulse Doppler, monopulse, phased array, and synthetic aperture; the first three were known and even used during wartime developments, but were matured later.

- Pulse-Doppler radar (often known as moving target indication or MTI), uses the Doppler-shifted signals from targets to better detect moving targets in the presence of clutter.
- Monopulse radar (also called simultaneous lobing) was conceived by Robert Page at the NRL in 1943. With this, the system derives error-angle information from a single pulse, greatly improving the tracking accuracy.
- Phased-array radar has the many segments of a large antenna separately controlled, allowing the beam to be quickly directed. This greatly reduces the time necessary to change the beam direction from one point to another, allowing almost simultaneous tracking of multiple targets while maintaining overall surveillance.
- Synthetic-aperture radar (SAR), was invented in the early 1950s at Goodyear Aircraft Corporation. Using a single, relatively small antenna carried on an aircraft, a SAR combines the returns from each pulse to produce a high-resolution image of the terrain comparable to that obtained by a much larger antenna. SAR has wide applications, particularly in mapping and remote sensing.

One of the early applications of digital computers was in switching the signal phase in elements of large phased-array antennas. As smaller computers came into being, these were quickly applied to digital signal processing using algorithms for improving radar performance.

Other advances in radar systems and applications in the decades following WWII are far too many to be included herein. The following sections are intended to provide representative samples.

Military Radars

In the United States, the Rad Lab at MIT officially closed at the end of 1945. The Naval Research Laboratory (NRL) and the Army's Evans Signal Laboratory continued with new activities in centimeter radar development. The United States Air Force (USAF) – separated from the Army in 1946 – concentrated radar research at their Cambridge Research Center (CRC) at Hanscom Field, Massachusetts. In 1951, MIT opened the Lincoln Laboratory for joint developments with the CRC. While the Bell Telephone Laboratories embarked on major communications upgrades, they continued with the Army in radar for their ongoing Nike air-defense program

In Britain, the RAF's Telecommunications Research Establishment (TRE) and the Army's Radar Research and Development Establishment (RRDE) both continued at reduced levels at Malvern, Worcestershire, then in 1953 were combined to form the Radar Research Establishment. In 1948, all of the Royal Navy's radio and radar R&D activities were combined to form the Admiralty Signal and Radar Establishment, located near Portsmouth, Hampshire. The USSR, although devastated by the war, immediately embarked on the development of new weapons, including radars.

During the Cold War period following WWII, the primary "axis" of combat shifted to lie between the United States and the Soviet Union. By 1949, both sides had nuclear weapons carried by bombers. To provide early warning of an attack, both deployed huge radar networks of increasing sophistication at ever-more remote locations. In the West, the first such system was the Pinetree Line, deployed across Canada in the early 1950s, backed up with radar pickets on ships and oil platforms off the east and west coasts.

The Pinetree Line initially used vintage pulsed radars and was soon supplemented with the Mid-Canada Line (MCL). Soviet technology improvements made these Lines inadequate and, in a construction project involving 25,000 persons, the Distant Early Warning Line (DEW Line) was completed in 1957. Stretching from Alaska to Baffin Island and covering over 6,000 miles (9,700 km), the DEW Line consisted of 63 stations with AN/FPS-19 high-power, pulsed, L-Band radars, most augmented by AN/FPS-23 pulse-Doppler systems. The Soviet Unit tested its first Intercontinental Ballistic Missile (ICBM) in August 1957, and in a few years the early-warning role was passed almost entirely to the more capable DEW Line.

Both the U.S. and the Soviet Union then had ICBMs with nuclear warheads, and each began the development of a major anti-ballistic missile (ABM) system. In the USSR, this was the Fakel V-1000, and for this they developed powerful radar systems. This was eventually deployed around Moscow as the A-35 anti-ballistic missile system, supported by radars designated by NATO as the Cat House, Dog House, and Hen House.

In 1957, the U.S. Army initiated an ABM system first called Nike-X; this passed through several names, eventually becoming the Safeguard Program. For this, there was a long-range Perimeter Acquisition Radar (PAR) and a shorter-range, more precise Missile Site Radar (MSR).

The PAR was housed in a 128-foot (39 m)-high nuclear-hardened building with one face sloping 25 degrees facing north. This contained 6,888 antenna elements separated in transmitting and receiving phased arrays. The L-Band transmitter used 128 long-life traveling-wave tubes (TWTs), having a combined power in the megawatt range The PAR could detect incoming missiles outside the atmosphere at distances up to 1,800 miles (2,900 km).

The MSR had an 80-foot (24 m), truncated pyramid structure, with each face holding a phased-array antenna 13 feet (4.0 m) in diameter and containing 5,001 array elements used for both transmitting and receiving. Operating in the S-Band, the transmitter used two klystrons functioning in parallel, each with megawatt-level power. The MSR could search for targets from all directions, acquiring them at up to 300 miles (480 km) range.

One Safeguard site, intended to defend Minuteman ICBM missile silos near the Grand Forks AFB in North Dakota, was finally completed in October 1975, but the U.S. Congress withdrew all funding after it was operational but a single day. During the following decades, the U.S. Army and the U.S. Air Force developed a variety of large radar systems, but the long-serving BTL gave up military development work in the 1970s.

A modern radar developed by of the U.S. Navy that should be noted is the AN/SPY-1. First fielded in 1973, this S-Band, 6 MW system has gone through a number of variants and is a major component of the Aegis Combat System. An automatic detect-and-track system, it is computer controlled using four complementary three-dimensional passive electronically scanned array antennas to provide hemispherical coverage.

Radar signals, traveling with line-of-sight propagation, normally have a range to ground targets limited by the visible horizon, or less than about 10 miles (16 km). Airborne targets can be detected by ground-level radars at greater ranges, but, at best, several hundred miles. Since the beginning of radio, it had been known that signals of appropriate frequencies (3 to 30 MHz) could be "bounced" from the ionosphere and received at considerable distances. As long-range bombers and missiles came into being, there was a need to have radars give early warnings at great ranges. In the early 1950s, a team at the Naval Research Laboratory came up with the Over-the-Horizon (OTH) radar for this purpose.

To distinguish targets from other reflections, it was necessary to use a phase-Doppler system. Very sensitive receivers with low-noise amplifiers had to be developed. Since the signal going to the target and returning had a propagation loss proportional to the range raised to the fourth power, a powerful transmitter and large antennas were required. A digital computer with considerable capability (new at that time) was necessary for analyzing the data. In 1950, their first experimental system was able to detect rocket launches 600 miles (970 km) away at Cape Canaveral, and the cloud from a nuclear explosion in Nevada 1,700 miles (2,700 km) distant.

In the early 1970s, a joint American-British project, code named Cobra Mist, used a 10-MW OTH radar at Orfordness (the birthplace of British radar), England, in an attempt to detect aircraft and missile launchings over the Western USSR. Because of US-USSR ABM agreements, this was abandoned within two years. In the same time period, the Soviets were developing a similar system; this successfully detected a missile launch at 2,500 km (1,600 mi). By 1976, this had matured into an operational system named *Duga* ("Arc" in English), but known to western intelligence as Steel Yard and called Woodpecker by radio amateurs and others who suffered from its interference – the transmitter was estimated to have a power of 10 MW.Australia, Canada, and France also developed OTH radar systems.

With the advent of satellites with early-warning capabilities, the military lost most of its interest in OTH radars. However, in recent years, this technology has been reactivated for detecting and tracking ocean shipping in applications such as maritime reconnaissance and drug enforcement.

Systems using an alternate technology have also been developed for over-the-horizon detection. Due to diffraction, electromagnetic surface waves are scattered to the rear of objects, and these signals can be detected in a direction opposite from high-powered transmissions. Called OTH-SW (SW for Surface Wave), Russia is using such a system to monitor the Sea of Japan, and Canada has a system for coastal surveillance.

Civil Aviation Radars

The post-war years saw the beginning of a revolutionary development in Air Traffic Control (ATC) – the introduction of radar. In 1946, the Civil Aeronautics Administration (CAA) unveiled an experimental radar-equipped tower for control of civil flights. By 1952, the CAA had begun its first routine use of radar for approach and departure control. Four years later, it placed a large order for long-range radars for use in *en route* ATC; these had the capability, at higher altitudes, to see aircraft within 200 nautical miles (370 km). In 1960, it became required for aircraft flying in certain areas to carry a radar transponder that identified the aircraft and helped improve radar performance. Since 1966, the responsible agency has been called the Federal Aviation Administration (FAA).

A Terminal Radar Approach Control (TRACON) is an ATC facility usually located within the vicinity of a large airport. In the US Air Force it is known as RAPCON (Radar Approach Control), and in the US Navy as a RATCF (Radar Air Traffic Control Facility). Typically, the TRACON controls aircraft within a 30 to 50 nautical mile (56 to 93 km) radius of the airport at an altitude between 10,000 and 15,000 feet (3,000 to 4,600 m). This uses one or more Airport Surveillance Radars (ASR-7, 8, & 9), sweeping the sky once every few seconds.

The Digital Airport Surveillance Radar (DASR) is a newer TRACON radar system, replacing the old analog systems with digital technology. The civilian nomenclature for this radar is the ASR-11, and AN/GPN-30 is used by the military. Two radar systems are included. The primary is an S-Band (~2.8 GHz) system with 25 kW pulse power. It provides 3-D tracking of target aircraft and also measures rainfall intensity. The secondary is a P-Band (~1.05 GHz) system with a peak-power of about 25 kW. It uses a transponder set to interrogate aircraft and receive operational data. The antennas for both systems rotate atop a tall tower.

Weather Radar

David Atlas

During World War II, military radar operators noticed noise in returned echoes due to weather elements like rain, snow, and sleet. Just after the war, military scientists returned to civilian life or continued in the Armed Forces and pursued their work in developing a use for those echoes. In the United States, David Atlas, for the Air Force group at first, and later for MIT, developed the first operational weather radars. In Canada, J.S. Marshall and R.H. Douglas formed the "Stormy Weather Group " in Montreal. Marshall and his doctoral student Walter Palmer are well known for their work on the drop size distribution in mid-latitude rain that led to understanding of the Z-R relation, which correlates a given radar reflectivity with the rate at which water is falling on the ground. In the United Kingdom, research continued to study the radar echo patterns and weather elements such as stratiform rain and convective clouds, and experiments were done to evaluate the potential of different wavelengths from 1 to 10 centimetres.

Between 1950 and 1980, reflectivity radars, which measure position and intensity of precipitation, were built by weather services around the world. In United States, the U.S. Weather Bureau, established in 1870 with the specific mission of to provide meteorological observations and giving notice of approaching storms, developed the WSR-1 (Weather Surveillance Radar-1), one of the first weather radars. This was a modified version of the AN/APS-2F radar, which the Weather Bureau acquired from the Navy. The WSR-1A, WSR-3, and WSR-4 were also variants of this radar. This was followed by the WSR-57 (Weather Surveillance Radar – 1957) was the first weather radar designed specifically for a national warning network. Using WWII technology based on vacuum tubes, it gave only coarse reflectivity data and no velocity information. Operating at 2.89 GHz (S-Band), it had a peak-power of 410 kW and a maximum range of about 580 mi (930 km). AN/FPS-41 was the military designation for the WSR-57.

The early meteorologists had to watch a cathode ray tube. During the 1970s, radars began to be standardized and organized into larger networks. The next significant change in the United States was the WSR-74 series, beginning operations in 1974. There were two types: the WSR-74S, for replacements and filling gaps in the WSR-57 national network, and the WSR-74C, primarily for local use. Both were transistor-based, and their primary technical difference was indicated by the letter, S band (better suited for long range) and C band, respectively. Until the 1990s, there were 128 of the WSR-57 and WSR-74 model radars were spread across that country. The first devices to capture radar images were developed during the same period. The number of scanned angles was increased to get a three-dimensional view of the precipitation, so that horizontal cross-sections (CAPPI) and vertical ones could be performed. Studies of the organization of thunderstorms were then possible for the Alberta Hail Project in Canada and National Severe Storms Laboratory (NSSL) in the US in particular. The NSSL, created in 1964, began experimentation on dual polarization signals and on Doppler effect uses. In May 1973, a tornado devastated Union City, Oklahoma, just west of Oklahoma City. For the first time, a Dopplerized 10-cm wavelength radar from NSSL documented the entire life cycle of the tornado. The researchers discovered a mesoscale rotation in the cloud aloft before the tornado touched the ground : the tornadic vortex signature. NSSL's research helped convince the National Weather Service that Doppler radar was a crucial forecasting tool.

Between 1980 and 2000, weather radar networks became the norm in North America, Europe, Japan and other developed countries. Conventional radars were replaced by Doppler radars, which in addition to position and intensity of could track the relative velocity of the particles in the air. In the United States, the construction of a network consisting of 10 cm (4 in) wavelength radars, called NEXRAD or WSR-88D (Weather Service Radar 1988 Doppler), was started in 1988 following NSSL's research. In Canada, Environment Canada constructed the King City station, with a five centimeter research Doppler radar, by 1985; McGill University dopplerized its radar (J. S. Marshall Radar Observatory) in 1993. This led to a complete Canadian Doppler network between 1998 and 2004. France and other European countries switched to Doppler network by the end of the 1990s to early 2000s. Meanwhile, rapid advances in computer technology led to algorithms to detect signs of severe weather and a plethora of "products" for media outlets and researchers.

After 2000, research on dual polarization technology moved into operational use, increasing the amount of information available on precipitation type (e.g. rain vs. snow). "Dual polarization" means that microwave radiation which is polarized both horizontally and vertically (with respect to the ground) is emitted. Wide-scale deployment is expected by the end of the decade in some countries such as the United States, France, and Canada.

Since 2003, the U.S. National Oceanic and Atmospheric Administration has been experimenting with phased-array radar as a replacement for conventional parabolic antenna to provide more time resolution in atmospheric sounding. This would be very important in severe thunderstorms as their evolution can be better evaluated with more timely data.

Also in 2003, the National Science Foundation established the Engineering Research Center for Collaborative Adaptive Sensing of the Atmosphere, "CASA", a multidisciplinary, multi-university collaboration of engineers, computer scientists, meteorologists, and sociologists to conduct fundamental research, develop enabling technology, and deploy prototype engineering systems designed to augment existing radar systems by sampling the generally undersampled lower troposphere with inexpensive, fast scanning, dual polarization, mechanically scanned and phased array radars.

Mapping Radar

The plan position indicator, dating from the early days of radar and still the most common type of display, provides a map of the targets surrounding the radar location. If the radar antenna on an aircraft is aimed downward, a map of the terrain is generated, and the larger the antenna, the greater the image resolution. After centimeter radar came into being, downward-looking radars – the H2S (L-Band) and H2X (C-Band) – provided real-time maps used by the U.S. and Britain in bombing runs over Europe at night and through dense clouds.

In 1951, Carl Wiley led a team at Goodyear Aircraft Corporation (later Goodyear Aerospace) in developing a technique for greatly expanding and improving the resolution of radar-generated images. Called synthetic aperture radar (SAR), an ordinary-sized antenna fixed to the side of an aircraft is used with highly complex signal processing to give an image that would otherwise require a much larger, scanning antenna; thus, the name synthetic aperture. As each pulse is emitted, it is radiated over a lateral band onto the terrain. The return is spread in time, due to reflections from features at different distances. Motion of the vehicle along the flight path gives the horizontal increments. The amplitude and phase of returns are combined by the signal processor using Fourier transform techniques in forming the image. The overall technique is closely akin to optical holography. Through the years, many variations of the SAR have been made with diversified applications resulting. In initial systems, the signal processing was too complex for on-board operation; the signals were recorded and processed later. Processors using optical techniques were then tried for generating real-time images, but advances in high-speed electronics now allow on-board processes for most applications. Early systems gave a resolution in tens of meters, but more recent airborne systems provide resolutions to about 10 cm. Current ultra-wideband systems have resolutions of a few millimeters.

Other Radars and Applications

There are many other post-war radar systems and applications. Only a few will be noted.

Radar gun

The most widespread radar device today is undoubtedly the radar gun. This is a small, usually hand-held, Doppler radar that is used to detect the speed of objects, especially trucks and automobiles in regulating traffic, as well as pitched baseballs, runners, or other moving objects in sports. This device can also be used to measure the surface speed of water and continuously manufactured materials. A radar gun does not return information regarding the object's position; it uses the Doppler effect to measure the speed of a target. First developed in 1954, most radar guns operate with very low power in the X or Ku Bands. Some use infrared radiation or laser light; these are usually called LIDAR. A related technology for velocity measurements in flowing liquids or gasses is called laser Doppler velocimetry; this technology dates from the mid-1960s.

Impulse Radar

As pulsed radars were initially being developed, the use of very narrow pulses was examined. The pulse length governs the accuracy of distance measurement by radar – the shorter the pulse, the greater the precision. Also, for a given pulse repetition frequency (PRF), a shorter pulse results in a higher peak power. Harmonic analysis shows that the narrower the pulse, the wider the band of frequencies that contain the energy, leading to such systems also being called wide-band radars. In the early days, the electronics for generating and receiving these pulses was not available; thus, essentially no applications of this were initially made. By the 1970s, advances in electronics led to renewed interest in what was often called short-pulse radar. With further advances, it became practical to generate pulses having a width on the same order as the period of the RF carrier (T = 1/f). This is now generally called impulse radar.

The first significant application of this technology was in ground-penetrating radar (GPR). Developed in the 1970s, GPR is now used for structural foundation analysis, archeological mapping, treasure hunting, unexploded ordnance identification, and other shallow investigations. This is possible because impulse radar can concisely locate the boundaries between the general media (the soil) and the desired target. The results, however, are non-unique and are highly dependent upon the skill of the operator and the subsequent interpretation of the data.

In dry or otherwise favorable soil and rock, penetration up to 300 feet (91 m) is often possible. For distance measurements at these short ranges, the transmitted pulse is usually only one radio-frequency cycle in duration; With a 100 MHz carrier and a PRF of 10 kHz (typical parameters), the pulse duration is only 10 ns (nanosecond). leading to the "impulse" designation. A variety of GPR systems are commercially available in back-pack and wheeled-cart versions with pulse-power up to a kilowatt. With continued development of electronics, systems with pulse durations measured in picoseconds became possible. Applications are as varied as security and motion sensors, building stud-finders, collision-warning devices, and cardiac-dynamics monitors. Some of these devices are matchbox sized, including a long-life power source.

Radar Astronomy

As radar was being developed, astronomers considered its application in making observations of the Moon and other near-by extraterrestrial objects. In 1944, Zoltán Lajos Bay had this as a major objective as he developed a radar in Hungary. His radar telescope was taken away by the conquering Soviet army and had to be rebuilt, thus delaying the experiment. Under Project Diana conducted by the Army's Evans Signal Laboratory in New Jersey, a modified SCR-271 radar (the fixed-position version of the SCR-270) operating at 110 MHz with 3 kW peak-power, was used in receiving echoes from the Moon on January 10, 1946. Zoltán Bay accomplished this on the following February 6. Radio astronomy also had its start following WWII, and many scientists involved in radar development then entered this field. A number of radio observatories were constructed during the following years; however, because of the additional cost and complexity of involving transmitters and associated receiving equipment, very few were dedicated to radar astronomy. In fact, essentially all major radar astronomy activities have been conducted as adjuncts to radio astronomy observatories.

The radio telescope at the Arecibo Observatory, opened in 1963, is the largest in the world. Owned by the U.S. National Science Foundation and contractor operated, it is used primarily for radio astronomy, but equipment is available for radar astronomy. This includes transmitters operating at 47 MHz, 439 MHz, and 2.38 GHz, all with very-high pulse power. It has a 305-m (1,000-ft) primary reflector fixed in position; the secondary reflector is on tracks to allow precise pointing to different parts of the sky. Many significant scientific discoveries have been made using the Arecibo radar telescope, including mapping of surface roughness of Mars and observations of Saturns and its largest moon, Titan. In 1989, the observatory radar-imaged an asteroid for the first time in history.

Several spacecraft orbiting the Moon, Mercury, Venus, Mars, and Saturn have carried radars for surface mapping; a ground-penetration radar was carried on the Mars Express mission. Radar systems on a number of aircraft and orbiting spacecraft have mapped the entire Earth for various purposes; on the Shuttle Radar Topography Mission, the entire planet was mapped at a 30-m resolution.

The Jodrell Bank Observatory, an operation of the University of Manchester in Britain, was originally started by Bernard Lovell to be a radar astronomy facility. It initially used a war-surplus GL-II radar system operating at 71 MHz (4.2 m). The first observations were of ionized trails in the Geminids meteor shower during December 1945. While the facility soon evolved to become the third largest radio observatory in the world, some radar astronomy continued. The largest (250-ft or 76-m in diameter) of their three fully steerable radio telescopes became operational just in time to radar track *Sputnik 1*, the first artificial satellite, in October 1957.

COMPUTERS

A computer is a programmable machine that receives input, stores and manipulates data, and provides output in a useful format.

Although mechanical examples of computers have existed through much of recorded human history, the first electronic computers were developed in the mid-20th century (1940–1945). These were the size of a large room, consuming as much power as several hundred modern personal computers (PCs). Modern computers based on integrated circuits are millions to billions of times more capable than the early machines, and occupy a fraction of the space. Simple computers are small enough to fit into small pocket devices, and can be powered by a small battery. Personal computers in their various forms are icons of the Information Age and are what most people think of as "computers". However, the embedded computers found in many devices from MP3 players to fighter aircraft and from toys to industrial robots are the most numerous.

The ability to store and execute lists of instructions called programs makes computers extremely versatile, distinguishing them from calculators. The Church–Turing thesis is a mathematical statement of this versatility: any computer with a certain minimum capability is, in principle, capable of performing the same tasks that any other computer can perform. Therefore, computers ranging from a netbook to a supercomputer are all able to perform the same computational tasks, given enough time and storage capacity.

MICROPROCESSORS

In 1969, Ted Hoff conceived the commercial microprocessor at Intel and thus ignited the development of the personal computer. Hoff's invention was part of an order by a Japanese company for a desktop programmable electronic calculator, which Hoff wanted to build as cheaply as possible. The first realization of the microprocessor was the Intel 4004, a 4-bit processor, in 1969, but only in 1973 did the Intel 8080, an 8-bit processor, make the building of the first personal computer, the MITS Altair 8800, possible. The first PC was announced to the general public on the cover of the January 1975 issue of Popular Electronics.

Many electronics engineers today specialize in the development of programs for microprocessor based electronic systems, known as embedded systems. Hybrid specializations such as Computer Engineering have emerged due to the detailed knowledge of the hardware that is required for working on such systems. Software engineers typically do not study microprocessors at the same level as computer and electronics engineers. Engineers who exclusively carry out the role of programming embedded systems or microprocessors are referred to as "embedded systems engineers", or "firmware engineers".

MICROPROCESSOR CHRONOLOGY

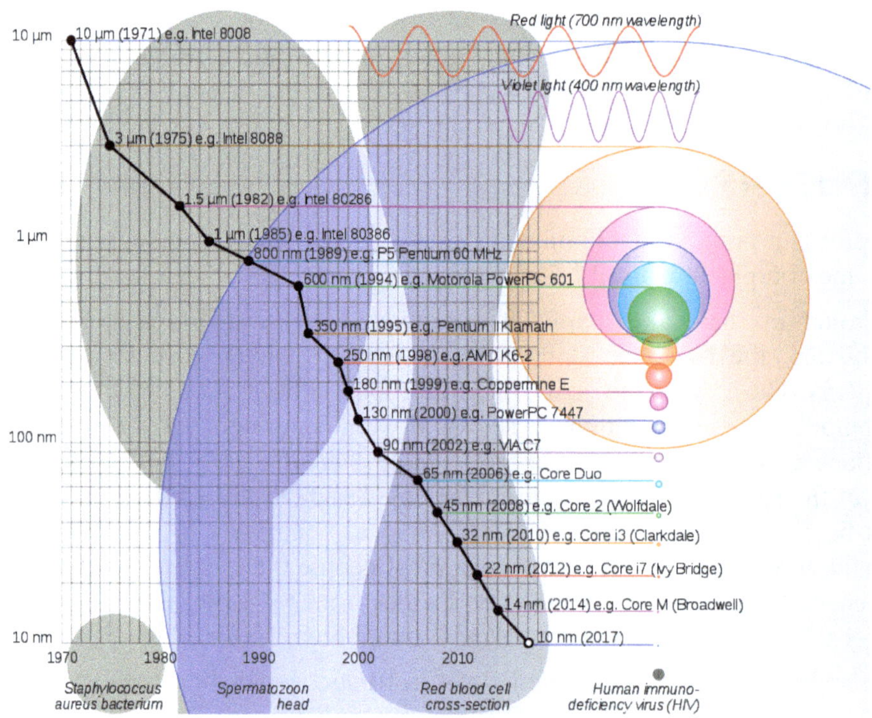

Progress of miniaturisation, and comparison of sizes of semiconductor manufacturing process nodes with some microscopic objects and visible light wavelengths.

1970s

The first microprocessors were manufactured in the 1970s. Designers predominantly used NMOS logic and they experimented with various word lengths. Early on, 4-bit processors were common (e.g. Intel 4004). Later in the decade, 8-bit processors such as the Motorola 6502 superseded the 4-bit chips. 16-bit processors emerged by the decade's end. Some unusual word lengths were tried, including 12-bit and 20-bit. The 20-bit MP944, designed for the U.S. Navy's F-14 Tomcat fighter, is considered by its designer to be the first microprocessor. It was classified by the Navy until 1998, meaning that Intel's 4004 was widely regarded as the first-ever microprocessor.

Date	Name	Developer	Max clock (first version)	Word size (bits)	Process	Transistors
1970	MP944	Garrett AiResearch	375 kHz	20		Multi-chip, pMOS Enhanced Mode
1971	4004	Intel	740 kHz	4	10 µm	2,250 pMOS
1972	PPS-25	Fairchild	400 kHz	4		Multi-chip, pMOS
1972	8008	Intel	500 kHz	8	10 µm	3,500 pMOS
1972	PPS-4	Rockwell	200 kHz	4		pMOS
1973	µCOM 4	NEC	1 MHz	4		2,500 NMOS
1973	IMP-16	National	715 kHz	16		Multi-chip, pMOS
1973	Mini-D	Burroughs	1 MHz	8		pMOS
1974	IMP-8	National	715 kHz	8		Multi-chip, pMOS
1974	8080	Intel	2 MHz	8	6 µm	6,000 NMOS
1974	5065	Mostek	1.4 MHz	8		pMOS
1974	TLCS-12	Toshiba	1 MHz	12		NMOS
1974	CP1600	General Instrument	3.3 MHz	16		NMOS
1974	IMP-4	National	500 kHz	4		Multi-chip, pMOS
1974	4040	Intel	740 kHz	4	10 µm	3,000 pMOS
1974	6800	Motorola	1 MHz	8	-	4,100 NMOS
1974	TMS 1000	Texas Instruments	400 kHz	4	8 µm	8,000
1974	PACE	National		16		pMOS
1974	ISP-8A/500 (SC/MP)	National	1 MHz	8		P Channel MOS technology
1975	6100	Intersil	4 MHz	12	-	4,000 CMOS
1975	2650	Signetics	1.2 MHz	8		NMOS
1975	PPS-8	Rockwell	256 kHz	8		pMOS
1975	F-8	Fairchild	2 MHz	8		NMOS
1975	CDP 1801	RCA	2 MHz	8	5 µm	5,000 CMOS two-chip

1975	6502	MOS Technology	1 MHz	8	-	3,510 dynamic NMOS
1975	BPC	Hewlett Packard	10 MHz	16	-	6,000 + ROM
1975	MCP-1600	Western Digital	3.3 MHz	16	-	Multi-chip, NMOS
1976	CDP 1802	RCA	6.4 MHz	8		CMOS
1976	Z-80	Zilog	2.5 MHz	8	4 μm	8,500 NMOS
1976	TMS9900	Texas Instruments	3.3 MHz	16	-	8,000
1976	8x300	Signetics	8 MHz	8		Bipolar
1977	8085	Intel	3.0 MHz	8	3 μm	6,500
1978	6809	Motorola	1 MHz	8	5 μm	40,000
1978	8086	Intel	5 MHz	16	3 μm	29,000
1978	6801	Motorola	-	8	5 μm	35,000
1979	Z8000	Zilog	-	16	-	17,500
1979	8088	Intel	5 MHz	8/16	3 μm	29,000 NMOS HMOS
1979	68000	Motorola	8 MHz	16/32	3.5 μm	68,000 NMOS HMOS

This list is incomplete; you can help by expanding it.

1980s

In the 1980s, 16-bit and 32-bit microprocessors were common among new designs, and CMOS technology overtook NMOS. Transistor count increased dramatically during the decade.

The home computers of the 1980s predominantly used processors that were introduced in the 1970s. Versions of the Motorola 6502, first released in 1975, and the Zilog Z80 (1976), were at the core of many of the home computers, such as the Commodore 64 and the ZX Spectrum. Even the first-generation IBM PC used a processor from the 1970s, the Intel 8088.

It was not until Intel's 80286 (in the IBM PC AT), and later the 80386, that processors designed in the 1980s drove the computers of the 1980s. These processors offered higher clock speeds and 32-bit word length as well as new operating modes, such as protected mode, that were not available in earlier chips. Critically, protected mode allowed the use of virtual memory and brought the graphical user interface to business computers, beginning with Microsoft Windows 2.0.

Date	**Name**	**Developer**	**Clock**	**Word size (bits)**	**Process**	**Transistors**
1980	16032	National Semiconductor	-	16/32	-	60,000
1981	6120	Harris Corporation	10 MHz	12	-	20,000 CMOS
1981	ROMP	IBM	10 MHz	32	2 μm	45,000
1981	T-11	DEC	2.5 MHz	16	5 μm	17,000 NMOS

1982	RISC-I	UC Berkeley	1 MHz	-	5 µm	44,420 NMOS
1982	FOCUS	Hewlett Packard	18 MHz	32	1.5 µm	450,000
1982	80186	Intel	6 MHz	16	-	55,000
?	80C186	Intel	6 MHz	16	-	? CMOS
1982	80188	Intel	8 MHz	8/16	-	29,000
1982	80286	Intel	6 MHz	16	1.5 µm	134,000
1983	RISC-II	UC Berkeley	3 MHz	-	3 µm	40,760 NMOS
1983	MIPS	Stanford University	2 MHz	32	3 µm	25,000
1984	68020	Motorola	16 MHz	32	2 µm	190,000
1984	32032	National Semiconductor	-	32	-	70,000
1984	V20	NEC	5 MHz	8/16	-	63,000
1985	80386	Intel	16–40 MHz	32	1.5 µm	275,000
1985	MicroVax II 78032	DEC	5 MHz	32	3.0 µm	125,000
1985	R2000	MIPS	8 MHz	32	2 µm	115,000
1985	Novix NC4016	Harris Corporation	8 MHz	16	3 µm	16,000
1986	Z80000	Zilog	-	32	-	91,000
1986	SPARC	Sun	40 MHz	32	0.8 µm	800,000
1986	V60	NEC	16 MHz	16/32	1.5 µm	375,000
1987	CVAX 78034	DEC	12.5 MHz	32	2.0 µm	134,000
1987	ARM2	Acorn	8 MHz	32	2 µm	25,000
1987	Gmicro/200	Hitachi	-	-	1.0 µm	730,000
1987	68030	Motorola	16 MHz	32	1.3 µm	273,000
1987	V70	NEC	20 MHz	16/32	1.5 µm	385,000
1988	R3000	MIPS	12 MHz	32	1.2 µm	120,000
1988	80386SX	Intel	12–33 MHz	16/32	-	-
1988	i960	Intel	10 MHz	33/32	1.5 µm	250,000
1989	VAX DC520 "Rigel"	DEC	35 MHz	32	1.5 µm	320,000
1989	80486	Intel	25 MHz	32	1 µm	1,180,000
1989	i860	Intel	25 MHz	32	1 µm	1,000,000

This list is incomplete; you can help by expanding it.

1990s

The 32-bit microprocessor dominated the consumer market in the 1990s. Processor clock speeds increased by more than tenfold between 1990 and 1999, and 64-bit processors began to emerge later in the decade. In the 1990s, microprocessors no longer used the same clock speed for the processor and the RAM. Processors began to have a front-side bus (FSB) clock speed used in communication with RAM and other components.

Typically, the processor itself ran at a clock speed that was a multiple of the FSB clock speed. Intel's Pentium III, for example, had an internal clock speed of 450–600 MHz and a FSB speed of 100–133 MHz. Only the processor's internal clock speed is shown here.

Date	**Name**	**Developer**	**Clock**	**Word size (bits)**	**Process**	**Transistors (M)**	**threads per core**
1990	68040	Motorola	40 MHz	32	-	1.2	
1990	POWER1	IBM	20–30 MHz	32	1.0 µm	6.9	
1991	R4000	MIPS Computer Systems	100 MHz	64	0.8 µm	1.35	
1991	NVAX	DEC	62.5–90.91 MHz	-	0.75 µm	1.3	
1991	RSC	IBM	33 MHz	32	0.8 µm	1.0	
1992	Alpha 21064	DEC	100–200 MHz	64	0.75 µm	1.68	
1992	microSPARC I	Sun	40–50 MHz	32	0.8 µm	0.8	
1992	PA-7100	Hewlett Packard	100 MHz	32	0.80 µm	0.85	
1993	PowerPC 601	IBM, Motorola	50–80 MHz	32	0.6 µm	2.8	
1993	Pentium	Intel	60–66 MHz	32	0.8 µm	3.1	
1993	POWER2	IBM	55–71.5 MHz	32	0.72 µm	23	
1994	68060	Motorola	50 MHz	32	0.6 µm	2.5	
1994	Alpha 21064A	DEC	200–300 MHz	64	0.5 µm	2.85	
1994	R4600	QED	100–125 MHz	64	0.65 µm	2.2	
1994	PA-7200	Hewlett Packard	125 MHz	32	0.55 µm	1.26	
1994	PowerPC 603	IBM, Motorola	60–120 MHz	32	0.5 µm	1.6	
1994	PowerPC 604	IBM, Motorola	100–180 MHz	32	0.5 µm	3.6	
1994	PA-7100LC	Hewlett Packard	100 MHz	32	0.75 µm	0.90	
1995	Alpha 21164	DEC	266–333 MHz	64	0.5 µm	9.3	
1995	UltraSPARC	Sun	143–167 MHz	64	0.47 µm	5.2	
1995	SPARC64	HAL Computer Systems	101–118 MHz	64	0.40 µm	-	
1995	Pentium Pro	Intel	150–200 MHz	32	0.35 µm	5.5	
1996	Alpha 21164A	DEC	400–500 MHz	64	0.35 µm	9.7	
1996	K5	AMD	75–100 MHz	32	0.5 µm	4.3	
1996	R10000	MTI	150–250 MHz	64	0.35 µm	6.7	
1996	R5000	QED	180–250 MHz	-	0.35 µm	3.7	
1996	SPARC64 II	HAL Computer Systems	141–161 MHz	64	0.35 µm	-	
1996	PA-8000	Hewlett-Packard	160–180 MHz	64	0.50 µm	3.8	
1996	P2SC	IBM	150 MHz	32	0.29 µm	15	
1997	RS64	IBM	125 MHz	64	? nm	?	
1997	Pentium II	Intel	233–300 MHz	32	0.35 µm	7.5	
1997	PowerPC 620	IBM, Motorola	120–150 MHz	64	0.35 µm	6.9	

1997	UltraSPARC IIs	Sun	250–400 MHz	64	0.35 μm	5.4	
1997	S/390 G4	IBM	370 MHz	32	0.5 μm	7.8	
1997	PowerPC 750	IBM, Motorola	233–366 MHz	32	0.26 μm	6.35	
1997	K6	AMD	166–233 MHz	32	0.35 μm	8.8	
1998	RS64-II	IBM	262 MHz	64	350 nm	12.5	
1998	Alpha 21264	DEC	450–600 MHz	64	0.35 μm	15.2	
1998	MIPSR12000	SGI	270–400 MHz	64	0.25 μm, 0.18 μm	6.9	
1998	RM7000	QED	250–300 MHz	-	0.25 μm	18	
1998	SPARC64 III	HAL Computer Systems	250–330 MHz	64	0.24 μm	17.6	
1998	S/390 G5	IBM	500 MHz	32	0.25 μm	25	
1998	PA-8500	Hewlett Packard	300–440 MHz	64	0.25 μm	140	
1998	POWER3	IBM	200 MHz	64	0.25 μm	15	
1999	Pentium III	Intel	450–600 MHz	32	0.25 μm	9.5	
1999	RS64-III	IBM	450 MHz	64	220 nm	34	2
1999	PowerPC 7400	Motorola	350–500 MHz	32	200–130 nm	10.5	
1999	Athlon	AMD	500–1000 MHz	32	0.25 μm	22	

This list is incomplete; you can help by expanding it.

2000s

64-bit processors became mainstream in the 2000s. Microprocessor clock speeds reached a ceiling because of the *heat dissipation barrier*. Instead of implementing expensive and impractical cooling systems, manufacturers turned to parallel computing in the form of the multi-core processor. Overclocking had its roots in the 1990s, but came into its own in the 2000s. Off-the-shelf cooling systems designed for overclocked processors became common, and the gaming PC had its advent as well. Over the decade, transistor counts increased by about an order of magnitude, a trend continued from previous decades. Process sizes decreased about fourfold, from 180 nm to 45 nm.

Date	Name	Developer	Clock	Process	Transistors (M)	Cores per die / Dies per module
2000	Athlon XP	AMD	1.33–1.73 GHz	180 nm	37.5	1 / 1
2000	Duron	AMD	550 MHz–1.3 GHz	180 nm	25	1 / 1
2000	RS64-IV	IBM	600–750 MHz	180 nm	44	1 / 2
2000	Pentium 4	Intel	1.3–2 GHz	180–130 nm	42	1 / 1
2000	SPARC64 IV	Fujitsu	450–810 MHz	130 nm	-	1 / 1

2000	z900	IBM	918 MHz	180 nm	47	1 / 12, 20
2001	MIPS R14000	SGI	500–600 MHz	130 nm	7.2	1 / 1
2001	POWER4	IBM	1.1–1.4 GHz	180–130 nm	174	2 / 1, 4
2001	UltraSPARC III	Sun	750–1200 MHz	130 nm	29	1 / 1
2001	Itanium	Intel	733–800 MHz	180 nm	25	1 / 1
2001	PowerPC 7450	Motorola	733–800 MHz	180–130 nm	33	1 / 1
2002	SPARC64 V	Fujitsu	1.1–1.35 GHz	130 nm	190	1 / 1
2002	Itanium 2	Intel	0.9–1 GHz	180 nm	410	1 / 1
2003	PowerPC 970	IBM	1.6–2.0 GHz	130–90 nm	52	1 / 1
2003	Pentium M	Intel	0.9–1.7 GHz	130–90 nm	77	1 / 1
2003	Opteron	AMD	1.4–2.4 GHz	130 nm	106	1 / 1
2004	POWER5	IBM	1.65–1.9 GHz	130–90 nm	276	2 / 1, 2, 4
2004	PowerPC BGL	IBM	700 MHz	130 nm	95	2 / 1
2005	Opteron "Athens"	AMD	1.6–3.0 GHz	90 nm	114	1 / 1
2005	Pentium D	Intel	2.8–3.2 GHz	90 nm	115	1 / 2
2005	Athlon 64 X2	AMD	2–2.4 GHz	90 nm	243	2 / 1
2005	PowerPC 970MP	IBM	1.2–2.5 GHz	90 nm	183	2 / 1
2005	UltraSPARC IV	Sun	1.05–1.35 GHz	130 nm	66	2 / 1
2005	UltraSPARC T1	Sun	1–1.4 GHz	90 nm	300	8 / 1
2005	Xenon	IBM	3.2 GHz	90–45 nm	165	3 / 1
2006	Core Duo	Intel	1.1–2.33 GHz	90–65 nm	151	2 / 1
2006	Core 2	Intel	1.06–2.67 GHz	65–45 nm	291	2 / 1, 2
2006	Cell/B.E.	IBM, Sony, Toshiba	3.2–4.6 GHz	90–45 nm	241	1+8 / 1
2006	Itanium "Montecito"	Intel	1.4–1.6 GHz	90 nm	1720	2 / 1
2007	POWER6	IBM	3.5–4.7 GHz	65 nm	790	2 / 1
2007	SPARC64 VI	Fujitsu	2.15–2.4 GHz	90 nm	543	2 / 1
2007	UltraSPARC T2	Sun	1–1.4 GHz	65 nm	503	8 / 1
2007	TILE64	Tilera	600–900 MHz	90–45 nm	?	64 / 1
2007	Opteron "Barcelona"	AMD	1.8–3.2 GHz	65 nm	463	4 / 1
2007	PowerPC BGP	IBM	850 MHz	90 nm	208	4 / 1
2008	Phenom	AMD	1.8–2.6 GHz	65 nm	450	2, 3, 4 / 1
2008	z10	IBM	4.4 GHz	65 nm	993	4 / 7
2008	PowerXCell 8i	IBM	2.8–4.0 GHz	65 nm	250	1+8 / 1
2008	SPARC64 VII	Fujitsu	2.4–2.88 GHz	65 nm	600	4 / 1

2008	Atom	Intel	0.8–1.6 GHz	65–45 nm	47	1 / 1
2008	Core i7	Intel	2.66–3.2 GHz	45–32 nm	730	2, 4, 6 / 1
2008	TILEPro64	Tilera	600–866 MHz	90–45 nm	?	64 / 1
2008	Opteron "Shanghai"	AMD	2.3–2.9 GHz	45 nm	751	4 / 1
2009	Phenom II	AMD	2.5–3.2 GHz	45 nm	758	2, 3, 4, 6 / 1
2009	Opteron "Istanbul"	AMD	2.2–2.8 GHz	45 nm	904	6 / 1

2010s

Date	Name	Developer	Clock	Process	Transistors (M)	Cores per die / Dies per module	threads per core
2010	POWER7	IBM	3–4.14 GHz	45 nm	1200	4, 6, 8 / 1, 4	4
2010	Itanium "Tukwila"	Intel	2 GHz	65 nm	2000	2, 4 / 1	2
2010	Opteron "Magny-cours"	AMD	1.7–2.4 GHz	45 nm	1810	4, 6 / 2	1
2010	Xeon "Nehalem-EX"	Intel	1.73–2.66 GHz	45 nm	2300	4, 6, 8 / 1	2
2010	z196	IBM	3.8–5.2 GHz	45 nm	1400	4 / 1, 6	1
2010	SPARC T3	Sun	1.6 GHz	45 nm	2000	16 / 1	8
2010	SPARC64 VII+	Fujitsu	2.66–3.0 GHz	45 nm	?	4 / 1	2
2010	Intel "Westmere"	Intel	1.86–3.33 GHz	32 nm	1170	4–6 / 1	2
2011	Intel "Sandy Bridge"	Intel	1.6–3.4 GHz	32 nm	995	2, 4 / 1	(1,) 2
2011	AMD Llano	AMD	1.0–1.6 GHz	40 nm	380	1, 2 / 1	1
2011	Xeon E7	Intel	1.73–2.67 GHz	32 nm	2600	4, 6, 8, 10 / 1	1–2
2011	PowerPC BGQ	IBM	1.6 GHz	45 nm	1470	18 / 1	4
2011	SPARC64 VIIIfx	Fujitsu	2.0 GHz	45 nm	760	8 / 1	2
2011	FX "Bulldozer" Interlagos	AMD	3.1–3.6 GHz	32 nm	1200	4–8 / 2	1
2011	SPARC T4	Oracle	2.8–3 GHz	40 nm	855	8 / 1	8
2012	SPARC64 IXfx	Fujitsu	1.848 GHz	40 nm	1870	16 / 1	2
2012	zEC12	IBM	5.5 GHz	32 nm	2750	6 / 6	1
2012	POWER7+	IBM	3.1–5.3 GHz	32 nm	2100	8 / 1, 2	4
2012	Itanium "Poulson"	Intel	1.73–2.53 GHz	32 nm	3100	8 / 1	2
2013	Intel "Haswell"	Intel	1.9–4.4 GHz	22 nm	1400	4 / 1	2
2013	SPARC64 X	Fujitsu	2.8–3 GHz	28 nm	2950	16 / 1	2
2013	SPARC T5	Oracle	3.6 GHz	28 nm	1500	16 / 1	8

2014	POWER8	IBM	2.5–5 GHz	22 nm	4200	6, 12 / 1, 2	8

REFERENCES

1. Raymond C. Watson, Jr.; Radar Origins Worldwide", Trafford Publishing, 2009.
2. Part 4 - America Between The Wars; http://www.armysignalocs.com/index_oct_13.html
3. «L'histoire du "radar", les faits". Archived from the original on October 5, 2007. Le principe fondamental du radar appartient au patrimoine commun des physiciens : ce qui demeure en fin de compte au crédit réel des techniciens se mesure à la réalisation effective de matériels opérationnels
4. van Keuren, D.K. (1997). "Science Goes to War: The Radiation Laboratory, Radar, and Their Technological Consequences". Reviews in American History. 25: 643–647. doi:10.1353/rah.1997.0150.
5. Buderi, Robert; The Invention that Changed the World, Simon & Schuster, 1996
6. Wald, Matthew L. (June 22, 1997). "Jam Sessions". New York Times.
7. Such experiments were conducted by Oliver Lodge, Jagadish Chandra Bose, and Alexander Stepanovich Popov.
8. Marconi, Guglielmo (1922). "Radio Telegraphy". Proc. IRE. 10 *(4): 215–238. doi:10.1109/JRPROC.1922.219820.*
9. «Development of A Monopulse Radar System", Kirkpatrick, George M., letter to IEEE Transactions on Aerospace and Electronic Systems, vol. 45, no. 2 (April 2009).
10. James B. Campbell, Randolph H. Wynne, Introduction to Remote Sensing, Fifth Edition, Guilford Press - 2011, page 207
11. Christian Hülsmeyer in Radar World
12. Raymond C. Watson, Jr., "Radar Origins Worldwide", Trafford Publishing, 2009, p. 45.
13. Bowen, E. G.; Radar Days, Inst. of Physics Publishing, 1987, p. 16
14. Latham, Colin, and Anne Stobbs (2011). The Birth of British Radar: The memoirs of Arnold ‹Skip' Wilkins, Second Edition, Radio Society of Great Britain, ISBN 9781-9050-8675-7
15. Judkins, Phil. "Making Vision into Power", International Journal of Engineering and Technology, Vol 82, No 1 (January 2012), pp. 103-104.
16. Judkins, p.113.
17. Judkins, pp.113-114.
18. Judkins, p.109.
19. Judkins, p.114.

20. Judkins, p.116.

21. Judkins, p.117.

22. Judkins, p.116: quoting from Jones, R. V. Most Secret War (London: Hamish Hamilton, 1978), pp.155-156.

23. Judkins, p.115.

24. Braham, J. R. D., Wing Commander, RAF. Night Fighter (Bantam, 1984).

25. Judkins, pp.114, 116, 118, & 119-120.

26. Butement, W. A. S., and P. E. Pollard; "Coastal Defense Apparatus", Recorded in the Inventions Book of the Royal Engineers Board, Jan. 1931

27. Coales, J. F., and J. D. S. Rawlinson; "The Development of Naval Radar 1935-1945", J. Naval Science, vol. 13, nos. 2-3, 1987.

28. Kummritz, Herbert; "On the Development of Radar Technologies in Germany up to 1945", in Tracking the History of Radar, ed. by Oskar Blumtritt et al., IEEE-Rutgers, 1994

29. Kroge, Harry von; GEMA: Birthplace of German Radar and Sonar, translated by Louis Brown, Inst. of Physics Publishing, 2000

30. "Telefunken firm in Berlin reveals details of a 'mystery ray' system capable of locating position of aircraft through fog, smoke and clouds." Electronics, September 1935

31. Runge. W.; "A personal reminiscence", in Radar Development to 1945, edited by Russell Burns, Peter Peregrinus Ltd, 1988, p.227

32. Brown, Louis; A Radar History of World War II; Inst. of Physics Publishing, 1999, p.43

33. Hyland, L.A., A.H. Taylor, and L.C. Young; "System for detecting objects by radio", U.S. Patent No. 1981884, 27 Nov. 1934

34. Breit, Gregory, and Merle A. Tuve; "A Radio Method for Estimating the Height of the Conducting Layer", Nature, vol. 116, 1925, p. 116

35. Page, Robert Morris; The Origin of Radar, Doubleday, 1962, p. 66.

36. Coulton, Roger B.; "Radar in the U.S. Army", Proc. IRE, vol. 33, 1947, pp. 740-753

37. Erickson, J.; "The air defense problem and the Soviet radar programme 1934/35-1945", in Radar Development to 1945, ed. by Russell Burns, Peter Peregrinus Ltd., 1988, pp. 227-234

38. Ioffe, A. F.; "Contemporary problems of the development of the technology of air defense", Sbornik PVO, February 1934 (in Russian)

39. Shembel, B. K.; At the Origin of Radar in USSR, Sovetskoye Radio, 1977 (in Russian)

40. Slutzkin, A. A., and D. S. Shteinberg, "Die Erzeugung von kurzwelligen ungedämpften Schwingungen bei Anwendung des Magnetfeldes" ["The generation of undamped shortwave oscillations by application of a magnetic field"], Annalen

der Physik, vol. 393, no. 5, pages 658-670 (May 1929)

41. Siddiqi, Asif A.; "Rockets Red Glare: Technology, Conflict, and Terror in the Soviet Union"; Technology & Culture, vol. 44, 2003, p.470
42. Lobanov, M. M.; The Beginning of Soviet Radar, Sovetskoye Radio, 1975 (in Russian)
43. Watson, Raymond C. (2009).Radar Origins Worldwide. Trafford Publishing, p. 306. ISBN 1-4269-2110-1
44. Kostenko, Alexei A., Alexander I, Nosich, and Irina A. Tishchenko; "Development of the First Soviet Three-Coordinate L-Band Pulsed Radar in Kharkov Before WWII", IEEE Antennas and Propagation Magazine, vol. 43, June 2001, pp. 29-48; http://www.ire.kharkov.ua/apm2001-radar.pdf
45. Chernyak, V. S., I. Ya. Immoreev, and B. M. Vovshin; "Radar in the Soviet Union and Russia: A Brief Historical Outline", IEEE AES Magazine, vol. 19, December 2003, p. 8
46. Yagi, H., "Beam Transmission of Ultra Short Waves", Proc. of the IRE, vol. 16, June 1928
47. Nakajima, S., "The history of Japanese radar development to 1945", in Russell Burns, Radar Development to 1945, Peter Peregrinus Ltd, 1988
48. Wilkinson, Roger I.; "Short survey of Japanese radar – Part I", Trans. AIEE, vol. 65, 1946, p. 370
49. Nakajima, S.; "Japanese radar development prior to 1945", IEEE Antennas and Propagation Magazine, vol. 34, Dec., 1992, pp. 17-22
50. Le Pair, C. (Kees); "Radar in the Dutch Knowledge Network", Telecommunication and Radar Conference, EUMW98, Amsterdam, 1998; http://www.clepair.net/radar-web.htm
51. Posthumus, K; "Oscillations in a Split-Anode Magnetron, Mechanism of Generation", Wireless Engineer, vol. 12, 1935, pp. 126-13
52. Staal, M., and J.L.C. Weiller; "Radar Development in the Netherlands before the war", in Radar Development to 1945, ed. by Russell Burns, Peter Peregrinus, 1988, pp. 235-237
53. "Measurements Building" http://www.museumwaalsdorp.nl/en/meetgben.html
54. Swords, S. S.; Technical history of the beginnings of radar, Peter Peregrinus Ltd, 1986, pp. 142-144
55. French patent (no. 788.795, "New system of location of obstacles and its applications")
56. Molyneux-Berry, R. B.; "Henri Gutton, French radar pioneer", in Radar Development to 1945, ed. by Russell Burns, Peter Peregrinus, 1988, pp. 45-52
57. "System for Object Detection and Distance Measurement" http://www.freepatentsonline.com/2433838.html
58. David, Pierre; Le Radar (The Radar), Presses Universitaires de France, 1949 (in French)

59. Megaw, Eric C. S.; "The High-Power Magnetron: A Review of Early Developments", J. of the IEE, vol. 93, 1946, p. 928, see copy at http://ieeexplore.ieee.org/xpls/abs_all.jsp?arnumber=5299357
60. Paul A. Redhead, The invention of the cavity magnetron and its introduction into Canada and the U.S.A., PHYSICS IN CANADA, November/December 2001, http://www.cap.ca/wyp/profiles/Redhead-Nov01.PDF
61. Calamia, M., and R. Palandri; "The History of the Italian Radio Detector Telemetro", in Radar Development to 1945, ed. by Russell Burns, Peter Peregrinus, 1988, pp. 97-105
62. Carrara, N.; "The detection of microwaves", Proc. of the IRE, vol. 20, Oct. 1932, pp. 1615-1625
63. Tiberio, U.; "Some historical data concerning the first Italian naval radar", IEEE Trans. AES, vol. 15, Sept., 1979, p. 733
64. Sinnott, D. H.; "Radar Development in Australia: 1939 to Present", Proc. of IEEE 2005 International Radar Conference, 9–12 May, pp. 5-9
65. Lamb, James B. (1987). On the triangle run. Toronto: Totem Books. pp. 26–28. ISBN 0-00-217909-1.
66. Moorcroft, Don; "Origins of Radar-based Research in Canada", Univ. Western Ontario, 2002; http://www.physics.uwo.ca/~drm/history/radar/radar_history.html
67. Unwin, R. S.; "Development of Radar in New Zealand in World War II", IEEE Antennas and Propagation Magazine, vol. 34, June, pp.31-33, 1992
68. Hewitt, F. J.; "South Africa's Role in the Development and Use of Radar in World War II", Military History Journal, vol. 3, no, 3, June 1975; http://samilitaryhistory.org/vol033fh.html
69. Renner, Peter; "The Role of the Hungarian Engineers in the Development of Radar Systems", Periodica Polytechnica Ser. Soc. Man. Sci, Vol. 12, p. 277, 2004; http://www.pp.bme.hu/so/2004_2/pdf/so2004_2_12.pdf
70. Angela Hind (February 5, 2007). "Briefcase ‹that changed the world'". BBC News. Retrieved 2007-08-16.
71. James Phinney Baxter III (Official Historian of the Office of Scientific Research and Development), Scientists Against Time (Boston: Little, Brown, and Co., 1946), page 142.
72. Barlow, E. J.; "Doppler Radar", Proc. IRE, vol. 37, pp. 340-355, April 1949
73. Page, R. M.; "Monopulse Radar", op. cet.
74. Von Aulock, W. H.; "Properties of Phased Arrays", Proc. IRE, vol. 48, pp. 1715-1727, Oct., 1960
75. "Airborne Synthetic Aperture Radar"; http://airsar.jpl.nasa.gov/
76. "ABM Research and Development at Bell Laboratories", http://srmsc.org/pdf/004438p0.pdf
77. "Cobra Mist"; http://www.users.zetnet.co.uk/Ray.Flint/cobra/cm.htm

78. "Mystery Signals Of The Short Wave", Wireless World, Feb. 1977; http://www.brogers.dsl.pipex.com/Wpecker2.htm
79. "Airport Surveillance Radars"; http://www.faa.gov/air_traffic/technology/asr-11/
80. David Atlas, "Radar in Meteorology", published by American Meteorological Society
81. «Stormy Weather Group". McGill University. 2000. Retrieved 2006-05-21.
82. Whiton, Roger C., et al. "History of Operational Use of Weather Radar by U.S. Weather Services. Part I: The Pre-NEXRAD Era"; Weather and Forecasting, vol. 13, no. 2, pp. 219–243, 19 Feb. 1998; http://ams.allenpress.com/amsonline/?request=get-document&doi=10.1175%2F1520-0434(1998)013%3C0219:HOOUOW%3E2.0.CO%3B2
83. Susan Cobb (October 29, 2004). "Weather radar development highlight of the National Severe Storms Laboratory first 40 years". NOAA Magazine. NOAA. Retrieved 2009-03-07.
84. Crozier, C.L.; P.I. Joe; J.W. Scott; H.N. Herscovitch; T.R. Nichols (1990). "The King City Operational Doppler Radar: Development, All-Season Applications and Forecasting (PDF)" (PDF). Canadian Meteorological and Oceanographic Society. Archived from the original (PDF) on 2006-10-02. Retrieved 2006-05-24.
85. «Information about Canadian radar network". The National Radar Program. Environment Canada. 2002. Retrieved 2006-06-14.
86. Parent du Châtelet, Jacques; et al. (2005). "The PANTHERE project and the evolution of the French operational radar network and products: Rain estimation, Doppler winds, and dual polarization" (PDF). Météo-France. 32nd Radar Conference of the AMS, Albuquerque, NM. Retrieved 2006-06-24.
87. Daniels, Jeffrey J.; "Ground Penetrating Radar Fundamentals"; http://www.clu-in.org/download/char/GPR_ohio_stateBASICS.pdf
88. "Micropower Impulse Radar"; https://ipo.llnl.gov/?q=technologies-micropower_impulse_radar
89. Mofenson, Jack; "Radio Echoes From the Moon", Electronics, April 1946; http://www.campevans.org/_CE/html/diamof.html
90. Bay, Z.; "Reflection of microwaves from the moon", Hung. Acta Phys., vol. 1, pp. 1-22, April 1946.
91. Lovell, Bernard; Story of Jodrell Bank, Oxford U. Press, 1968

CHAPTER 3

Semiconductor Devices

Chapter Contents

- **Introduction**
- **Transistor**
- **History of Semiconductor Device Development**

A semiconductor is used to manipulate the flow of energy within an electronic setting. Semiconductor devices are the many manipulations of this technology to provide a specific control over the energy. These devices are found in many forms and within many types of electronics.

Semiconductor devices are comprised of material that can be manipulated easily to control how power reaches an electronic device. Semiconductors can be altered by using an induction of an electric field, exposure to light, pressure and heat. These devices are comprised of material with holes in it, which are positive, or p-type, and material with extra electrons, which are negative, or n-type.

INTRODUCTION

Semiconductor devices are electronic components that exploit the electronic properties of semiconductor materials, principally silicon, germanium, and gallium arsenide, as well as organic semiconductors. Semiconductor devices have replaced thermionic devices (vacuum tubes) in most applications. They use electronic conduction in the solid state as opposed to the gaseous state or thermionic emission in a high vacuum.

Semiconductor devices are manufactured both as single discrete devices and as *integrated circuits* (ICs), which consist of a number—from a few (as low as two) to billions—of devices manufactured and interconnected on a single semiconductor substrate, or wafer. Semiconductor materials are useful because their behavior can be easily manipulated by the addition of impurities, known as doping. Semiconductor conductivity can be controlled by the introduction of an electric or magnetic field, by exposure to light or heat, or by the mechanical deformation of a doped monocrystalline grid; thus, semiconductors can make excellent sensors. Current conduction in a semiconductor occurs via mobile or "free" *electrons* and *holes*, collectively known as *charge carriers*. Doping a semiconductor such as silicon with a small amount of impurity atoms, such as phosphorus or boron, greatly increases the number of free electrons or holes within the semiconductor. When a doped semiconductor contains excess holes it is called "p-type", and when it contains excess free electrons it is known as "n-type", where *p* (positive for holes) or *n* (negative for electrons) is the sign of the charge of the majority mobile charge carriers. The semiconductor material used in devices is doped under highly controlled conditions in a fabrication facility, or *fab*, to control precisely the location and concentration of p- and n-type dopants. The junctions which form where n-type and p-type semiconductors join together are called p–n junctions.

TRANSISTOR

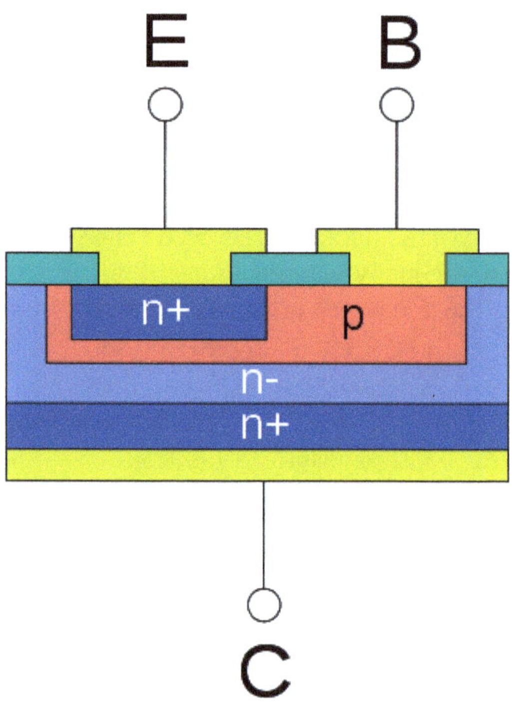

An n–p–n bipolar junction transistor structure.

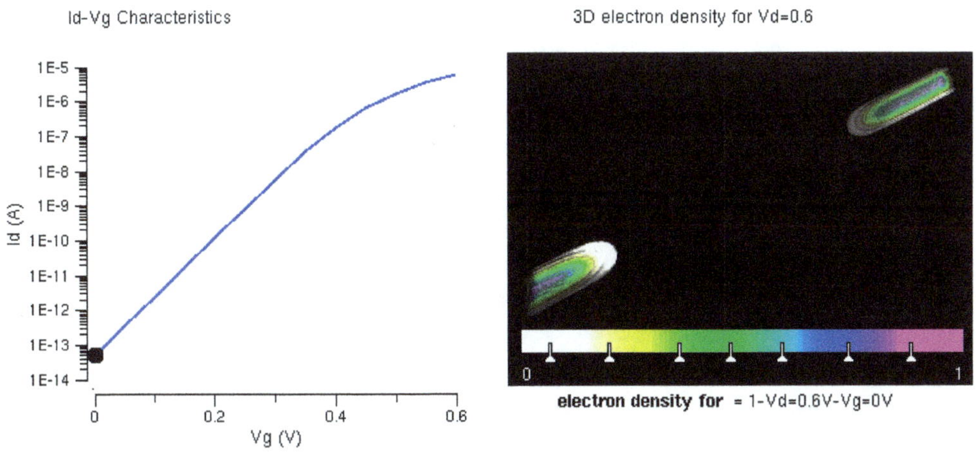

Operation of a MOSFET and its Id-Vg curve. At first, when no gate voltage is applied. There is no inversion electron in the channel, the device is OFF. As gate voltage increase, inversion electron density in the channel increase, current increase, the device turns on.

Bipolar Junction Transistor

Bipolar junction transistors are formed from two p–n junctions, in either n–p–n or p–n–p configuration. The middle, or *base*, region between the junctions is typically very narrow. The other regions, and their associated terminals, are known as the *emitter* and the *collector*. A small current injected through the junction between the base and the emitter changes the properties of the base-collector junction so that it can conduct current even though it is reverse biased. This creates a much larger current between the collector and emitter, controlled by the base-emitter current.

Field-Effect Transistor

Another type of transistor, the field-effect transistor, operates on the principle that semiconductor conductivity can be increased or decreased by the presence of an electric field. An electric field can increase the number of free electrons and holes in a semiconductor, thereby changing its conductivity. The field may be applied by a reverse-biased p–n junction, forming a *junction field-effect transistor* (JFET) or by an electrode insulated from the bulk material by an oxide layer, forming a *metal–oxide–semiconductor field-effect transistor* (MOSFET).

The MOSFET, a solid-state device, is the most used semiconductor device today. The *gate* electrode is charged to produce an electric field that controls the conductivity of a "channel" between two terminals, called the *source* and *drain*. Depending on the type of carrier in the channel, the device may be an *n-channel* (for electrons) or a *p-channel* (for holes) MOSFET. Although the MOSFET is named in part for its "metal" gate, in modern devices polysilicon is typically used instead.

Semiconductor Device Materials

By far, silicon (Si) is the most widely used material in semiconductor devices. Its combination of low raw material cost, relatively simple processing, and a useful temperature range makes it currently the best compromise among the various competing materials. Silicon used in semiconductor device manufacturing is currently fabricated into boules that are large enough in diameter to allow the production of 300 mm (12 in.) wafers.

Germanium (Ge) was a widely used early semiconductor material but its thermal sensitivity makes it less useful than silicon. Today, germanium is often alloyed with silicon for use in very-high-speed SiGe devices; IBM is a major producer of such devices.

Gallium arsenide (GaAs) is also widely used in high-speed devices but so far, it has been difficult to form large-diameter boules of this material, limiting the wafer diameter to sizes significantly smaller than silicon wafers thus making mass production of GaAs devices significantly more expensive than silicon.

Other less common materials are also in use or under investigation.

Silicon carbide (SiC) has found some application as the raw material for blue light-emitting diodes (LEDs) and is being investigated for use in semiconductor devices that could withstand very high operating temperatures and environments with the presence of significant levels of ionizing radiation. IMPATT diodes have also been fabricated from SiC.

Various indium compounds (indium arsenide, indium antimonide, and indium phosphide) are also being used in LEDs and solid state laser diodes. Selenium sulfide is being studied in the manufacture of photovoltaic solar cells.

The most common use for organic semiconductors is Organic light-emitting diodes.

List of Common Semiconductor Devices

Two-terminal Devices

- DIAC
- Diode (rectifier diode)
- Gunn diode
- IMPATT diode
- Laser diode
- Light-emitting diode (LED)
- Photocell
- Phototransistor
- PIN diode
- Schottky diode
- Solar cell
- Transient-voltage-suppression diode
- Tunnel diode

- VCSEL
- Zener diode

Three-terminal devices:

- Bipolar transistor
- Darlington transistor
- Field-effect transistor
- Insulated-gate bipolar transistor (IGBT)
- Silicon-controlled rectifier
- Thyristor
- TRIAC
- Unijunction transistor

Four-terminal devices:

- Hall effect sensor (magnetic field sensor)
- Photocoupler (Optocoupler)

Semiconductor Device Applications

All transistor types can be used as the building blocks of logic gates, which are fundamental in the design of digital circuits. In digital circuits like microprocessors, transistors act as on-off switches; in the MOSFET, for instance, the voltage applied to the gate determines whether the switch is on or off.

Transistors used for analog circuits do not act as on-off switches; rather, they respond to a continuous range of inputs with a continuous range of outputs. Common analog circuits include amplifiers and oscillators.

Circuits that interface or translate between digital circuits and analog circuits are known as mixed-signal circuits.

Power semiconductor devices are discrete devices or integrated circuits intended for high current or high voltage applications. Power integrated circuits combine IC technology with power semiconductor technology, these are sometimes referred to as "smart" power devices. Several companies specialize in manufacturing power semiconductors.

Component Identifiers

The type designators of semiconductor devices are often manufacturer specific. Nevertheless, there have been attempts at creating standards for type codes, and a subset of devices follow those. For discrete devices, for example, there are three standards: JEDEC JESD370B in United States, Pro Electron in Europe and Japanese Industrial Standards (JIS) in Japan.

HISTORY OF SEMICONDUCTOR DEVICE DEVELOPMENT

Cat's-whisker Detector

Semiconductors had been used in the electronics field for some time before the invention of the transistor. Around the turn of the 20th century they were quite common as detectors in radios, used in a device called a "cat's whisker" developed by Jagadish Chandra Bose and others. These detectors were somewhat troublesome, however, requiring the operator to move a small tungsten filament (the whisker) around the surface of a galena (lead sulfide) or carborundum (silicon carbide) crystal until it suddenly started working. Then, over a period of a few hours or days, the cat's whisker would slowly stop working and the process would have to be repeated. At the time their operation was completely mysterious. After the introduction of the more reliable and amplified vacuum tube based radios, the cat's whisker systems quickly disappeared. The "cat's whisker" is a primitive example of a special type of diode still popular today, called a Schottky diode.

Metal Rectifier

Another early type of semiconductor device is the metal rectifier in which the semiconductor is copper oxide or selenium. Westinghouse Electric (1886) was a major manufacturer of these rectifiers.

World War II

During World War II, radar research quickly pushed radar receivers to operate at ever higher frequencies and the traditional tube based radio receivers no longer worked well. The introduction of the cavity magnetron from Britain to the United States in 1940 during the Tizard Mission resulted in a pressing need for a practical high-frequency amplifier.

On a whim, Russell Ohl of Bell Laboratories decided to try a cat's whisker. By this point they had not been in use for a number of years, and no one at the labs had one. After hunting one down at a used radio store in Manhattan, he found that it worked much better than tube-based systems.

Ohl investigated why the cat's whisker functioned so well. He spent most of 1939 trying to grow more pure versions of the crystals. He soon found that with higher quality crystals their finicky behaviour went away, but so did their ability to operate as a radio detector. One day he found one of his purest crystals nevertheless worked well, and interestingly, it had a clearly visible crack near the middle. However as he moved about the room trying to test it, the detector would mysteriously work, and then stop again. After some study he found that the behaviour was controlled by the light in the room–more light caused more conductance in the crystal. He invited several other people to see this crystal, and Walter Brattain immediately realized there was some sort of junction at the crack.

Further research cleared up the remaining mystery. The crystal had cracked because either side contained very slightly different amounts of the impurities Ohl could not remove–about 0.2%. One side of the crystal had impurities that added extra electrons (the carriers of electric current) and made it a “conductor”. The other had impurities that wanted to bind to these electrons, making it (what he called) an “insulator”. Because the two parts of the crystal were in contact with each other, the electrons could be pushed out of the conductive side which had extra electrons (soon to be known as the *emitter*) and replaced by new ones being provided (from a battery, for instance) where they would flow into the insulating portion and be collected by the whisker filament (named the *collector*). However, when the voltage was reversed the electrons being pushed into the collector would quickly fill up the “holes” (the electron-needy impurities), and conduction would stop almost instantly. This junction of the two crystals (or parts of one crystal) created a solid-state diode, and the concept soon became known as semiconduction. The mechanism of action when the diode is off has to do with the separation of charge carriers around the junction. This is called a “depletion region”.

Development of the Diode

Armed with the knowledge of how these new diodes worked, a vigorous effort began to learn how to build them on demand. Teams at Purdue University, Bell Labs, MIT, and the University of Chicago all joined forces to build better crystals. Within a year germanium production had been perfected to the point where military-grade diodes were being used in most radar sets.

Development of the Transistor

After the war, William Shockley decided to attempt the building of a triode-like semiconductor device. He secured funding and lab space, and went to work on the problem with Brattain and John Bardeen.

The key to the development of the transistor was the further understanding of the process of the electron mobility in a semiconductor. It was realized that if there were some way to control the flow of the electrons from the emitter to the collector of this newly discovered diode, an amplifier could be built. For instance, if contacts are placed on both sides of a single type of crystal, current will not flow between them through the crystal. However if a third contact could then “inject” electrons or holes into the material, current would flow.

Actually doing this appeared to be very difficult. If the crystal were of any reasonable size, the number of electrons (or holes) required to be injected would have to be very large, making it less than useful as an amplifier because it would require a large injection current to start with. That said, the whole idea of the crystal diode was that the crystal itself could provide the electrons over a very small distance, the depletion region. The key appeared to be to place the input and output contacts very close together on the surface of the crystal on either side of this region.

Brattain started working on building such a device, and tantalizing hints of amplification continued to appear as the team worked on the problem. Sometimes the system would work but then stop working unexpectedly. In one instance a non-working system started working when placed in water. Ohl and Brattain eventually developed a new branch of quantum mechanics, which became known as surface physics, to account for the behaviour. The electrons in any one piece of the crystal would migrate about due to nearby charges. Electrons in the emitters, or the "holes" in the collectors, would cluster at the surface of the crystal where they could find their opposite charge "floating around" in the air (or water). Yet they could be pushed away from the surface with the application of a small amount of charge from any other location on the crystal. Instead of needing a large supply of injected electrons, a very small number in the right place on the crystal would accomplish the same thing.

Their understanding solved the problem of needing a very small control area to some degree. Instead of needing two separate semiconductors connected by a common, but tiny, region, a single larger surface would serve. The electron-emitting and collecting leads would both be placed very close together on the top, with the control lead placed on the base of the crystal. When current flowed through this "base" lead, the electrons or holes would be pushed out, across the block of semiconductor, and collect on the far surface. As long as the emitter and collector were very close together, this should allow enough electrons or holes between them to allow conduction to start.

The First Transistor

A stylized replica of the first transistor.

The Bell team made many attempts to build such a system with various tools, but generally failed. Setups where the contacts were close enough were invariably as fragile as the original cat's whisker detectors had been, and would work briefly, if at all. Eventually they had a practical breakthrough. A piece of gold foil was glued to the edge of a plastic wedge, and then the foil was sliced with a razor at the tip of the triangle. The result was two very closely spaced contacts of gold. When the wedge was pushed down onto the surface of a crystal and voltage applied to the other side (on the base of the crystal), current started to flow from one contact to the other as the base voltage pushed the electrons away from the base towards the other side near the contacts. The point-contact transistor had been invented.

While the device was constructed a week earlier, Brattain's notes describe the first demonstration to higher-ups at Bell Labs on the afternoon of 23 December 1947, often given as the birthdate of the transistor. what is now known as the "p–n–p point-contact germanium transistor" operated as a speech amplifier with a power gain of 18 in that trial. John Bardeen, Walter Houser Brattain, and William Bradford Shockley were awarded the 1956 Nobel Prize in physics for their work.

Origin of the term "Transistor"

Bell Telephone Laboratories needed a generic name for their new invention: "Semiconductor Triode", "Solid Triode", "Surface States Triode" [*sic*], "Crystal Triode" and "Iotatron" were all considered, but "transistor", coined by John R. Pierce, won an internal ballot. The rationale for the name is described in the following extract from the company's Technical Memoranda (May 28, 1948) calling for votes:

Transistor. This is an abbreviated combination of the words "transconductance" or "transfer", and "varistor". The device logically belongs in the varistor family, and has the transconductance or transfer impedance of a device having gain, so that this combination is descriptive.

Improvements in Transistor Design

Shockley was upset about the device being credited to Brattain and Bardeen, who he felt had built it "behind his back" to take the glory. Matters became worse when Bell Labs lawyers found that some of Shockley's own writings on the transistor were close enough to those of an earlier 1925 patent by Julius Edgar Lilienfeld that they thought it best that his name be left off the patent application.

Shockley was incensed, and decided to demonstrate who was the real brains of the operation. A few months later he invented an entirely new, considerably more robust, type of transistor with a layer or ‹sandwich' structure. This structure went on to be used for the vast majority of all transistors into the 1960s, and evolved into the bipolar junction transistor.

With the fragility problems solved, a remaining problem was purity. Making germanium of the required purity was proving to be a serious problem, and limited the yield of transistors that actually worked from a given batch of material. Germanium's sensitivity to temperature also limited its usefulness. Scientists theorized that silicon would be easier to fabricate, but few investigated this possibility. Gordon K. Teal was the first to develop a working silicon transistor, and his company, the nascent Texas Instruments, profited from its technological edge. From the late 1960s most transistors were silicon-based. Within a few years transistor-based products, most notably easily portable radios, were appearing on the market.

A major improvement in manufacturing yield came when a chemist advised the companies fabricating semiconductors to use distilled rather than tap water: calcium ions present in tap water were the cause of the poor yields. "Zone melting", a technique using a band of molten material moving through the crystal, further increased crystal purity.

REFERENCES

1. Ernest Braun & Stuart MacDonald (1982). Revolution in Miniature: The History and Impact of Semiconductor Electronics. Cambridge University Press. pp. 11–13.

CHAPTER 4

Diodes

Chapter Contents

In simple language, a diode is like one-way valve which allows electric current to flow in one direction but generally does not allow it to flow in the opposite direction. The direction of the electric current in the diode may be reversed. However, even if it is, the flow will still be one directional.

A diode contains two electrodes that act in much the same manner as semiconductors. The positive or p-type is usually the anode and the negative or n-type is the cathode. In other words, the cathode is negatively charged as compared to the anode. If the cathode is charged at the same or very similar voltage to the anode, current will not flow.

INTRODUCTION

In electronics, a **diode** is a two-terminal electronic component that conducts primarily in one direction (asymmetric conductance); it has low (ideally zero) resistance to the flow of current in one direction, and high (ideally infinite) resistance in the other. A **semiconductor diode**, the most common type today, is a crystalline piece of semiconductor material with a p–n junction connected to two electrical terminals. A vacuum tube diode has two electrodes, a plate (anode) and a heated cathode. Semiconductor diodes were the first semiconductor electronic devices. The discovery of crystals' rectifying abilities was made by German physicist Ferdinand Braun in 1874. The first semiconductor diodes, called cat's whisker diodes, developed around 1906, were made of mineral crystals such as galena. Today, most diodes are made of silicon, but other semiconductors such as selenium or germanium are sometimes used.

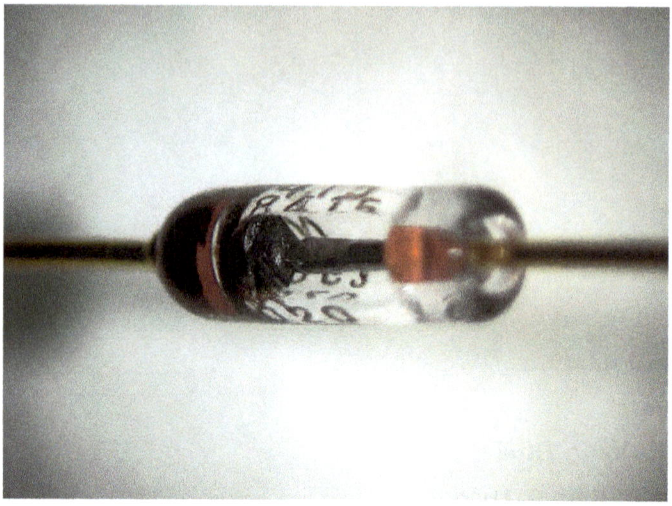

Closeup of a diode, showing the square-shaped semiconductor crystal (black object on left).

Extreme macro photo of a Chinese diode of the seventies.

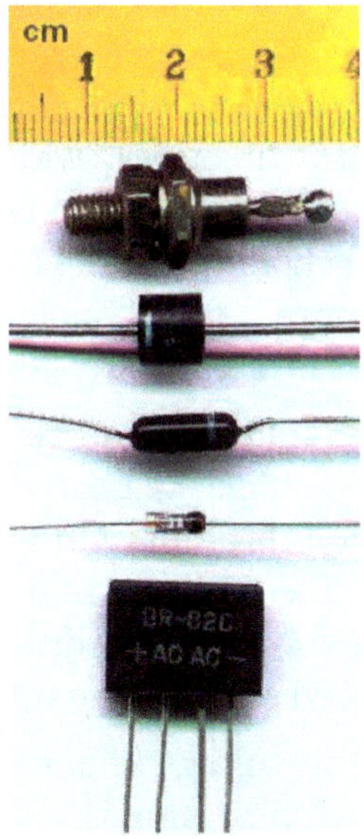

Various semiconductor diodes. Bottom: A bridge rectifier. In most diodes, a white or black painted band identifies the cathode into which electrons will flow when the diode is conducting. Electron flow is the reverse of conventional current flow.

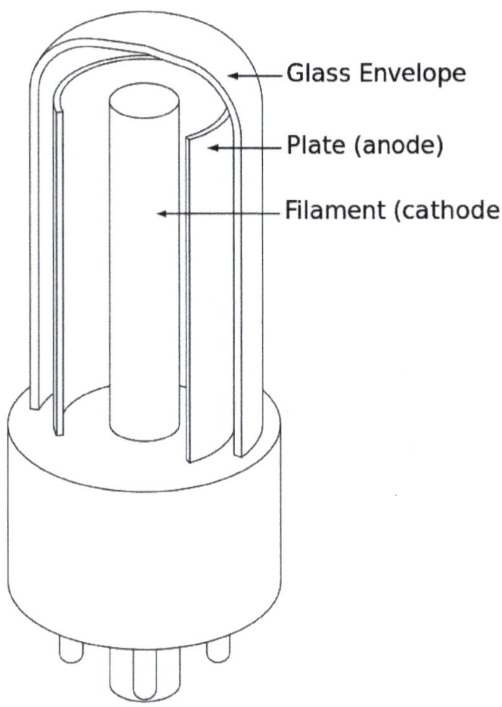

Structure of a vacuum tube diode. The filament may be bare, or more commonly (as shown here), embedded within and insulated from an enclosing cathode.

MAIN FUNCTIONS

The most common function of a diode is to allow an electric current to pass in one direction (called the diode's forward direction), while blocking current in the opposite direction (the reverse direction). Thus, the diode can be viewed as an electronic version of a check valve. This unidirectional behavior is called rectification, and is used to convert alternating current (AC) to direct current (DC), including extraction of modulation from radio signals in radio receivers—these diodes are forms of rectifiers.

However, diodes can have more complicated behavior than this simple on–off action, because of their nonlinear current-voltage characteristics. Semiconductor diodes begin conducting electricity only if a certain threshold voltage or cut-in voltage is present in the forward direction (a state in which the diode is said to be forward-biased). The voltage drop across a forward-biased diode varies only a little with the current, and is a function of temperature; this effect can be used as a temperature sensor or as a voltage reference.

A semiconductor diode's current–voltage characteristic can be tailored by selecting the semiconductor materials and the doping impurities introduced into the materials during manufacture. These techniques are used to create special-purpose diodes that perform many different functions. For example, diodes are used to regulate voltage (Zener diodes), to protect circuits from high voltage surges (avalanche diodes), to electronically tune radio and TV receivers (varactor diodes), to generate radio-frequency oscillations (tunnel diodes, Gunn diodes, IMPATT diodes), and to produce light (light-emitting diodes). Tunnel, Gunn and IMPATT diodes exhibit negative resistance, which is useful in microwave and switching circuits.

Diodes, both vacuum and semiconductor, can be used as shot-noise generators.

HISTORY

Thermionic (vacuum tube) diodes and solid state (semiconductor) diodes were developed separately, at approximately the same time, in the early 1900s, as radio receiver detectors. Until the 1950s vacuum tube diodes were used more frequently in radios because the early point-contact type semiconductor diodes were less stable. In addition, most receiving sets had vacuum tubes for amplification that could easily have the thermionic diodes included in the tube (for example the 12SQ7double diode triode), and vacuum tube rectifiers and gas-filled rectifiers were capable of handling some high voltage/high current rectification tasks better than the semiconductor diodes (such as selenium rectifiers) which were available at that time.

Vacuum Tube Diodes

In 1873, Frederick Guthrie discovered the basic principle of operation of thermionic diodes. Guthrie discovered that a positively charged electroscope could be discharged by bringing a grounded piece of white-hot metal close to it (but not actually touching it). The same did not apply to a negatively charged electroscope, indicating that the current flow was only possible in one direction.

Thomas Edison independently rediscovered the principle on February 13, 1880. At the time, Edison was investigating why the filaments of his carbon-filament light bulbs nearly always burned out at the positive-connected end. He had a special bulb made with a metal plate sealed into the glass envelope. Using this device, he confirmed that an invisible current flowed from the glowing filament through the vacuum to the metal plate, but only when the plate was connected to the positive supply.

Edison devised a circuit where his modified light bulb effectively replaced the resistor in a DC voltmeter. Edison was awarded a patent for this invention in 1884. Since there was no apparent practical use for such a device at the time, the patent application was most likely simply a precaution in case someone else did find a use for the so-called Edison effect.

About 20 years later, John Ambrose Fleming (scientific adviser to the Marconi Company and former Edison employee) realized that the Edison effect could be used as a precision radio detector. Fleming patented the first true thermionic diode, the Fleming valve, in Britain on November 16, 1904 (followed by U.S. Patent 803,684 in November 1905).

Solid-State Diodes

In 1874 German scientist Karl Ferdinand Braun discovered the "unilateral conduction" of crystals. Braun patented the crystal rectifier in 1899. Copper oxide and selenium rectifiers were developed for power applications in the 1930s.

Indian scientist Jagadish Chandra Bose was the first to use a crystal for detecting radio waves in 1894. The crystal detector was developed into a practical device for wireless telegraphy by Greenleaf Whittier Pickard, who invented a silicon crystal detector in 1903 and received a patent for it on November 20, 1906. Other experimenters tried a variety of other substances, of which the most widely used was the mineral galena (lead sulfide). Other substances offered slightly better performance, but galena was most widely used because it had the advantage of being cheap and easy to obtain. The crystal detector in these early crystal radio sets consisted of an adjustable wire point-contact, often made of gold or platinum because of their incorrodible nature (the so-called "cat's whisker"), which could be manually moved over the face of the crystal in search of a portion of that mineral with rectifying qualties. This troublesome device was superseded by thermionic diodes (vacuum tubes) by the 1920s, but after high purity semiconductor materials became available, the crystal detector returned to dominant use with the advent, in the 1950s, of inexpensive fixed-germanium diodes. Bell Labs also developed a germanium diode for microwave reception, and AT&T used these in their microwave towers that criss-crossed the nation starting in the late 1940s, carrying telephone and network television signals. Bell Labs did not develop a satisfactory thermionic diode for microwave reception.

Etymology

At the time of their invention, such devices were known as rectifiers. In 1919, the year tetrodes were invented, William Henry Eccles coined the term ***diode*** from the Greek roots di (from *δί*), meaning ‹two', and ode (from *ὁδός*), meaning ‹path'. (However, the word diode itself, as well as triode, tetrode, pentode, hexode, were already in use as terms of multiplex telegraphy; see, for example, The telegraphic journal and electrical review, September 10, 1886, p. 252).

Rectifiers

Although all diodes rectify, the term 'rectifier' is normally reserved for higher currents and voltages than would normally be found in the rectification of lower power signals; examples include:

- Power supply rectifiers (half-wave, full-wave, bridge)
- Flyback diodes

World's Smallest Diode

Researchers from the University of Georgia and Ben-Gurion University of the Negev (BGU) have developed a diode made from a molecule of DNA. Professor Bingqian Xu from the College of Engineering at the University of Georgia and his team took a single DNA molecule made from 11 base pairs and connected it to an electronic circuit a few nanometers in size. When layers of coralyne were inserted between layers of DNA, the current jumped up to 15 times larger negative versus positive, which is necessary for a nano diode.

THERMIONIC DIODES

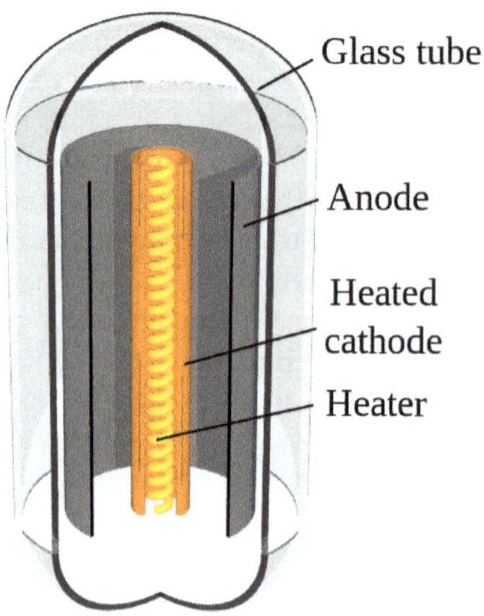

Diode vacuum tube construction.

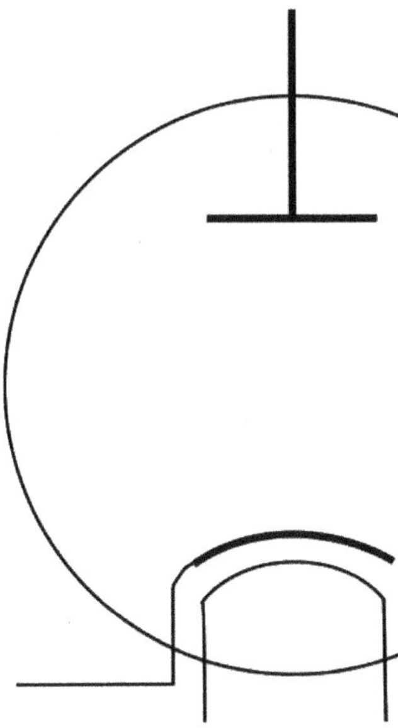

The symbol for an indirect heated vacuum-tube diode. From top to bottom, the components are the anode, the cathode, and the heater filament.

A thermionic diode is a thermionic-valve device (also known as a vacuum tube, tube, or valve), consisting of a sealed evacuated glass envelope containing two electrodes: a cathode heated by a filament, and a plate (anode). Early examples were fairly similar in appearance to incandescent light bulbs.

In operation, a current flows through the filament (heater)—a high resistance wire made of nichrome—and heats the cathode red hot (800–1000 °C). This causes the cathode to releases electrons into the vacuum, a process called thermionic emission. (Some valves use direct heating, in which a tungsten filament acts as both heater and cathode.) The alternating voltage to be rectified is applied between the cathode and the concentric plate electrode. When the plate has a positive voltage with respect to the cathode, it electrostatically attracts the electrons from the cathode, so a current of electrons flows through the tube from cathode to plate. However, when the polarity is reversed and the plate has a negative voltage, no current flows, because the cathode electrons are not attracted to it. The plate, being unheated, does not emit any electrons. So electrons can only flow through the tube in one direction, from the cathode to the anode plate.

The cathode is coated with oxides of alkaline earth metals, such as barium and strontium oxides. These have a low work function, meaning that they more readily emit electrons than would the uncoated cathode.

In a mercury-arc valve, an arc forms between a refractory conductive anode and a pool of liquid mercury acting as cathode. Such units were made with ratings up to hundreds of kilowatts, and were important in the development of HVDC power transmission. Some types of smaller thermionic rectifiers had mercury vapor fill to reduce their forward voltage drop and to increase current rating over thermionic hard-vacuum devices.

Throughout the vacuum tube era, valve diodes were used in analog signal applications and as rectifiers in DC power supplies in consumer electronics such as radios, televisions, and sound systems. They were replaced in power supplies beginning in the 1940s by selenium rectifiers and then by semiconductor diodes by the 1960s. Today they are still used in a few high power applications where their ability to withstand transient voltages and their robustness gives them an advantage over semiconductor devices. The recent (2012) resurgence of interest among audiophiles and recording studios in old valve audio gear such as guitar amplifiers and home audio systems has provided a market for the legacy consumer diode valves.

SEMICONDUCTOR DIODES

Electronic Symbols

The symbol used for a semiconductor diode in a circuit diagram specifies the type of diode. There are alternative symbols for some types of diodes, though the differences are minor. The triangle in the symbols points to the forward direction.

Diode

Light-emitting diode (LED)

Photodiode

Schottky diode

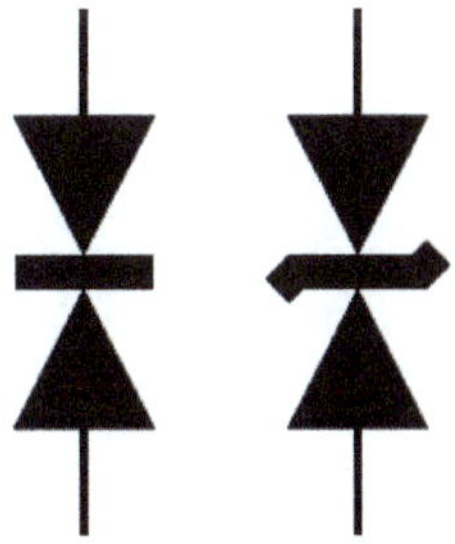

Transient-voltage-suppression diode (TVS)

Tunnel diode

Varicap

Zener diode

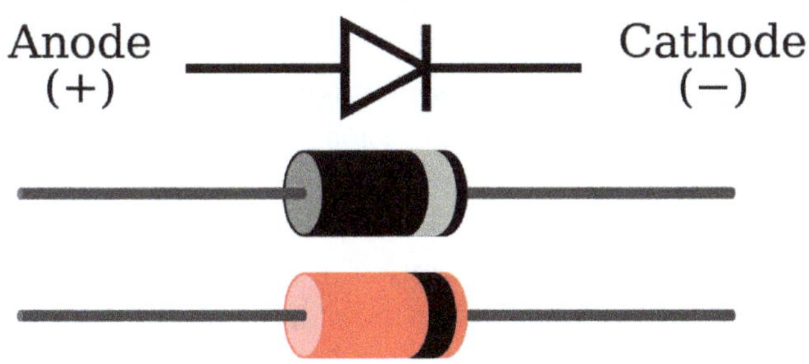

Typical diode packages in same alignment as diode symbol. Thin bar depicts the cathode.

A galena cat's-whisker detector, a point-contact diode.

Point-Contact Diodes

A **point-contact diode** works the same as the junction diodes described below, but its construction is simpler. A pointed metal wire is placed in contact with an n-type semiconductor. Some metal migrates into the semiconductor to make a small p-type region around the contact. The 1N34 germanium version is still used in radio receivers as a detector and occasionally in specialized analog electronics.

Junction Diodes

p–n Junction Diode

A p–n junction diode is made of a crystal of semiconductor, usually silicon, but germanium and gallium arsenide are also used. Impurities are added to it to create a region on one side that contains negative charge carriers (electrons), called an n-type semiconductor, and a region on the other side that contains positive charge carriers (holes), called a p-type semiconductor. When the n-type and p-type materials are attached together, a momentary flow of electrons occur from the n to the p side resulting in a third region between the two where no charge carriers are present. This region is called the depletion region because there are no charge carriers (neither electrons nor holes) in it. The diode's terminals are attached to the n-type and p-regions. The boundary between these two regions, called a p–n junction, is where the action of the diode takes place. When a sufficiently higher electrical potential is applied to the P side (the anode) than to the N side (the cathode), it allows electrons to flow through the depletion region from the N-type side to the P-type side. The junction does not allow the flow of electrons in the opposite direction when the potential is applied in reverse, creating, in a sense, an electrical check valve.

Schottky Diode

Another type of junction diode, the Schottky diode, is formed from a metal–semiconductor junction rather than a p–n junction, which reduces capacitance and increases switching speed.

Current–Voltage Characteristic

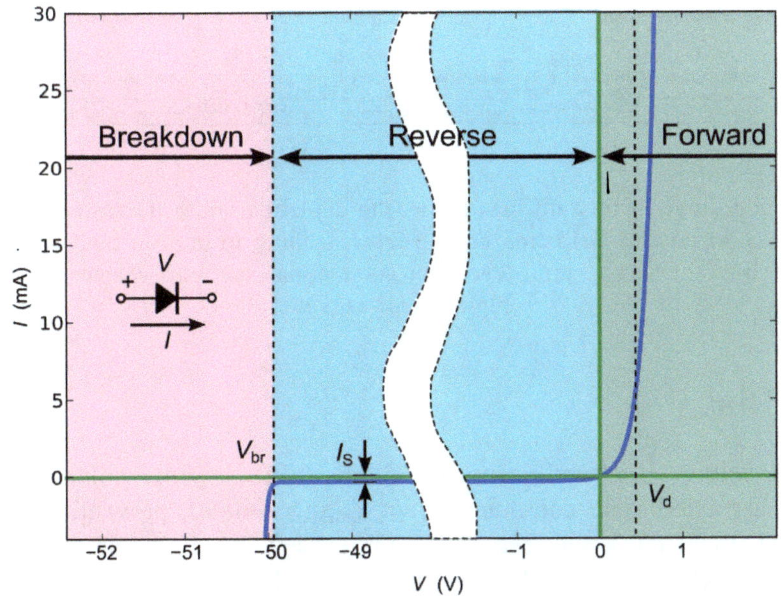

I–V (current vs. voltage) characteristics of a p–n junction diode.

A semiconductor diode's behavior in a circuit is given by its current–voltage characteristic, or I–V graph (see graph below). The shape of the curve is determined by the transport of charge carriers through the so-called depletion layer or depletion region that exists at the p–n junction between differing semiconductors. When a p–n junction is first created, conduction-band (mobile) electrons from the N-doped region diffuse into the P-doped region where there is a large population of holes (vacant places for electrons) with which the electrons "recombine". When a mobile electron recombines with a hole, both hole and electron vanish, leaving behind an immobile positively charged donor (dopant) on the N side and negatively charged acceptor (dopant) on the P side. The region around the p–n junction becomes depleted of charge carriers and thus behaves as an insulator.

However, the width of the depletion region (called the depletion width) cannot grow without limit. For each electron–hole pair recombination made, a positively charged dopant ion is left behind in the N-doped region, and a negatively charged dopant ion is created in the P-doped region. As recombination proceeds and more ions are created, an increasing electric field develops through the depletion zone that acts to slow and then finally stop recombination. At this point, there is a "built-in" potential across the depletion zone.

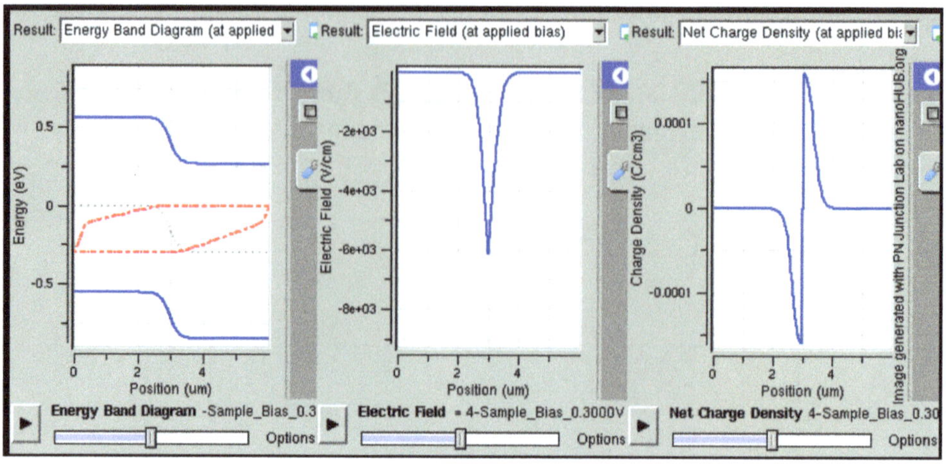

A PN junction diode in forward bias mode, the depletion width decreases. Both p and n junctions are doped at a 1e15/cm3 doping level, leading to built-in potential of ~0.59V. Observe the different Quasi Fermi levels for conduction band and valence band in n and p regions (red curves).

Reverse Bias

If an external voltage is placed across the diode with the same polarity as the built-in potential, the depletion zone continues to act as an insulator, preventing any significant electric current flow (unless electron–hole pairs are actively being created in the junction by, for instance, light; see photodiode). This is called the reverse bias phenomenon.

Forward Bias

However, if the polarity of the external voltage opposes the built-in potential, recombination can once again proceed, resulting in a substantial electric current through the p–n junction (i.e. substantial numbers of electrons and holes recombine at the junction). For silicon diodes, the built-in potential is approximately 0.7 V (0.3 V for germanium and 0.2 V for Schottky). Thus, if an external voltage greater than and opposite to the built-in voltage is applied, a current will flow and the diode is said to be "turned on" as it has been given an external forward bias. The diode is commonly said to have a forward "threshold" voltage, above which it conducts and below which conduction stops. However, this is only an approximation as the forward characteristic is according to the Shockley equation absolutely smooth (see graph below).

A diode's I–V characteristic can be approximated by four regions of operation:

- At very large reverse bias, beyond the peak inverse voltage or PIV, a process called reverse breakdown occurs that causes a large increase in current (i.e., a large number of electrons and holes are created at, and move away from the p–n junction) that usually damages the device permanently. The avalanche diode is deliberately designed for use in that manner. In the Zener diode, the concept of PIV is not applicable. A Zener diode contains a heavily doped p–n junction allowing electrons to tunnel from the valence band of the p-type material to the conduction band of the n-type material, such that the reverse voltage is "clamped" to a known value (called the Zener voltage), and avalanche does not occur. Both devices, however, do have a limit to the maximum current and power they can withstand in the clamped reverse-voltage region. Also, following the end of forward conduction in any diode, there is reverse current for a short time. The device does not attain its full blocking capability until the reverse current ceases.

- For a bias less than the PIV, the reverse current is very small. For a normal P–N rectifier diode, the reverse current through the device in the micro-ampere (μA) range is very low. However, this is temperature dependent, and at sufficiently high temperatures, a substantial amount of reverse current can be observed (mA or more).
- With a small forward bias, where only a small forward current is conducted, the current–voltage curve is exponential in accordance with the ideal diode equation. There is a definite forward voltage at which the diode starts to conduct significantly. This is called the knee voltage or cut-in voltage and is equal to the barrier potential of the p-n junction. This is a feature of the exponential curve, and appears sharper on a current scale more compressed than in the diagram shown here.
- At larger forward currents the current-voltage curve starts to be dominated by the ohmic resistance of the bulk semiconductor. The curve is no longer exponential, it is asymptotic to a straight line whose slope is the bulk resistance. This region is particularly important for power diodes. The diode can be modeled as an ideal diode in series with a fixed resistor.

In a small silicon diode operating at its rated currents, the voltage drop is about 0.6 to 0.7 volts. The value is different for other diode types—Schottky diodes can be rated as low as 0.2 V, germanium diodes 0.25 to 0.3 V, and red or blue light-emitting diodes (LEDs) can have values of 1.4 V and 4.0 V respectively.

At higher currents the forward voltage drop of the diode increases. A drop of 1 V to 1.5 V is typical at full rated current for power diodes.

Shockley Diode Equation

The Shockley ideal diode equation or the diode law (named after transistor co-inventor William Bradford Shockley) gives the I–V characteristic of an ideal diode in either forward or reverse bias (or no bias). The following equation is called the Shockley ideal diode equation when n, the ideality factor, is set equal to 1 :

$$I = I_S \left(e^{\frac{V_D}{nV_T}} - 1 \right)$$

where

I is the diode current,

I_S is the reverse bias saturation current (or scale current),

V_D is the voltage across the diode,

V_T is the thermal voltage, and

n is the ideality factor, also known as the quality factor or sometimes emission coefficient. The ideality factor n typically varies from 1 to 2 (though can in some cases be higher), depending on the fabrication process and semiconductor material and is set equal to 1 for the case of an "ideal" diode (thus the n is sometimes omitted). The ideality factor was added to account for imperfect junctions as observed in real transistors. The factor mainly accounts for carrier recombination as the charge carriers cross the depletion region.

The thermal voltage V_T is approximately 25.85 mV at 300 K, a temperature close to "room temperature" commonly used in device simulation software. At any temperature it is a known constant defined by:

$$V_T = \frac{kT}{q},$$

where k is the Boltzmann constant, T is the absolute temperature of the p–n junction, and q is the magnitude of charge of an electron (the elementary charge).

The reverse saturation current, I_S, is not constant for a given device, but varies with temperature; usually more significantly than V_T, so that V_D typically decreases as T increases.

The Shockley ideal diode equation or the diode law is derived with the assumption that the only processes giving rise to the current in the diode are drift (due to electrical field), diffusion, and thermal recombination–generation (R–G) (this equation is derived by setting n = 1 above). It also assumes that the R–G current in the depletion region is insignificant. This means that the Shockley ideal diode equation doesn't account for the processes involved in reverse breakdown and photon-assisted R–G. Additionally, it doesn't describe the "leveling off" of the I–V curve at high forward bias due to internal resistance. Introducing the ideality factor, n, accounts for recombination and generation of carriers.

Under reverse bias voltages the exponential in the diode equation is negligible, and the current is a constant (negative) reverse current value of $-I_S$. The reverse breakdown region is not modeled by the Shockley diode equation.

For even rather small forward bias voltages the exponential is very large, since the thermal voltage is very small in comparison. The subtracted '1' in the diode equation is then negligible and the forward diode current can be approximated by

$$I = I_S e^{\frac{V_D}{nV_T}}$$

The use of the diode equation in circuit problems is illustrated in the article on diode modeling.

Small-signal Behavior

For circuit design, a small-signal model of the diode behavior often proves useful. A specific example of diode modeling is discussed in the article on small-signal circuits.

Reverse-Recovery Effect

Following the end of forward conduction in a p–n type diode, a reverse current can flow for a short time. The device does not attain its blocking capability until the mobile charge in the junction is depleted.

The effect can be significant when switching large currents very quickly. A certain amount of "reverse recovery time" t_r (on the order of tens of nanoseconds to a few microseconds) may be required to remove the reverse recovery charge Q_r from the diode. During this recovery time, the diode can actually conduct in the reverse direction. This might give rise to a large constant current in the reverse direction for a short time while the diode is reverse biased. The magnitude of such a reverse current is determined by the operating circuit (i.e., the series resistance) and the diode is said to be in the storage-phase. In certain real-world cases it is important to consider the losses that are incurred by this non-ideal diode effect. However, when the slew rate of the current is not so severe (e.g. Line frequency) the effect can be safely ignored. For most applications, the effect is also negligible for Schottky diodes.

The reverse current ceases abruptly when the stored charge is depleted; this abrupt stop is exploited in step recovery diodes for generation of extremely short pulses.

Types of Semiconductor Diode

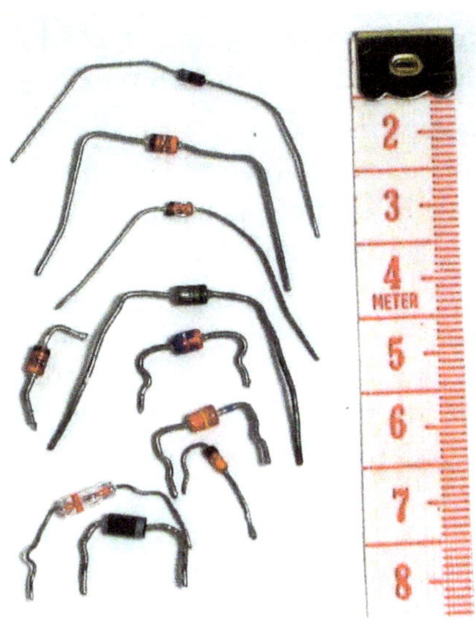

Several types of diodes. The scale is centimeters.

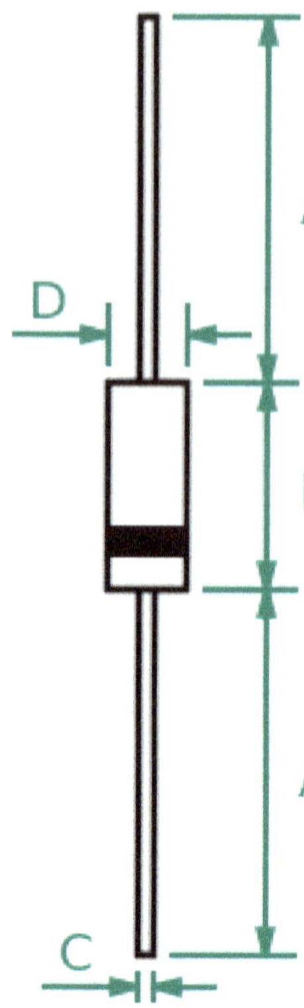

Typical datasheet drawing showing the dimensions of a DO-41 diode package.

There are several types of p–n junction diodes, which emphasize either a different physical aspect of a diode often by geometric scaling, doping level, choosing the right electrodes, are just an application of a diode in a special circuit, or are really different devices like the Gunn and laser diode and the MOSFET:

Normal (p–n) diodes, which operate as described above, are usually made of doped silicon or, more rarely, germanium. Before the development of silicon power rectifier diodes, cuprous oxide and later selenium was used. Their low efficiency required a much higher forward voltage to be applied (typically 1.4 to 1.7 V per "cell", with multiple cells stacked so as to increase the peak inverse voltage rating for application in high voltage rectifiers), and required a large heat sink (often an extension of the diode's metal substrate), much larger than the later silicon diode of the same current ratings would require. The vast majority of all diodes are the p–n diodes found in CMOS integrated circuits, which include two diodes per pin and many other internal diodes.

Avalanche Diodes

These are diodes that conduct in the reverse direction when the reverse bias voltage exceeds the breakdown voltage. These are electrically very similar to Zener diodes (and are often mistakenly called Zener diodes), but break down by a different mechanism: the avalanche effect. This occurs when the reverse electric field applied across the p–n junction causes a wave of ionization, reminiscent of an avalanche, leading to a large current. Avalanche diodes are designed to break down at a well-defined reverse voltage without being destroyed. The difference between the avalanche diode (which has a reverse breakdown above about 6.2 V) and the Zener is that the channel length of the former exceeds the mean free path of the electrons, resulting in many collisions between them on the way through the channel. The only practical difference between the two types is they have temperature coefficients of opposite polarities.

Cat's Whisker or Crystal Diodes

These are a type of point-contact diode. The cat's whisker diode consists of a thin or sharpened metal wire pressed against a semiconducting crystal, typically galena or a piece of coal. The wire forms the anode and the crystal forms the cathode. Cat's whisker diodes were also called crystal diodes and found application in the earliest radios called crystal radio receivers. Cat's whisker diodes are generally obsolete, but may be available from a few manufacturers.

Constant Current Diodes

These are actually JFETs with the gate shorted to the source, and function like a two-terminal current-limiting analog to the voltage-limiting Zener diode. They allow a current through them to rise to a certain value, and then level off at a specific value. Also called CLDs, constant-current diodes, diode-connected transistors, or current-regulating diodes.

Esaki or Tunnel Diodes

These have a region of operation showing negative resistance caused by quantum tunneling, allowing amplification of signals and very simple bistable circuits. Because of the high carrier concentration, tunnel diodes are very fast, may be used at low (mK) temperatures, high magnetic fields, and in high radiation environments. Because of these properties, they are often used in spacecraft.

These are similar to tunnel diodes in that they are made of materials such as GaAs or InP that exhibit a region of negative differential resistance. With appropriate biasing, dipole domains form and travel across the diode, allowing high frequency microwave oscillators to be built.

When an LED-like structure is contained in a resonant cavity formed by polishing the parallel end faces, a laser can be formed. Laser diodes are commonly used in optical storage devices and for high speed optical communication.

Thermal Diodes

This term is used both for conventional p–n diodes used to monitor temperature because of their varying forward voltage with temperature, and for Peltier heat pumps for thermoelectric heating and cooling. Peltier heat pumps may be made from semiconductor, though they do not have any rectifying junctions, they use the differing behaviour of charge carriers in N and P type semiconductor to move heat.

Perun's Diodes

This is a special type of voltage-surge protection diode. It is characterized by the symmetrical voltage-current characteristic, similar to DIAC. It has much faster response time however, that's why it is used in demanding applications.

Photodiodes

All semiconductors are subject to optical charge carrier generation. This is typically an undesired effect, so most semiconductors are packaged in light blocking material. Photodiodes are intended to sense light(photodetector), so they are packaged in materials that allow light to pass, and are usually PIN (the kind of diode most sensitive to light). A photodiode can be used in solar cells, in photometry, or in optical communications. Multiple photodiodes may be packaged in a single device, either as a linear array or as a two-dimensional array. These arrays should not be confused with charge-coupled devices.

PIN Diodes

A PIN diode has a central un-doped, or intrinsic, layer, forming a p-type/intrinsic/n-type structure. They are used as radio frequency switches and attenuators. They are also used as large-volume, ionizing-radiation detectors and as photodetectors. PIN diodes are also used in power electronics, as their central layer can withstand high voltages. Furthermore, the PIN structure can be found in many power semiconductor devices, such as IGBTs, power MOSFETs, and thyristors.

Schottky Diodes

Schottky diodes are constructed from a metal to semiconductor contact. They have a lower forward voltage drop than p–n junction diodes. Their forward voltage drop at forward currents of about 1 mA is in the range 0.15 V to 0.45 V, which makes them useful in voltage clamping applications and prevention of transistor saturation. They can also be used as low loss rectifiers, although their reverse leakage current is in general higher than that of other diodes. Schottky diodes are majority carrier devices and so do not suffer from minority carrier storage problems that slow down many other diodes—so they have a faster reverse recovery than p–n junction diodes. They also tend to have much lower junction capacitance than p–n diodes, which provides for high switching speeds and their use in high-speed circuitry and RF devices such as switched-mode power supply, mixers, and detectors.

Super Barrier Diodes

Super barrier diodes are rectifier diodes that incorporate the low forward voltage drop of the Schottky diode with the surge-handling capability and low reverse leakage current of a normal p–n junction diode.

Gold-Doped Diodes

As a dopant, gold (or platinum) acts as recombination centers, which helps a fast recombination of minority carriers. This allows the diode to operate at signal frequencies, at the expense of a higher forward voltage drop. Gold-doped diodes are faster than other p–n diodes (but not as fast as Schottky diodes). They also have less reverse-current leakage than Schottky diodes (but not as good as other p–n diodes). A typical example is the 1N914.

Snap-off or Step Recovery Diodes

The term step recovery relates to the form of the reverse recovery characteristic of these devices. After a forward current has been passing in an SRD and the current is interrupted or reversed, the reverse conduction will cease very abruptly (as in a step waveform). SRDs can, therefore, provide very fast voltage transitions by the very sudden disappearance of the charge carriers.

Stabistors or *Forward Reference Diodes*

The term stabistor refers to a special type of diodes featuring extremely stable forward voltage characteristics. These devices are specially designed for low-voltage stabilization applications requiring a guaranteed voltage over a wide current range and highly stable over temperature.

Transient Voltage Suppression Diode (TVS)

These are avalanche diodes designed specifically to protect other semiconductor devices from high-voltage transients.Their p–n junctions have a much larger cross-sectional area than those of a normal diode, allowing them to conduct large currents to ground without sustaining damage.

Varicap or Varactor Diodes

These are used as voltage-controlled capacitors. These are important in PLL (phase-locked loop) and FLL (frequency-locked loop) circuits, allowing tuning circuits, such as those in television receivers, to lock quickly on to the frequency. They also enabled tunable oscillators in early discrete tuning of radios, where a cheap and stable, but fixed-frequency, crystal oscillator provided the reference frequency for a voltage-controlled oscillator.

Zener Diodes

These can be made to conduct in reverse bias (backward), and are correctly termed reverse breakdown diodes. This effect, called Zener breakdown, occurs at a precisely defined voltage, allowing the diode to be used as a precision voltage reference. The term Zener diode is colloquially applied to several types of breakdown diodes, but strictly speaking Zener diodes have a breakdown voltage of below 5 volts, whilst avalanche diodes are used for breakdown voltages above that value. In practical voltage reference circuits, Zener and switching diodes are connected in series and opposite directions to balance the temperature coefficient response of the diodes to near-zero. Some devices labeled as high-voltage Zener diodes are actually avalanche diodes (see above). Two (equivalent) Zeners in series and in reverse order, in the same package, constitute a transient absorber (or Transorb, a registered trademark).

Other uses for semiconductor diodes include the sensing of temperature, and computing analog logarithms (see Operational amplifier applications#Logarithmic output).

Numbering and coding schemes

There are a number of common, standard and manufacturer-driven numbering and coding schemes for diodes; the two most common being the EIA/JEDEC standard and the European Pro Electron standard:

EIA/JEDEC

The standardized 1N-series numbering EIA370 system was introduced in the US by EIA/JEDEC (Joint Electron Device Engineering Council) about 1960. Most diodes have a 1-prefix designation (e.g., 1N4003). Among the most popular in this series were: 1N34A/1N270 (germanium signal), 1N914/1N4148 (silicon signal), 1N400x (silicon 1A power rectifier), and 1N580x (silicon 3A power rectifier).

JIS

The JIS semiconductor designation system has all semiconductor diode designations starting with “1S”.

Pro Electron

The European Pro Electron coding system for active components was introduced in 1966 and comprises two letters followed by the part code. The first letter represents the semiconductor material used for the component (A = germanium and B = silicon) and the second letter represents the general function of the part (for diodes, A = low-power/

signal, B = variable capacitance, X = multiplier, Y = rectifier and Z = voltage reference); for example:

- AA-series germanium low-power/signal diodes (e.g., AA119)
- BA-series silicon low-power/signal diodes (e.g., BAT18 silicon RF switching diode)
- BY-series silicon rectifier diodes (e.g., BY127 1250V, 1A rectifier diode)
- BZ-series silicon Zener diodes (e.g., BZY88C4V7 4.7V Zener diode)

Other common numbering / coding systems (generally manufacturer-driven) include:

- GD-series germanium diodes (e.g., GD9) – this is a very old coding system
- OA-series germanium diodes (e.g., OA47) – a coding sequence developed by Mullard, a UK company

As well as these common codes, many manufacturers or organisations have their own systems too – for example:

- HP diode 1901-0044 = JEDEC 1N4148
- UK military diode CV448 = Mullard type OA81 = GEC type GEX23

Related devices

- Rectifier
- Transistor
- Thyristor or silicon controlled rectifier (SCR)
- TRIAC
- DIAC
- Varistor

In optics, an equivalent device for the diode but with laser light would be the Optical isolator, also known as an Optical Diode, that allows light to only pass in one direction. It uses a Faraday rotator as the main component.

APPLICATIONS

Radio Demodulation

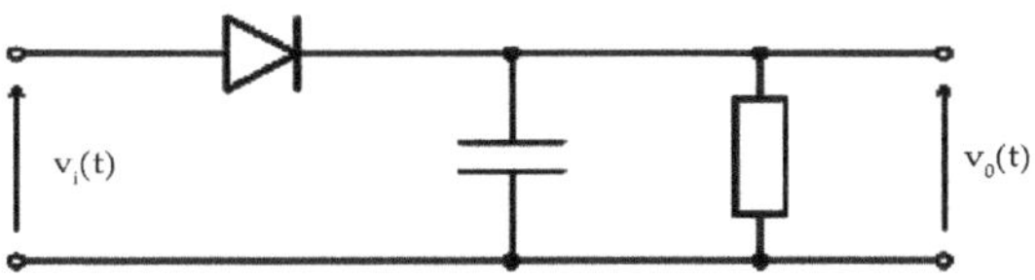

A simple envelope demodulator circuit.

The first use for the diode was the demodulation of amplitude modulated (AM) radio broadcasts. The history of this discovery is treated in depth in the radio article. In summary, an AM signal consists of alternating positive and negative peaks of a radio carrier wave, whose amplitude or envelope is proportional to the original audio signal. The diode (originally a crystal diode) rectifies the AM radio frequency signal, leaving only the positive peaks of the carrier wave. The audio is then extracted from the rectified carrier wave using a simple filter and fed into an audio amplifier or transducer, which generates sound waves.

Power Conversion

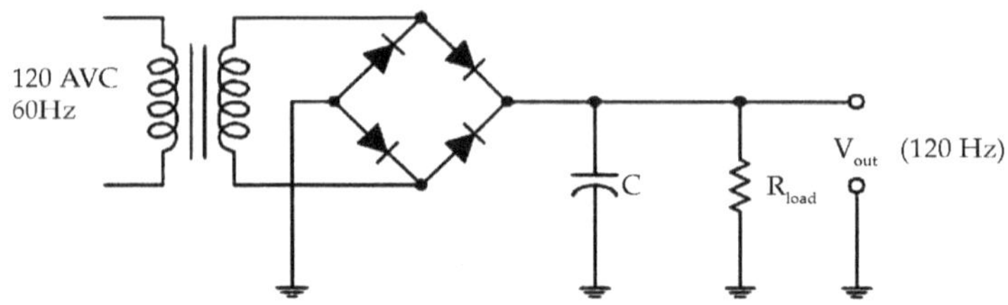

Schematic of basic AC-to-DC power supply.

Rectifiers are constructed from diodes, where they are used to convert alternating current (AC) electricity into direct current (DC). Automotive alternators are a common example, where the diode, which rectifies the AC into DC, provides better performance than the commutator or earlier, dynamo. Similarly, diodes are also used in Cockcroft–Walton voltage multipliers to convert AC into higher DC voltages.

Over-voltage Protection

Diodes are frequently used to conduct damaging high voltages away from sensitive electronic devices. They are usually reverse-biased (non-conducting) under normal circumstances. When the voltage rises above the normal range, the diodes become forward-biased (conducting). For example, diodes are used in (stepper motor and H-bridge) motor controller and relay circuits to de-energize coils rapidly without the damaging voltage spikes that would otherwise occur. (A diode used in such an application is called a flyback diode). Many integrated circuits also incorporate diodes on the connection pins to prevent external voltages from damaging their sensitive transistors. Specialized diodes are used to protect from over-voltages at higher power.

Logic Gates

Diodes can be combined with other components to construct AND and OR logic gates. This is referred to as diode logic.

Ionizing Radiation Detectors

In addition to light, mentioned above, semiconductor diodes are sensitive to more energetic radiation. In electronics, cosmic rays and other sources of ionizing radiation cause noise pulses and single and multiple bit errors. This effect is sometimes exploited by particle detectors to detect radiation. A single particle of radiation, with thousands or millions of electron volts of energy, generates many charge carrier pairs, as its energy is deposited in the semiconductor material. If the depletion layer is large enough to catch the whole shower or to stop a heavy particle, a fairly accurate measurement of the particle's energy can be made, simply by measuring the charge conducted and without the complexity of a magnetic spectrometer, etc. These semiconductor radiation detectors need efficient and uniform charge collection and low leakage current. They are often cooled by liquid nitrogen. For longer-range (about a centimetre) particles, they need a very large depletion depth and large area. For short-range particles, they need any contact or un-depleted semiconductor on at least one surface to be very thin. The back-bias voltages are near breakdown (around a thousand volts per centimetre). Germanium and silicon are common materials. Some of these detectors sense position as well as energy. They have a finite life, especially when detecting heavy particles, because of radiation damage. Silicon and germanium are quite different in their ability to convert gamma rays to electron showers.

Semiconductor detectors for high-energy particles are used in large numbers. Because of energy loss fluctuations, accurate measurement of the energy deposited is of less use.

Temperature Measurements

A diode can be used as a temperature measuring device, since the forward voltage drop across the diode depends on temperature, as in a silicon bandgap temperature sensor. From the Shockley ideal diode equation given above, it might appear that the voltage has a positive temperature coefficient (at a constant current), but usually the variation of the reverse saturation current term is more significant than the variation in the thermal voltage term. Most diodes therefore have a negative temperature coefficient, typically −2 mV/°C for silicon diodes. The temperature coefficient is approximately constant for temperatures above about 20 kelvins. Some graphs are given for 1N400x series, and CY7 cryogenic temperature sensor.

Current Steering

Diodes will prevent currents in unintended directions. To supply power to an electrical circuit during a power failure, the circuit can draw current from a battery. An uninterruptible power supply may use diodes in this way to ensure that current is only drawn from the battery when necessary. Likewise, small boats typically have two circuits each with their own battery/batteries: one used for engine starting; one used for domestics. Normally, both are charged from a single alternator, and a heavy-duty split-charge diode is used to prevent the higher-charge battery (typically the engine battery) from discharging through the lower-charge battery when the alternator is not running.

Diodes are also used in electronic musical keyboards. To reduce the amount of wiring needed in electronic musical keyboards, these instruments often use keyboard matrix circuits. The keyboard controller scans the rows and columns to determine which note the player has pressed. The problem with matrix circuits is that, when several notes are pressed at once, the current can flow backwards through the circuit and trigger "phantom keys" that cause "ghost" notes to play. To avoid triggering unwanted notes, most keyboard matrix circuits have diodes soldered with the switch under each key of the musical keyboard. The same principle is also used for the switch matrix in solid-state pinball machines.

Waveform Clipper

Diodes can be used to limit the positive or negative excursion of a signal to a prescribed voltage.

Clamper

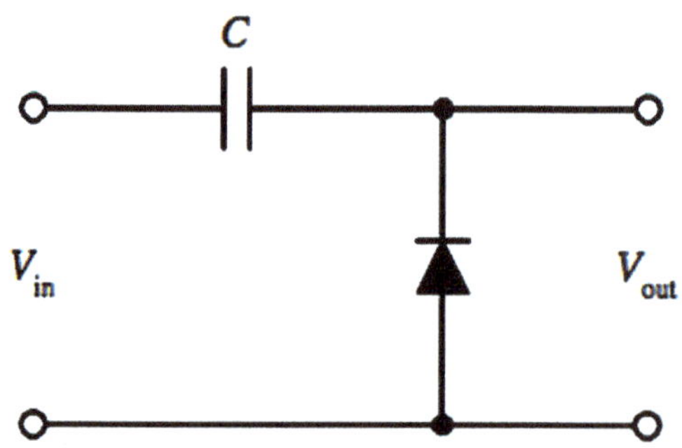

This simple diode clamp will clamp the negative peaks of the incoming waveform to the common rail voltage.

A diode clamp circuit can take a periodic alternating current signal that oscillates between positive and negative values, and vertically displace it such that either the positive, or the negative peaks occur at a prescribed level. The clamper does not restrict the peak-to-peak excursion of the signal, it moves the whole signal up or down so as to place the peaks at the reference level.

GUNN DIODE

A Gunn diode, also known as a transferred electron device (TED), is a form of diode, a two-terminal passive semiconductor electronic component, with negative resistance, used in high-frequency electronics. It is based on the "Gunn effect" discovered in 1962 by physicist J. B. Gunn. Its largest use is in electronic oscillators to generate microwaves, in applications such as radar speed guns, microwave relay data link transmitters, and automatic door openers.

A Russian-made Gunn diode.

Its internal construction is unlike other diodes in that it consists only of N-dopedsemiconductor material, whereas most diodes consist of both P and N-doped regions. It therefore does not conduct in only one direction and cannot rectify alternating current like other diodes, which is why some sources do not use the term diode but prefer TED. In the Gunn diode, three regions exist: two of those are heavily N-doped on each terminal, with a thin layer of lightly n-doped material between. When a voltage is applied to the device, the electrical gradient will be largest across the thin middle layer. If the voltage is increased, the current through the layer will first increase, but eventually, at higher field values, the conductive properties of the middle layer are altered, increasing its resistivity, and causing the current to fall. This means a Gunn diode has a region of negative differential resistance in its current-voltage characteristic curve, in which an increase of applied voltage, causes a decrease in current. This property allows it to amplify, functioning as a radio frequency amplifier, or to become unstable and oscillate when it is biased with a DC voltage.

Gunn Diode Oscillators

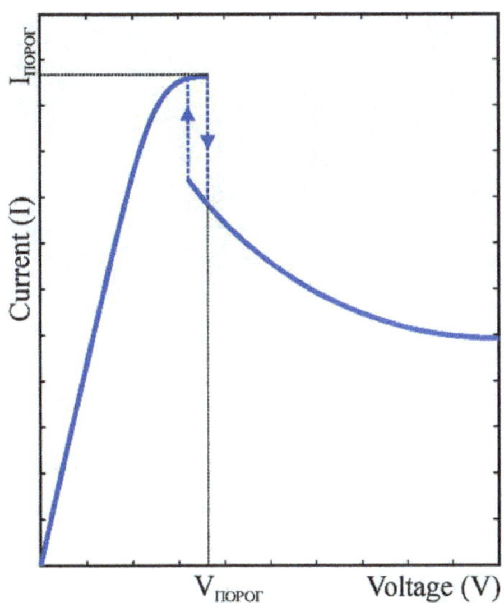

Current-voltage (IV) curve of a Gunn diode. It shows negative resistance above the threshold voltage ($V_{порог}$).

The negative differential resistance, combined with the timing properties of the intermediate layer, is responsible for the diode's largest use: in electronic oscillators at microwave frequencies and above. A microwave oscillator can be created simply by applying a DC voltage to bias the device into its negative resistance region. In effect, the negative differential resistance of the diode cancels the positive resistance of the load circuit, thus creating a circuit with zero differential resistance, which will produce spontaneous oscillations. The oscillation frequency is determined partly by the properties of the middle diode layer, but can be tuned by external factors. In practical oscillators an electronic resonator is usually added to control frequency, in the form of a waveguide, microwave cavity or YIG sphere. The diode is usually mounted inside the cavity. The diode cancels the loss resistance of the resonator, so it produces oscillations at its resonant frequency. The frequency can be tuned mechanically, by adjusting the size of the cavity, or in case of YIG spheres by changing the magnetic field. Gunn diodes are used to build oscillators in the 10 GHz to high (THz) frequency range.

Gallium arsenide Gunn diodes are made for frequencies up to 200 GHz, gallium nitride materials can reach up to 3 terahertz.

The Gunn diode is based on the Gunn effect, and both are named for the physicist J. B. Gunn who, at IBM in 1962, discovered the effect because he refused to accept inconsistent experimental results in gallium arsenide as "noise", and tracked down the cause. Alan Chynoweth, of Bell Telephone Laboratories, showed in June 1965 that only a transferred-electron mechanism could explain the experimental results. The interpretation refers to the Ridley-Watkins-Hilsum theory, whereby semiconductors display negative resistance, meaning that increasing the applied voltage causes the current to decrease.

NASA ERC scientist W. Deter Straub conducting an experiment with the Gunn effect.

The Gunn effect, and its relation to the Watkins-Ridley-Hilsum effect entered the monograph literature in the early 1970s, e.g. in books on transferred electron devices and, more recently on nonlinear wave methods for charge transport.Several other books that provided the same coverage were published in the intervening years, and can be found by searching library and bookseller catalogues on Gunn effect.

Russian Gunn diode oscillator. The diode is mounted inside the cavity (metal box), which functions as a resonator to determine the frequency. The negative resistance of the diode excites microwave oscillations in the cavity which radiate out the hole into a waveguide (not shown). The frequency can be adjusted by changing the size of the cavity using the thumbscrew.

How it Works

The electronic band structure of some semiconductor materials, including gallium arsenide (GaAs), have another energy band or sub-band in addition to the valence and conduction bands which are usually used in semiconductor devices. This third band is at a higher energy than the normal conduction band and is empty until energy is supplied to promote electrons to it. The energy comes from the kinetic energy of ballistic electrons, that is, electrons in the conduction band but moving with sufficient kinetic energy such that they are able to reach the third band.

These electrons either start out below the Fermi level and are given a sufficiently long mean free path to acquire the needed energy by applying a strong electric field, or they are injected by a cathode with the right energy. With forward voltage applied, the Fermi level in the cathode moves into the third band, and reflections of ballistic electrons starting around the Fermi level are minimized by matching the density of states and using the additional interface layers to let the reflected waves interfere destructively.

In GaAs the mobility or drift velocity in the third band is lower than that in the usual conduction band, so with a small increase in the forward voltage, more and more electrons can reach the third band and current decreases. This creates a region of negative incremental resistance in the voltage/current relationship.

When a high enough potential is applied to the diode, the charge carrier density along the cathode becomes unstable, and will develop small slices of low conductivity and high field strength which move from the cathode to the anode. It is not possible to balance the population in both bands, so there will always be thin slices of high field strength in a general background of low field strength. So in practice, with a small increase in forward voltage, a slice is created at the cathode, resistance increases, the slice takes off, and when it reaches the anode a new slice is created at the cathode to keep the total voltage constant. If the voltage is lowered, any existing slice is quenched and resistance decreases again.

The laboratory methods that are used to select materials for the manufacture of Gunn diodes include angle-resolved photoemission spectroscopy.

Applications

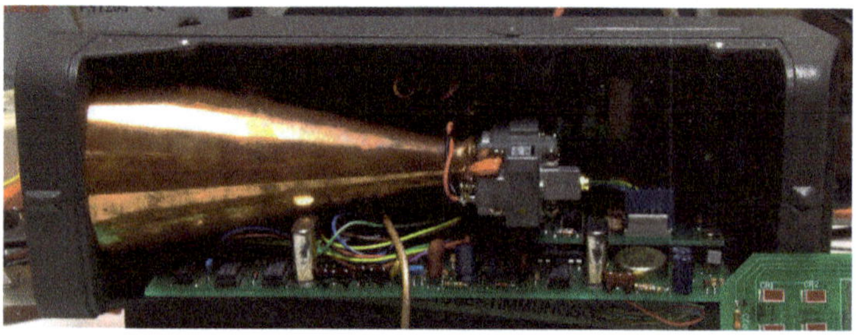

Disassembled radar speed gun. The grey assembly attached to the end of the copper-colored horn antenna is the Gunn diode oscillator which generates the microwaves.

Because of their high frequency capability, Gunn diodes are mainly used at microwave frequencies and above. They can produce some of the highest output power of any semiconductor devices at these frequencies. Their most common use is in oscillators, but they are also used in microwave amplifiers to amplify signals. Because the diode is a one-port (two terminal) device, an amplifier circuit must separate the outgoing amplified signal from the incoming input signal to prevent coupling. One common circuit is a reflection amplifier which uses a circulator to separate the signals. A bias tee is needed to isolate the bias current from the high frequency oscillations.

Sensors and Measuring Instruments

Gunn diode oscillators are used to generate microwave power for: airborne collision avoidance radar, anti-lock brakes, sensors for monitoring the flow of traffic, car radar detectors, pedestrian safety systems, "distance traveled" recorders, motion detectors, "slow-speed" sensors (to detect pedestrian and traffic movement up to 50 m.p.h), traffic signal controllers, automatic door openers, automatic traffic gates, process control equipment to monitor throughput, burglar alarms and equipment to detect trespassers, sensors to avoid derailment of trains, remote vibration detectors, rotational speed tachometers, moisture content monitors.

Radio Amateur Use

By virtue of their low voltage operation, Gunn diodes can serve as microwave frequency generators for very low powered (few-milliwatt) microwave transceivers called **Gunnplexers**. They were first used by British radio amateurs in the late 1970s, and many Gunnplexer designs have been published in journals. They typically consist of an approximately 3 inch waveguide into which the diode is mounted. A low voltage (less than 12 volt) direct current power supply, that can be modulated appropriately, is used to drive the diode. The waveguide is blocked at one end to form a resonant cavity and the other end usually feeds a horn antenna. An additional "mixer diode" is inserted into the waveguide, and it is often connected to a modified FM broadcast receiver to enable listening of other amateur stations. Gunnplexers are most commonly used in the 10 GHz and 24 GHz ham bands.

Some radio amateurs are now using these diodes for EmDrive as they are a lot safer than magnetrons and far more controllable which is useful as the resonance with a typical frustrum can be less than 50 Hz in some cases and even less with a superconducting frustrum.

Radio Astronomy

Gunn oscillators are used as local oscillators for millimeter-wave and submillimeter-wave radio astronomy receivers. The Gunn diode is mounted in a cavity tuned to resonate at twice the fundamental frequency of the diode. The cavity length is changed by a micrometer adjustment. Gunn oscillators capable of generating over 50 mW over a 50% tuning range (one waveguide band) are available.

The Gunn oscillator frequency is multiplied by a diode frequency multiplier for submillimeter-wave applications.

IMPATT DIODE

An **IMPATT diode** (**IMP**act ionization **A**valanche **T**ransit-**T**ime diode) is a form of high-power semiconductor diode used in high-frequency microwave electronics devices. They have negative resistance and are used as oscillators to generate microwaves as well as amplifiers. They operate at frequencies between about 3 and 100 GHz or more. A main advantage is their high-power capability. These diodes are used in a variety of applications from low-power radar systems to proximity alarms. A major drawback of using IMPATT diodes is the high level of phase noise they generate. This results from the statistical nature of the avalanche process.

Device Structure

The IMPATT diode family includes many different junctions and metal semiconductor devices. The first IMPATT oscillation was obtained from a simple silicon p-n junction diode biased into a reverse avalanche break down and mounted in a microwave cavity. Because of the strong dependence of the ionization coefficient on the electric field, most of the electron–hole pairs are generated in the high field region. The generated electron immediately moves into the N region, while the generated holes drift across the P region. The time required for the hole to reach the contact constitutes the transit time delay.

The original proposal for a microwave device of the IMPATT type was made by Read. The Read diode consists of two regions (i) The Avalanche region (a region with relatively high doping and high field) in which avalanche multiplication occurs and (ii) the drift region (a region with essentially intrinsic doping and constant field) in which the generated holes drift towards the contact. A similar device can be built with the configuration in which electrons generated from the avalanche multiplication drift through the intrinsic region.

An IMPATT diode generally is mounted in a microwave package. The diode is mounted with its high–field region close to a copper heatsink so that the heat generated at the diode junction can be readily dissipated. Similar microwave packages are used to house other microwave devices.

The IMPATT diode operates over a narrow frequency band, and diode internal dimensions must correlate with the desired operating frequency. An IMPATT oscillator can be tuned by adjusting the resonant frequency of the coupled circuit, and also by varying the current in the diode; this can be used for frequency modulation.

Principle of Operation

If a free electron with sufficient energy strikes a silicon atom, it can break the covalent bond of silicon and liberate an electron from the covalent bond. If the electron liberated gains energy by being in an electric field and liberates other electrons from other covalent bonds then this process can cascade very quickly into a chain reaction producing a large number of electrons and a large current flow. This phenomenon is called impact avalanche.

At breakdown, the n – region is punched through and forms the avalanche region of the diode. The high resistivity region is the drift zone through which the avalanche generated electrons move toward the anode.

Consider a dc bias V_B, just short of that required to cause breakdown, applied to the diode. Let an AC voltage of sufficiently large magnitude be **superimposed on the dc bias**, such that during the positive cycle of the AC voltage, the diode is driven deep into the avalanche breakdown. At t=0, the AC voltage is zero, and only a small pre-breakdown current flows through the diode. As t increases, the voltage goes above the breakdown voltage and secondary electron-hole pairs are produced by impact ionization. As long as the field in the avalanche region is maintained above the breakdown field, the electron-hole concentration grows exponentially with t. Similarly this concentration decays exponentially with time when the field is reduced below breakdown voltage during the negative swing of the AC voltage. The holes generated in the avalanche region disappear in the p+ region and are collected by the cathode. The electrons are injected into the i – zone where they drift toward the n+ region. Then, the field in the avalanche region reaches its maximum value and the population of the electron-hole pairs starts building up. At this time, the ionization coefficients have their maximum values. The generated electron concentration does not follow the electric field instantaneously because it also depends on the number of electron-hole pairs already present in the avalanche region. Hence, the electron concentration at this point will have a small value. Even after the field has passed its maximum value, the electron-hole concentration continues to grow because the secondary carrier generation rate still remains above its average value. For this reason, the electron concentration in the avalanche region attains its maximum value at, when the field has dropped to its average value. Thus, it is clear that the avalanche region introduces a 90° phase shift between the AC signal and the electron concentration in this region. With a further increase in t, the AC voltage becomes negative, and the field in the avalanche region drops below its critical value. The electrons in the avalanche region are then injected into the drift zone which induces a current in the external circuit which has a phase opposite to that of the AC voltage. The AC field, therefore, absorbs energy from the drifting electrons as they are decelerated by the decreasing field. It is clear that an ideal phase shift between the diode current and the AC signal is achieved if the thickness of the drift zone is such that the bunch of electron is collected at the n^+ - anode at the moment the AC voltage goes to zero. This condition is achieved by making the length of the drift region equal to the wavelength of the signal. This situation produces an additional phase shift of 90° between the AC voltage and the diode current.

Origins

In 1956 W.T. Read and Ralph L. Johnston of Bell Laboratories proposed that an avalanche diode that exhibited significant transit time delay might exhibit a negative resistance characteristic. The effect was soon demonstrated in ordinary silicon diodes and by the late 1960s oscillators at 340 GHz had been produced. Silicon IMPATT diodes can produce up to 3 kilowatts of power continuously, with higher power available in pulses.

LASER DIODE

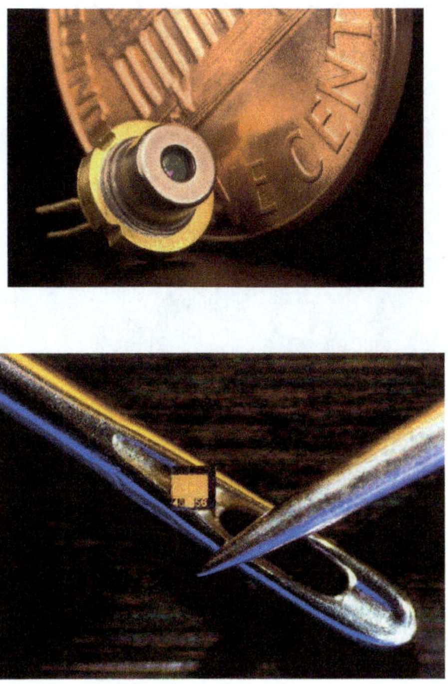

Top: a packaged laser diode shown with a penny for scale. Bottom: the laser diode chip is removed from the above package and placed on the eye of a needle for scale.

A laser diode with the case cut away. The actual laser diode chip is the small black chip at the front; a photodiode at the back is used to control output power.

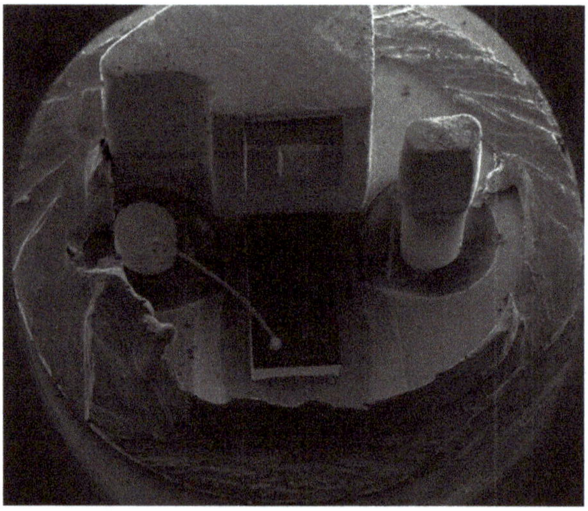

SEM (Scanning Electron Microscope) image of a commercial laser diode with its case and window cut away. The anode connection on the right has been accidentally broken by the case cut process.

A **laser diode**, or **LD** also known as **injection laser diode** or **ILD**, is an electrically pumped semiconductor laser in which the active laser medium is formed by a p-n junction of a semiconductor diode similar to that found in a light-emitting diode.

The laser diode is the most common type of laser produced with a wide range of uses that include fiber optic communications, barcode readers, laser pointers, CD/DVD/Blu-ray Disc reading and recording, laser printing, laser scanning and increasingly directional lighting sources.

Theory of Operation

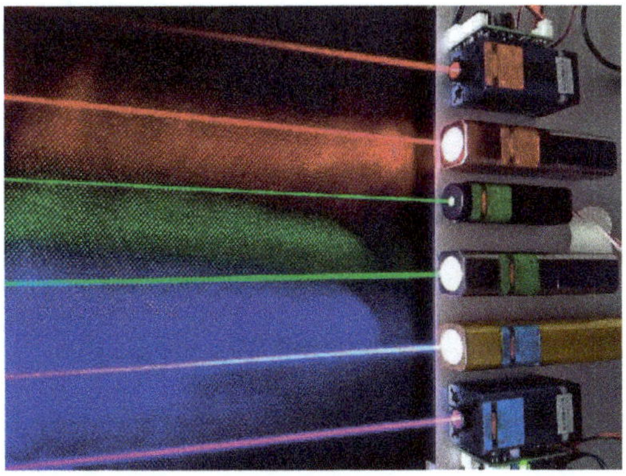

Semi-conductor lasers (660 nm, 635 nm, 532 nm, 520 nm, 445 nm, 405 nm).

A laser diode is electrically a PIN diode. The active region of the laser diode is in the intrinsic (I) region, and the carriers (electrons and holes) are pumped into that region from the N and P regions respectively. While initial diode laser research was conducted on simple P-N diodes, all modern lasers use the double-heterostructure implementation, where the carriers and the photons are confined in order to maximize their chances for recombination and light generation. Unlike a regular diode, the goal for a laser diode is to recombine all carriers in the I region, and produce light. Thus, laser diodes are fabricated using direct bandgap semiconductors. The laser diode epitaxial structure is grown using one of the crystal growth techniques, usually starting from an N doped substrate, and growing the I doped active layer, followed by the P doped cladding, and a contact layer. The active layer most often consists of quantum wells, which provide lower threshold current and higher efficiency.

Laser diodes form a subset of the larger classification of semiconductor p-n junction diodes. Forward electrical bias across the laser diode causes the two species of charge carrier – holes and electrons – to be "injected" from opposite sides of the p-n junction into the depletion region. Holes are injected from the p-doped, and electrons from the n-doped, semiconductor. (A depletion region, devoid of any charge carriers, forms as a result of the difference in electrical potential between n- and p-type semiconductors wherever they are in physical contact.) Due to the use of charge injection in powering most diode lasers, this class of lasers is sometimes termed "injection lasers," or "injection laser diode" (ILD). As diode lasers are semiconductor devices, they may also be classified as semiconductor lasers. Either designation distinguishes diode lasers from solid-state lasers.

Another method of powering some diode lasers is the use of optical pumping. Optically pumped semiconductor lasers (OPSL) use a III-V semiconductor chip as the gain medium, and another laser (often another diode laser) as the pump source. OPSL offer several advantages over ILDs, particularly in wavelength selection and lack of interference from internal electrode structures.

When an electron and a hole are present in the same region, they may recombine or "annihilate" producing a spontaneous emission — i.e., the electron may re-occupy the energy state of the hole, emitting a photon with energy equal to the difference between the electron's original state and hole's state. (In a conventional semiconductor junction diode, the energy released from the recombination of electrons and holes is carried away as phonons, i.e., lattice vibrations, rather than as photons.) Spontaneous emission below the lasing threshold produces similar properties to an LED. Spontaneous emission is necessary to initiate laser oscillation, but it is one among several sources of inefficiency once the laser is oscillating.

The difference between the photon-emitting semiconductor laser and a conventional phonon-emitting (non-light-emitting) semiconductor junction diode lies in the type of semiconductor used, one whose physical and atomic structure confers the possibility for photon emission. These photon-emitting semiconductors are the so-called "direct bandgap" semiconductors. The properties of silicon and germanium, which are single-element semiconductors, have bandgaps that do not align in the way needed to allow photon emission and are not considered "direct." Other materials, the so-called compound semiconductors, have virtually identical crystalline structures as silicon or germanium but use alternating arrangements of two different atomic species in a checkerboard-like pattern to break the symmetry. The transition between the materials in the alternating pattern creates the critical "direct bandgap" property. Gallium arsenide, indium phosphide, gallium antimonide, and gallium nitride are all examples of compound semiconductor materials that can be used to create junction diodes that emit light.

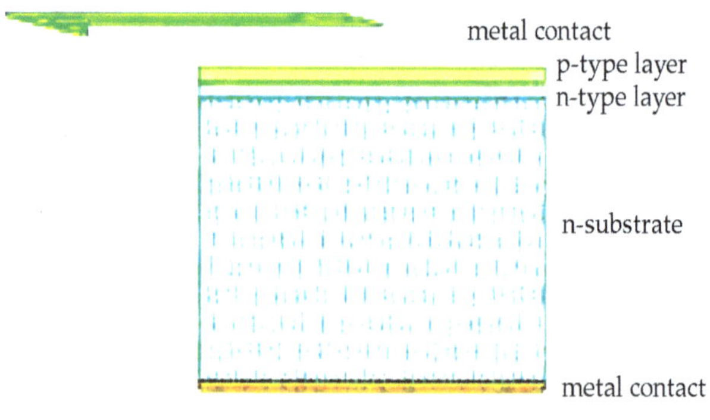

Diagram of a simple laser diode, such as shown above; not to scale.

In the absence of stimulated emission (e.g., lasing) conditions, electrons and holes may coexist in proximity to one another, without recombining, for a certain time, termed the "upper-state lifetime" or "recombination time" (about a nanosecond for typical diode laser materials), before they recombine. A nearby photon with energy equal to the recombination energy can cause recombination by stimulated emission. This generates another photon of the same frequency, polarization, and phase , travelling in the same direction as the first photon. This means that stimulated emission will cause gain in an optical wave (of the correct wavelength) in the injection region, and the gain increases as the number of electrons and holes injected across the junction increases. The spontaneous and stimulated emission processes are vastly more efficient in direct bandgap semiconductors than in indirect bandgap semiconductors; therefore silicon is not a common material for laser diodes.

As in other lasers, the gain region is surrounded with an optical cavity to form a laser. In the simplest form of laser diode, an optical waveguide is made on that crystal's surface, such that the light is confined to a relatively narrow line. The two ends of the crystal are cleaved to form perfectly smooth, parallel edges, forming a Fabry–Pérot resonator. Photons emitted into a mode of the waveguide will travel along the waveguide and be reflected several times from each end face before they exit. As a light wave passes through the cavity, it is amplified by stimulated emission, but light is also lost due to absorption and by incomplete reflection from the end facets. Finally, if there is more amplification than loss, the diode begins to "lase".

Some important properties of laser diodes are determined by the geometry of the optical cavity. Generally, the light is contained within a very thin layer, and the structure supports only a single optical mode in the direction perpendicular to the layers. In the transverse direction, if the waveguide is wide compared to the wavelength of light, then the waveguide can support multiple transverse optical modes, and the laser is known as "multi-mode". These transversely multi-mode lasers are adequate in cases where one needs a very large amount of power, but not a small diffraction-limited beam; for example in printing, activating chemicals, or pumping other types of lasers.

In applications where a small focused beam is needed, the waveguide must be made narrow, on the order of the optical wavelength. This way, only a single transverse mode is supported and one ends up with a diffraction-limited beam. Such single spatial mode devices are used for optical storage, laser pointers, and fiber optics. Note that these lasers may still support multiple longitudinal modes, and thus can lase at multiple wavelengths simultaneously. The wavelength emitted is a function of the band-gap of the semiconductor material and the modes of the optical cavity. In general, the maximum gain will occur for photons with energy slightly above the band-gap energy, and the modes nearest the peak of the gain curve will lase most strongly. The width of the gain curve will determine the number of additional "side modes" that may also lase, depending on the operating conditions. Single spatial mode lasers that can support multiple longitudinal modes are called Fabry Perot (FP) lasers. An FP laser will lase at multiple cavity modes within the gain bandwidth of the lasing medium. The number of lasing modes in an FP laser is usually unstable, and can fluctuate due to changes in current or temperature.

Single spatial mode diode lasers can be designed so as to operate on a single longitudinal mode. These single frequency diode lasers exhibit a high degree of stability, and are used in spectroscopy and metrology, and as frequency references. Single frequency diode lasers classed as either distributed feedback (DFB) lasers or distributed Bragg reflector (DBR) lasers.

Due to diffraction, the beam diverges (expands) rapidly after leaving the chip, typically at 30 degrees vertically by 10 degrees laterally. A lens must be used in order to form a collimated beam like that produced by a laser pointer. If a circular beam is required, cylindrical lenses and other optics are used. For single spatial mode lasers, using symmetrical lenses, the collimated beam ends up being elliptical in shape, due to the difference in the vertical and lateral divergences. This is easily observable with a red laser pointer.

The simple diode described above has been heavily modified in recent years to accommodate modern technology, resulting in a variety of types of laser diodes, as described below.

Types

The simple laser diode structure, described above, is extremely inefficient. Such devices require so much power that they can only achieve pulsed operation without damage. Although historically important and easy to explain, such devices are not practical.

Double Heterostructure Lasers

In these devices, a layer of low bandgap material is sandwiched between two high bandgap layers. One commonly-used pair of materials is gallium arsenide (GaAs) with aluminium gallium arsenide ($Al_xGa_{(1-x)}As$). Each of the junctions between different bandgap materials is called a heterostructure, hence the name "double heterostructure laser" or DH laser. The kind of laser diode described in the first part of the article may be referred to as a homojunction laser, for contrast with these more popular devices.

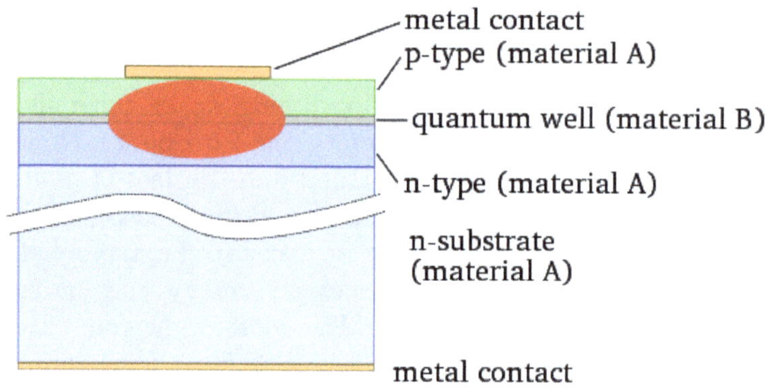

Diagram of front view of a double heterostructure laser diode; not to scale.

The advantage of a DH laser is that the region where free electrons and holes exist simultaneously—the active region—is confined to the thin middle layer. This means that many more of the electron-hole pairs can contribute to amplification—not so many are left out in the poorly amplifying periphery. In addition, light is reflected from the heterojunction; hence, the light is confined to the region where the amplification takes place.

Quantum Well Lasers

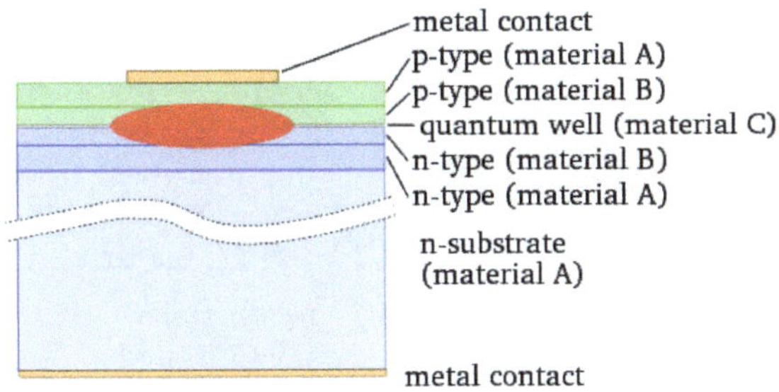

Diagram of front view of a simple quantum well laser diode; not to scale.

If the middle layer is made thin enough, it acts as a quantum well. This means that the vertical variation of the electron's wavefunction, and thus a component of its energy, is quantized. The efficiency of a quantum well laser is greater than that of a bulk laser because the density of states function of electrons in the quantum well system has an abrupt edge that concentrates electrons in energy states that contribute to laser action.

Lasers containing more than one quantum well layer are known as multiple quantum well lasers. Multiple quantum wells improve the overlap of the gain region with the optical waveguide mode.

Further improvements in the laser efficiency have also been demonstrated by reducing the quantum well layer to a quantum wire or to a "sea" of quantum dots.

Quantum Cascade Lasers

In a quantum cascade laser, the difference between quantum well energy levels is used for the laser transition instead of the bandgap. This enables laser action at relatively long wavelengths, which can be tuned simply by altering the thickness of the layer. They are heterojunction lasers.

Interband Cascade Lasers

A Interband cascade laser (ICL) is a type of laser diode that can produce coherent radiation over a large part of the mid-infrared region of the electromagnetic spectrum.

Separate Confinement Heterostructure Lasers

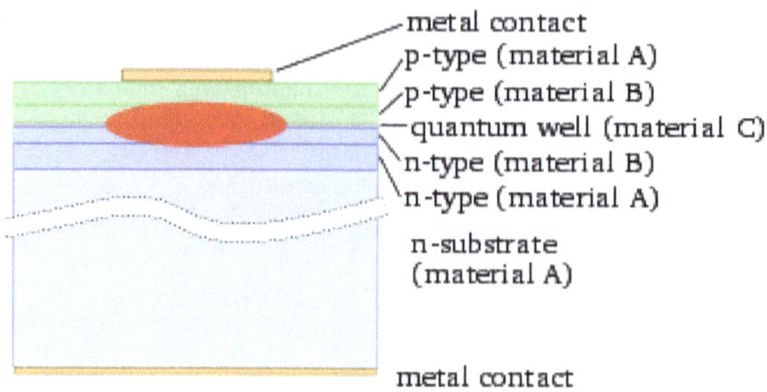

Diagram of front view of a separate confinement heterostructure quantum well laser diode; not to scale

The problem with the simple quantum well diode described above is that the thin layer is simply too small to effectively confine the light. To compensate, another two layers are added on, outside the first three. These layers have a lower refractive index than the centre layers, and hence confine the light effectively. Such a design is called a separate confinement heterostructure (SCH) laser diode.

Almost all commercial laser diodes since the 1990s have been SCH quantum well diodes.

Distributed Bragg Reflector lasers

A **distributed Bragg reflector laser** (**DBR**) is a type of single frequency laser diode. It is characterized by an optical cavity consisting of an electrically or optically pumped gain region between two mirrors to provide feedback. One of the mirrors is a broadband reflector and the other mirror is wavelength selective so that gain is favored on a single longitudinal mode, resulting in lasing at a single resonant frequency. The broadband mirror is usually coated with a low reflectivity coating to allow emission. The wavelength selective mirror is a periodically structured diffraction grating with high reflectivity. The diffraction grating is within a non-pumped, or passive region of the cavity . A DBR laser is a monolithic single chip device with the grating etched into the semiconductor. DBR lasers can be edge emitting lasers or VCSELs. Alternative hybrid architectures that share the same topology include extended cavity diode lasers and volume Bragg grating lasers, but these are not properly called DBR lasers.

Distributed Feedback Lasers

A distributed feedback laser (DFB) is a type of single frequency laser diode. DFBs are the most common transmitter type in DWDM-systems. To stabilize the lasing wavelength, a diffraction grating is etched close to the p-n junction of the diode. This grating acts like an optical filter, causing a single wavelength to be fed back to the gain region and lase. Since the grating provides the feedback that is required for lasing, reflection from the facets is not required. Thus, at least one facet of a DFB is anti-reflection coated. The DFB laser has a stable wavelength that is set during manufacturing by the pitch of the grating, and can only be tuned slightly with temperature. DFB lasers are widely used in optical communication applications where a precise and stable wavelength is critical.

The threshold current of this DFB laser, based on its static characteristic, is around 11 mA. The appropriate bias current in a linear regime could be taken in the middle of the static characteristic (50 mA).

VCSELs

Vertical-cavity surface-emitting lasers (VCSELs) have the optical cavity axis along the direction of current flow rather than perpendicular to the current flow as in conventional laser diodes. The active region length is very short compared with the lateral dimensions so that the radiation emerges from the surface of the cavity rather than from its edge as shown in the figure. The reflectors at the ends of the cavity are dielectric mirrors made from alternating high and low refractive index quarter-wave thick multilayer.

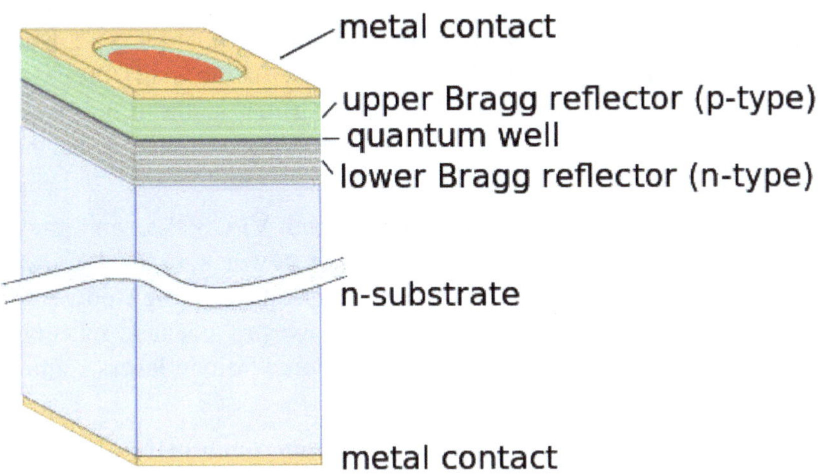

Diagram of a simple VCSEL structure; not to scale.

Such dielectric mirrors provide a high degree of wavelength-selective reflectance at the required free surface wavelength λ if the thicknesses of alternating layers d_1 and d_2 with refractive indices n_1 and n_2 are such that $n_1d_1 + n_2d_2 = \lambda/2$ which then leads to the constructive interference of all partially reflected waves at the interfaces. But there is a disadvantage: because of the high mirror reflectivities, VCSELs have lower output powers when compared to edge-emitting lasers.

There are several advantages to producing VCSELs when compared with the production process of edge-emitting lasers. Edge-emitters cannot be tested until the end of the production process. If the edge-emitter does not work, whether due to bad contacts or poor material growth quality, the production time and the processing materials have been wasted.

Additionally, because VCSELs emit the beam perpendicular to the active region of the laser as opposed to parallel as with an edge emitter, tens of thousands of VCSELs can be processed simultaneously on a three-inch gallium arsenide wafer. Furthermore, even though the VCSEL production process is more labor- and material-intensive, the yield can be controlled to a more predictable outcome. However, they normally show a lower power output level.

VECSELs

Vertical external-cavity surface-emitting lasers, or VECSELs, are similar to VCSELs. In VCSELs, the mirrors are typically grown epitaxially as part of the diode structure, or grown separately and bonded directly to the semiconductor containing the active region. VECSELs are distinguished by a construction in which one of the two mirrors is external to the diode structure. As a result, the cavity includes a free-space region. A typical distance from the diode to the external mirror would be 1 cm.

One of the most interesting features of any VECSEL is the small thickness of the semiconductor gain region in the direction of propagation, less than 100 nm. In contrast, a conventional in-plane semiconductor laser entails light propagation over distances of from 250 μm upward to 2 mm or longer. The significance of the short propagation distance is that it causes the effect of "antiguiding" nonlinearities in the diode laser gain region to be minimized. The result is a large-cross-section single-mode optical beam which is not attainable from in-plane ("edge-emitting") diode lasers.

Several workers demonstrated optically pumped VECSELs, and they continue to be developed for many applications including high power sources for use in industrial machining (cutting, punching, etc.) because of their unusually high power and efficiency when pumped by multi-mode diode laser bars. However, because of their lack of p-n junction, optically-pumped VECSELs are not considered "diode lasers", and are classified as semiconductor lasers.

Electrically pumped VECSELs have also been demonstrated. Applications for electrically pumped VECSELs include projection displays, served by frequency doubling of near-IR VECSEL emitters to produce blue and green light.

External-cavity Diode Lasers

External-cavity diode lasers are tunable lasers which use mainly double heterostructures diodes of the $Al_xGa_{(1-x)}As$ type. The first external-cavity diode lasers used intracavity etalons and simple tuning Littrow gratings. Other designs include gratings in grazing-incidence configuration and multiple-prism grating configurations.

Failure Mechanisms

Laser diodes have the same reliability and failure issues as light emitting diodes. In addition they are subject to catastrophic optical damage (COD) when operated at higher power.

Many of the advances in reliability of diode lasers in the last 20 years remain proprietary to their developers. The reliability of a laser diode can make or break a product line. Moreover, reverse engineering is not always able to reveal the differences between more-reliable and less-reliable diode laser products.

At the edge of a diode laser, where light is emitted, a mirror is traditionally formed by cleaving the semiconductor wafer to form a specularly reflecting plane. This approach is facilitated by the weakness of the crystallographic plane in III-V semiconductor crystals (such as GaAs, InP, GaSb, etc.) compared to other planes. A scratch made at the edge of the wafer and a slight bending force causes a nearly atomically perfect mirror-like cleavage plane to form and propagate in a straight line across the wafer.

But it so happens that the atomic states at the cleavage plane are altered (compared to their bulk properties within the crystal) by the termination of the perfectly periodic lattice at that plane. Surface states at the cleaved plane have energy levels within the (otherwise forbidden) bandgap of the semiconductor.

Essentially, as a result, when light propagates through the cleavage plane and transits to free space from within the semiconductor crystal, a fraction of the light energy is absorbed by the surface states where it is converted to heat by phonon-electron interactions. This heats the cleaved mirror. In addition, the mirror may heat simply because the edge of the diode laser—which is electrically pumped—is in less-than-perfect contact with the mount that provides a path for heat removal. The heating of the mirror causes the bandgap of the semiconductor to shrink in the warmer areas. The bandgap shrinkage brings more electronic band-to-band transitions into alignment with the photon energy causing yet more absorption. This is thermal runaway, a form of positive feedback, and the result can be melting of the facet, known as catastrophic optical damage, or COD.

In the 1970s, this problem, which is particularly nettlesome for GaAs-based lasers emitting between 0.630 μm and 1 μm wavelengths (less so for InP-based lasers used for long-haul telecommunications which emit between 1.3 μm and 2 μm), was identified. Michael Ettenberg, a researcher and later Vice President at RCA Laboratories' David Sarnoff Research Center in Princeton, New Jersey, devised a solution. A thin layer of aluminum oxide was deposited on the facet. If the aluminum oxide thickness is chosen correctly, it functions as an anti-reflective coating, reducing reflection at the surface. This alleviated the heating and COD at the facet.

Since then, various other refinements have been employed. One approach is to create a so-called non-absorbing mirror (NAM) such that the final 10 μm or so before the light emits from the cleaved facet are rendered non-absorbing at the wavelength of interest.

In the very early 1990s, SDL, Inc. began supplying high power diode lasers with good reliability characteristics. CEO Donald Scifres and CTO David Welch presented new reliability performance data at, e.g., SPIE Photonics West conferences of the era. The methods used by SDL to defeat COD were considered to be highly proprietary and were still undisclosed publicly as of June 2006.

In the mid-1990s, IBM Research (Ruschlikon, Switzerland) announced that it had devised its so-called "E2 process" which conferred extraordinary resistance to COD in GaAs-based lasers. This process, too, was undisclosed as of June 2006.

Reliability of high-power diode laser pump bars (used to pump solid-state lasers) remains a difficult problem in a variety of applications, in spite of these proprietary advances. Indeed, the physics of diode laser failure is still being worked out and research on this subject remains active, if proprietary.

Extension of the lifetime of laser diodes is critical to their continued adaptation to a wide variety of applications.

Uses

Laser diodes are numerically the most common laser type, with 2004 sales of approximately 733 million units, as compared to 131,000 of other types of lasers.

Laser diodes find wide use in telecommunication as easily modulated and easily coupled light sources for fiber optics communication. They are used in various measuring instruments, such as rangefinders. Another common use is in barcode readers. Visible lasers, typically red but later also green, are common as laser pointers. Both low and high-power diodes are used extensively in the printing industry both as light sources for scanning (input) of images and for very high-speed and high-resolution printing plate (output) manufacturing. Infrared and red laser diodes are common in CD players, CD-ROMs and DVD technology. Violet lasers are used in HD DVD and Blu-ray technology. Diode lasers have also found many applications in laser absorption spectrometry (LAS) for high-speed, low-cost assessment or monitoring of the concentration of various species in gas phase. High-power laser diodes are used in industrial applications such as heat treating, cladding, seam welding and for pumping other lasers, such as diode-pumped solid-state lasers.

Uses of laser diodes can be categorized in various ways. Most applications could be served by larger solid-state lasers or optical parametric oscillators, but the low cost of mass-produced diode lasers makes them essential for mass-market applications. Diode lasers can be used in a great many fields; since light has many different properties (power, wavelength, spectral and beam quality, polarization, etc.) it is useful to classify applications by these basic properties.

Many applications of diode lasers primarily make use of the "directed energy" property of an optical beam. In this category, one might include the laser printers, barcode readers, image scanning, illuminators, designators, optical data recording, combustion ignition, laser surgery, industrial sorting, industrial machining, and directed energy weaponry. Some of these applications are well-established while others are emerging.

Laser diodes can be arrayed to produce very high power outputs, continuous wave or pulsed. Such arrays may be used to efficiently pump solid-state lasers for high average power drilling, burning or for inertial confinement fusion.

Laser medicine: medicine and especially dentistry have found many new uses for diode lasers. The shrinking size and cost of the units and their increasing user friendliness makes them very attractive to clinicians for minor soft tissue procedures. Diode wavelengths range from 810 to 1,100 nm, are poorly absorbed by soft tissue, and are not used for cutting or ablation. Soft tissue is not cut by the laser's beam, but is instead cut by contact with a hot charred glass tip.The laser's irradiation is highly absorbed at the distal end of the tip and heats it up to 500 °C to 900 °C. Because the tip is so hot, it can be used to cut soft-tissue and can cause hemostasis through cauterization and carbonization. Diode lasers when used on soft tissue can cause extensive collateral thermal damage to surrounding tissue.

Uses which may make use of the coherence of diode-laser-generated light include interferometric distance measurement, holography, coherent communications, and coherent control of chemical reactions.

Uses which may make use of "narrow spectral" properties of diode lasers include range-finding, telecommunications, infra-red countermeasures, spectroscopic sensing, generation of radio-frequency or terahertz waves, atomic clock state preparation, quantum key cryptography, frequency doubling and conversion, water purification (in the UV), and photodynamic therapy (where a particular wavelength of light would cause a substance such as porphyrin to become chemically active as an anti-cancer agent only where the tissue is illuminated by light).

Uses where the desired quality of laser diodes is their ability to generate ultra-short pulses of light by the technique known as "mode-locking" include clock distribution for high-performance integrated circuits, high-peak-power sources for laser-induced breakdown spectroscopy sensing, arbitrary waveform generation for radio-frequency waves, photonic sampling for analog-to-digital conversion, and optical code-division-multiple-access systems for secure communication.

Common wavelengths

- **405 nm** – InGaN blue-violet laser, in Blu-ray Disc and HD DVD drives
- **445–465 nm** – InGaN blue laser multimode diode recently introduced (2010) for use in mercury-free high-brightness data projectors
- **510–525 nm** – Green diodes recently (2010) developed by Nichia and OSRAM for laser projectors.
- **635 nm** – AlGaInP better red laser pointers, same power subjectively twice as bright as 650 nm
- **650–660 nm** – GaInP/AlGaInP CDDVD, cheap red laser pointers
- **670 nm** – AlGaInP bar code readers, first diode laser pointers (now obsolete, replaced by brighter 650 nm and 671 nm DPSS)
- **760 nm** – AlGaInP gas sensing: O 2
- **785 nm** – GaAlAs Compact Disc drives
- **808 nm** – GaAlAs pumps in DPSS Nd:YAG lasers (e.g., in green laser pointers or as arrays in higher-powered lasers)
- **848 nm** – laser mice
- **980 nm** – InGaAs pump for optical amplifiers, for Yb:YAG DPSS lasers
- **1,064 nm** – AlGaAs fiber-optic communication, DPSS laser pump frequency
- **1,310 nm** – InGaAsP, InGaAsN fiber-optic communication
- **1,480 nm** – InGaAsP pump for optical amplifiers
- **1,512 nm** – InGaAsP gas sensing: NH_3
- **1,550 nm** – InGaAsP, InGaAsNSb fiber-optic communication
- **1,625 nm** – InGaAsP fiber-optic communication, service channel
- **1,654 nm** – InGaAsP gas sensing: CH_4
- **1,877 nm** – GaInAsSb gas sensing: H_2O
- **2,004 nm** – GaInAsSb gas sensing: CO_2
- **2,330 nm** – GaInAsSb gas sensing: CO_2**,680 nm** – GaInAsSb gas sensing: CO_2
- **3,030 nm** – GaInAsSb gas sensing: C_2H_2
- **3,330 nm** – GaInAsSb gas sensing: CH_4

Coherent light emission from a gallium arsenide (GaAs) semiconductor diode (the first laser diode) was demonstrated in 1962 by two US groups led by Robert N. Hall at the General Electric research center and by Marshall Nathan at the IBM T.J. Watson Research Center. There has been ongoing debate as to whether IBM or GE invented the first laser diode which was largely based on theoretical work by William P. Dumke at

IBM Labs in Yonkers, NY. The priority is given to General Electric group who have obtained and submitted their results earlier; they also went further and made a resonant cavity for their diode. It was initially speculated Ben Lax amongst and other leading physicists that silicon or germanium could be used to create a lasing effect, but William P. Dumke insisted that these materials would not work and instead suggested Gallium Arsenide would be a good candidate according to his theoretical work. The first visible wavelength GaAs laser diode was demonstrated by Nick Holonyak, Jr. later in 1962.

Other teams at MIT Lincoln Laboratory, Texas Instruments, and RCA Laboratories were also involved in and received credit for their historic initial demonstrations of efficient light emission and lasing in semiconductor diodes in 1962 and thereafter. GaAs lasers were also produced in early 1963 in the Soviet Union by the team led by Nikolay Basov.

In the early 1960s liquid phase epitaxy (LPE) was invented by Herbert Nelson of RCA Laboratories. By layering the highest quality crystals of varying compositions, it enabled the demonstration of the highest quality heterojunction semiconductor laser materials for many years. LPE was adopted by all the leading laboratories, worldwide and used for many years. It was finally supplanted in the 1970s by molecular beam epitaxy and organometallic chemical vapor deposition.

Diode lasers of that era operated with threshold current densities of 1000 A/cm^2 at 77 K temperatures. Such performance enabled continuous-lasing to be demonstrated in the earliest days. However, when operated at room temperature, about 300 K, threshold current densities were two orders of magnitude greater, or 100,000 A/cm^2 in the best devices. The dominant challenge for the remainder of the 1960s was to obtain low threshold current density at 300 K and thereby to demonstrate continuous-wave lasing at room temperature from a diode laser.

The first diode lasers were homojunction diodes. That is, the material (and thus the bandgap) of the waveguide core layer and that of the surrounding clad layers, were identical. It was recognized that there was an opportunity, particularly afforded by the use of liquid phase epitaxy using aluminum gallium arsenide, to introduce heterojunctions. Heterostructures consist of layers of semiconductor crystal having varying bandgap and refractive index. Heterojunctions (formed from heterostructures) had been recognized by Herbert Kroemer, while working at RCA Laboratories in the mid-1950s, as having unique advantages for several types of electronic and optoelectronic devices including diode lasers. LPE afforded the technology of making heterojunction diode lasers.

The first heterojunction diode lasers were single-heterojunction lasers. These lasers utilized aluminum gallium arsenide p-type injectors situated over n-type gallium arsenide layers grown on the substrate by LPE. An admixture of aluminum replaced gallium in the semiconductor crystal and raised the bandgap of the p-type injector over that of the n-type layers beneath. It worked; the 300 K threshold currents went down by 10× to 10,000 amperes per square centimeter. Unfortunately, this was still not in the needed range and these single-heterostructure diode lasers did not function in continuous wave operation at room temperature.

The innovation that met the room temperature challenge was the double heterostructure laser. The trick was to quickly move the wafer in the LPE apparatus between different "melts" of aluminum gallium arsenide (p- and n-type) and a third melt of gallium arsenide. It had to be done rapidly since the gallium arsenide core region needed to be significantly under 1 μm in thickness. The first laser diode to achieve continuous wave operation was a double heterostructure demonstrated in 1970 essentially simultaneously by Zhores Alferov and collaborators (including Dmitri Z. Garbuzov) of the Soviet Union, and Morton Panish and Izuo Hayashi working in the United States. However, it is widely accepted that Zhores I. Alferov and team reached the milestone first.

For their accomplishment and that of their co-workers, Alferov and Kroemer shared the 2000 Nobel Prize in Physics.

LIGHT-EMITTING DIODE

light-emitting diode (LED) is a two-lead semiconductor light source. It is a p–n junction diode, which emits light when activated. When a suitable voltage is applied to the leads, electrons are able to recombine with electron holes within the device, releasing energy in the form of photons. This effect is called electroluminescence, and the color of the light (corresponding to the energy of the photon) is determined by the energy band gap of the semiconductor.

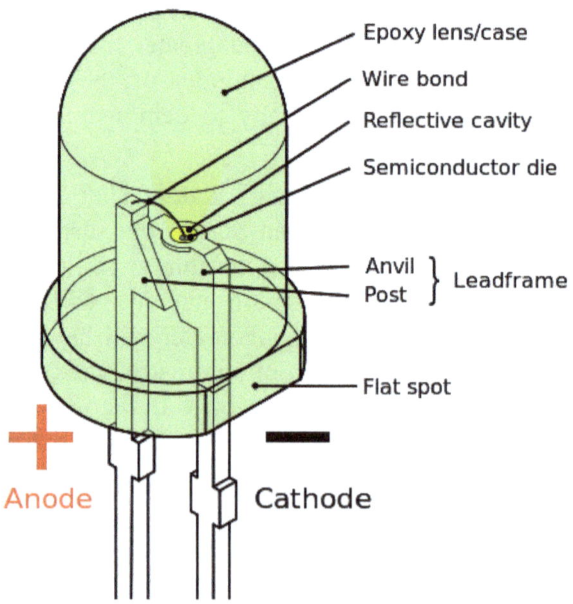

Parts of an LED. Although unlabeled, the flat bottom surfaces of the anvil and post embedded inside the epoxy act as anchors, to prevent the conductors from being forcefully pulled out via mechanical strain or vibration.

A bulb-shaped modern retrofit LED lamp with aluminium heat sink, a light diffusing dome and E27 screw base, using a built-in power supply working on mains voltage

An LED is often small in area (less than 1 mm^2) and integrated optical components may be used to shape its radiation pattern.

Appearing as practical electronic components in 1962, the earliest LEDs emitted low-intensity infrared light. Infrared LEDs are still frequently used as transmitting elements in remote-control circuits, such as those in remote controls for a wide variety of consumer electronics. The first visible-light LEDs were also of low intensity and limited to red. Modern LEDs are available across the visible, ultraviolet, and infrared wavelengths, with very high brightness.

Early LEDs were often used as indicator lamps for electronic devices, replacing small incandescent bulbs. They were soon packaged into numeric readouts in the form of seven-segment displays and were commonly seen in digital clocks.

Recent developments in LEDs permit them to be used in environmental and task lighting. LEDs have many advantages over incandescent light sources including lower energy consumption, longer lifetime, improved physical robustness, smaller size, and faster switching. Light-emitting diodes are now used in applications as diverse as aviation lighting, automotive headlamps, advertising, general lighting, traffic signals, camera flashes, and lighted wallpaper. As of 2016, LEDs powerful enough for room lighting remain somewhat more expensive, and require more precise current and heat management, than compact fluorescent lamp sources of comparable output. They are, however, significantly more energy efficient and, arguably, have fewer environmental concerns linked to their disposal.

LEDs have allowed new displays and sensors to be developed, while their high switching rates are also used in advanced communications technology.

Discoveries and Early Devices

Green electroluminescence from a point contact on a crystal of SiC recreates Round's original experiment from 1907.

Electroluminescence as a phenomenon was discovered in 1907 by the British experimenter H. J. Round of Marconi Labs, using a crystal of silicon carbide and a cat's-whisker detector. Russian inventor Oleg Losev reported creation of the first LED in 1927. His research was distributed in Soviet, German and British scientific journals, but no practical use was made of the discovery for several decades.Kurt Lehovec, Carl Accardo, and Edward Jamgochian explained these first light-emitting diodes in 1951 using an apparatus employing SiC crystals with a current source of battery or pulse generator and with a comparison to a variant, pure, crystal in 1953.

Rubin Braunstein of the Radio Corporation of America reported on infrared emission from gallium arsenide (GaAs) and other semiconductor alloys in 1955.Braunstein observed infrared emission generated by simple diode structures using gallium antimonide (GaSb), GaAs, indium phosphide (InP), and silicon-germanium (SiGe) alloys at room temperature and at 77 Kelvin.

In 1957, Braunstein further demonstrated that the rudimentary devices could be used for non-radio communication across a short distance. As noted by Kroemer Braunstein "… had set up a simple optical communications link: Music emerging from a record player was used via suitable electronics to modulate the forward current of a GaAs diode. The emitted light was detected by a PbS diode some distance away. This signal was fed into an audio amplifier and played back by a loudspeaker. Intercepting the beam stopped the music. We had a great deal of fun playing with this setup." This setup presaged the use of LEDs for optical communication applications.

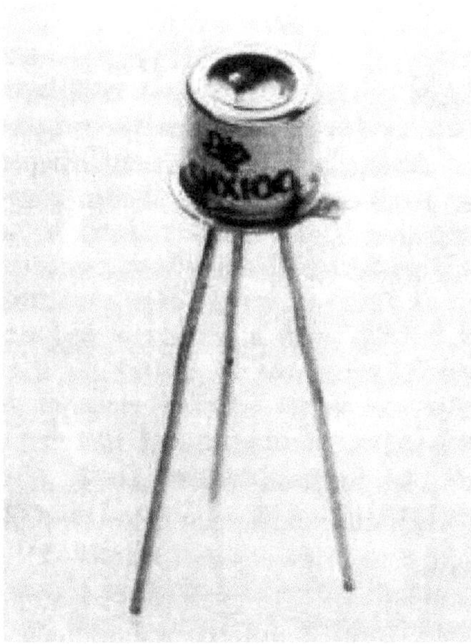

A Texas Instruments SNX-100 GaAs LED contained in a TO-18 transistor metal case.

In September 1961, while working at Texas Instruments in Dallas, Texas, James R. Biard and Gary Pittman discovered near-infrared (900 nm) light emission from a tunnel diode they had constructed on a GaAs substrate. By October 1961, they had demonstrated efficient light emission and signal coupling between a GaAs p-n junction light emitter and an electrically-isolated semiconductor photodetector. On August 8, 1962, Biard and Pittman filed a patent titled "Semiconductor Radiant Diode" based on their findings, which described a zinc diffused p–n junction LED with a spaced cathode contact to allow for efficient emission of infrared light under forward bias. After establishing the priority of their work based on engineering notebooks predating submissions from G.E. Labs, RCA Research Labs, IBM Research Labs, Bell Labs, and Lincoln Lab at MIT, the U.S. patent office issued the two inventors the patent for the GaAs infrared (IR) light-emitting diode (U.S. Patent US3293513), the first practical LED. Immediately after filing the patent, Texas Instruments (TI) began a project to manufacture infrared diodes. In October 1962, TI announced the first commercial LED product (the SNX-100), which employed a pure GaAs crystal to emit a 890 nm light output. In October 1963, TI announced the first commercial hemispherical LED, the SNX-110.

The first visible-spectrum (red) LED was developed in 1962 by Nick Holonyak, Jr. while working at General Electric. Holonyak first reported his LED in the journal Applied Physics Letters on December 1, 1962. M. George Craford, a former graduate student of Holonyak, invented the first yellow LED and improved the brightness of red and red-orange LEDs by a factor of ten in 1972. In 1976, T. P. Pearsall created the first high-brightness, high-efficiency LEDs for optical fiber telecommunications by inventing new semiconductor materials specifically adapted to optical fiber transmission wavelengths.

Initial Commercial Development

The first commercial LEDs were commonly used as replacements for incandescent and neon indicator lamps, and in seven-segment displays, first in expensive equipment such as laboratory and electronics test equipment, then later in such appliances as TVs, radios, telephones, calculators, as well as watches (see list of signal uses). Until 1968, visible and infrared LEDs were extremely costly, in the order of US$200 per unit, and so had little practical use. The Monsanto Company was the first organization to mass-produce visible LEDs, using gallium arsenide phosphide (GaAsP) in 1968 to produce red LEDs suitable for indicators. Hewlett Packard (HP) introduced LEDs in 1968, initially using GaAsP supplied by Monsanto. These red LEDs were bright enough only for use as indicators, as the light output was not enough to illuminate an area. Readouts in calculators were so small that plastic lenses were built over each digit to make them legible. Later, other colors became widely available and appeared in appliances and equipment. In the 1970s commercially successful LED devices at less than five cents each were produced by Fairchild Optoelectronics. These devices employed compound semiconductor chips fabricated with the planar process invented by Dr. Jean Hoerni at Fairchild Semiconductor. The combination of planar processing for chip fabrication and innovative packaging methods enabled the team at Fairchild led by optoelectronics pioneer Thomas Brandt to achieve the needed cost reductions. These methods continue to be used by LED producers.

LED display of a TI-30 scientific calculator (ca. 1978), which uses plastic lenses to increase the visible digit size.

Most LEDs were made in the very common 5 mm T1¾ and 3 mm T1 packages, but with rising power output, it has grown increasingly necessary to shed excess heat to maintain reliability, so more complex packages have been adapted for efficient heat dissipation. Packages for state-of-the-art high-power LEDs bear little resemblance to early LEDs.

Blue LED

Blue LEDs were first developed by Herbert Paul Maruska at RCA in 1972 using gallium nitride (GaN) on a sapphire substrate. SiC-types were first commercially sold in the United States by Cree in 1989. However, neither of these initial blue LEDs were very bright.

The first high-brightness blue LED was demonstrated by Shuji Nakamura of Nichia Corporation in 1994 and was based on InGaN. In parallel, Isamu Akasaki and Hiroshi Amano in Nagoya were working on developing the important GaN nucleation on sapphire substrates and the demonstration of p-type doping of GaN. Nakamura, Akasaki, and Amano were awarded the 2014 Nobel prize in physics for their work. In 1995, Alberto Barbieri at the Cardiff University Laboratory (GB) investigated the efficiency and reliability of high-brightness LEDs and demonstrated a "transparent contact" LED using indium tin oxide (ITO) on (AlGaInP/GaAs).

In 2001 and 2002, processes for growing gallium nitride (GaN) LEDs on silicon were successfully demonstrated. In January 2012, Osram demonstrated high-power InGaN LEDs grown on silicon substrates commercially.

White LEDs and the Illumination Breakthrough

The attainment of high efficiency in blue LEDs was quickly followed by the development of the first white LED. In this device a $Y_3Al_5O_{12}$:Ce (known as "YAG") phosphor coating on the emitter absorbs some of the blue emission and produces yellow light through fluorescence. The combination of that yellow with remaining blue light appears white to the eye. However, using different phosphors (fluorescent materials) it also became possible to instead produce green and red light through fluorescence. The resulting mixture of red, green and blue is not only perceived by humans as white light but is superior for illumination in terms of color rendering, whereas one cannot appreciate the color of red or green objects illuminated only by the yellow (and remaining blue) wavelengths from the YAG phosphor.

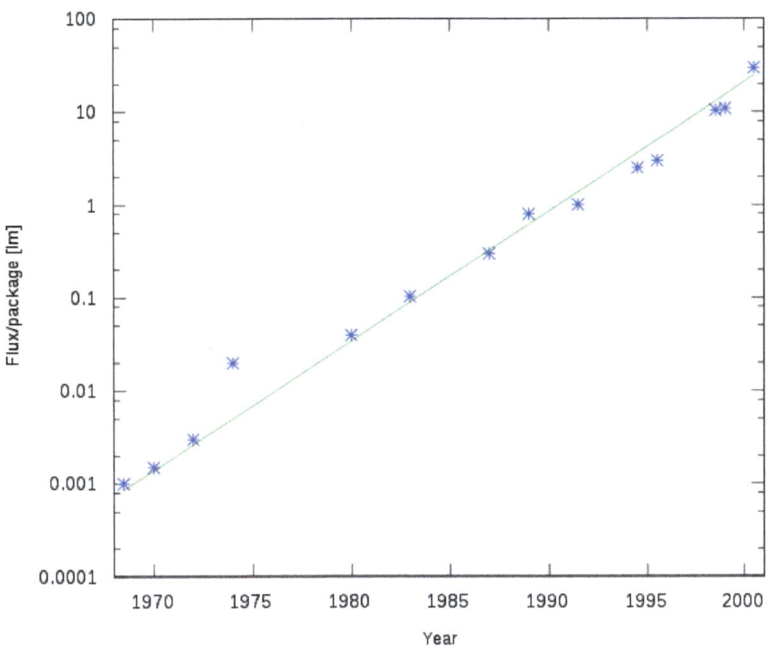

Illustration of Haitz's law, showing improvement in light output per LED over time, with a logarithmic scale on the vertical axis.

The first white LEDs were expensive and inefficient. However, the light output of LEDs has increased exponentially, with a doubling occurring approximately every 36 months since the 1960s (similar to Moore's law). This trend is generally attributed to the parallel development of other semiconductor technologies and advances in optics and materials science and has been called Haitz's law after Dr. Roland Haitz.

The light output and efficiency of blue and near-ultraviolet LEDs rose as the cost of reliable devices fell: this led to the use of (relatively) high-power white-light LEDs for the purpose of illumination which are replacing incandescent and fluorescent lighting.

Experimental white LEDs have been demonstrated to produce over 300 lumens per watt of electricity; some can last up to 100,000 hours. Compared to incandescent bulbs, this is not only a huge increase in electrical efficiency but – over time – a similar or lower cost per bulb.

Working Principle

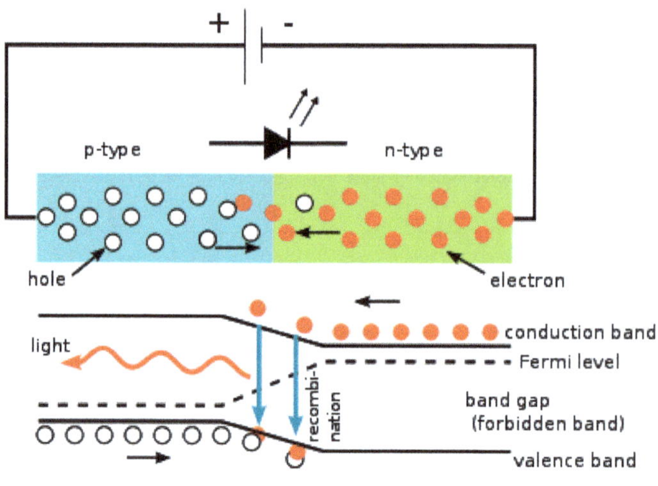

The inner workings of an LED, showing circuit (top) and band diagram (bottom).

A P-N junction can convert absorbed light energy into a proportional electric current. The same process is reversed here (i.e. the P-N junction emits light when electrical energy is applied to it). This phenomenon is generally called electroluminescence, which can be defined as the emission of light from a semi-conductor under the influence of an electric field. The charge carriers recombine in a forward-biased P-N junction as the electrons cross from the N-region and recombine with the holes existing in the P-region. Free electrons are in the conduction band of energy levels, while holes are in the valence energy band. Thus the energy level of the holes will be lesser than the energy levels of the electrons. Some portion of the energy must be dissipated in order to recombine the electrons and the holes. This energy is emitted in the form of heat and light.

The electrons dissipate energy in the form of heat for silicon and germanium diodes but in gallium arsenide phosphide (GaAsP) and gallium phosphide (GaP) semiconductors, the electrons dissipate energy by emitting photons. If the semiconductor is translucent, the junction becomes the source of light as it is emitted, thus becoming a light-emitting diode, but when the junction is reverse biased no light will be produced by the LED and, on the contrary, the device may also be damaged.

Technology

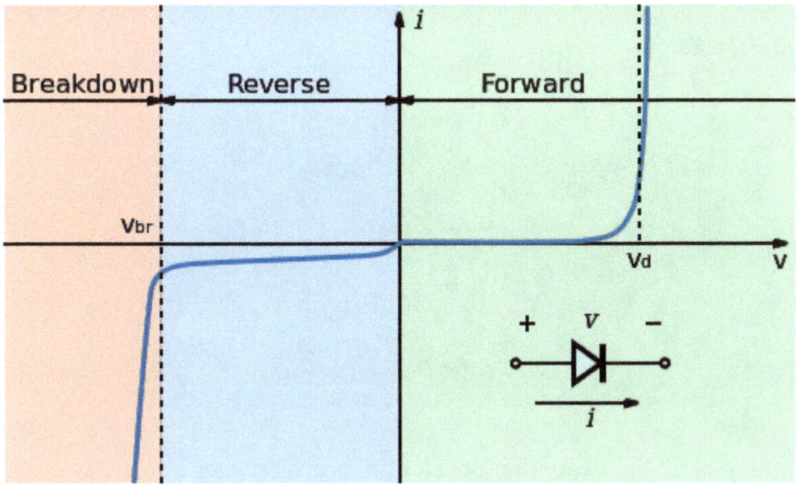

I-V diagram for a diode. An LED will begin to emit light when more than 2 or 3 volts is applied to it. The reverse bias region uses a different vertical scale from the forward bias region, in order to show that the leakage current is nearly constant with voltage until breakdown occurs. In forward bias, the current is small but increases exponentially with voltage.

Physics

The LED consists of a chip of semiconducting material doped with impurities to create a p-n junction. As in other diodes, current flows easily from the p-side, or anode, to the n-side, or cathode, but not in the reverse direction. Charge-carriers—electrons and holes—flow into the junction from electrodes with different voltages. When an electron meets a hole, it falls into a lower energy level and releases energy in the form of a photon.

The wavelength of the light emitted, and thus its color, depends on the band gap energy of the materials forming the p-n junction. In silicon or germanium diodes, the electrons and holes usually recombine by a non-radiative transition, which produces no optical emission, because these are indirect band gap materials. The materials used for the LED have a direct band gap with energies corresponding to near-infrared, visible, or near-ultraviolet light.

LED development began with infrared and red devices made with gallium arsenide. Advances in materials science have enabled making devices with ever-shorter wavelengths, emitting light in a variety of colors.

LEDs are usually built on an n-type substrate, with an electrode attached to the p-type layer deposited on its surface. P-type substrates, while less common, occur as well. Many commercial LEDs, especially GaN/InGaN, also use sapphire substrate.

Most materials used for LED production have very high refractive indices. This means that much of the light will be reflected back into the material at the material/air surface interface. Thus, light extraction in LEDs is an important aspect of LED production, subject to much research and development.

Refractive Index

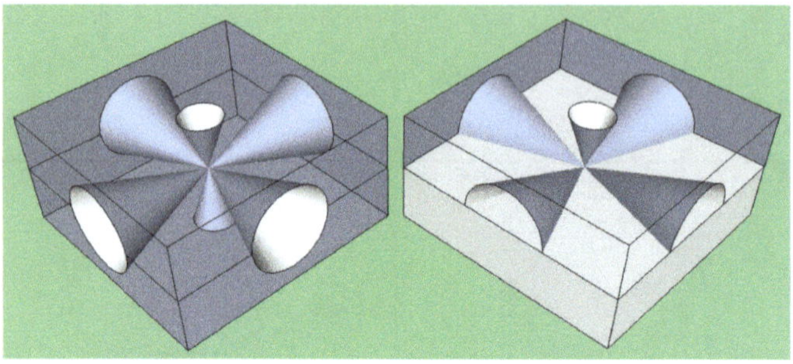

Idealized example of light emission cones in a semiconductor, for a single point-source emission zone. The left illustration is for a fully translucent wafer, while the right illustration shows the half-cones formed when the bottom layer is fully opaque. The light is actually emitted equally in all directions from the point-source, so the areas between the cones show the large amount of trapped light energy that is wasted as heat.

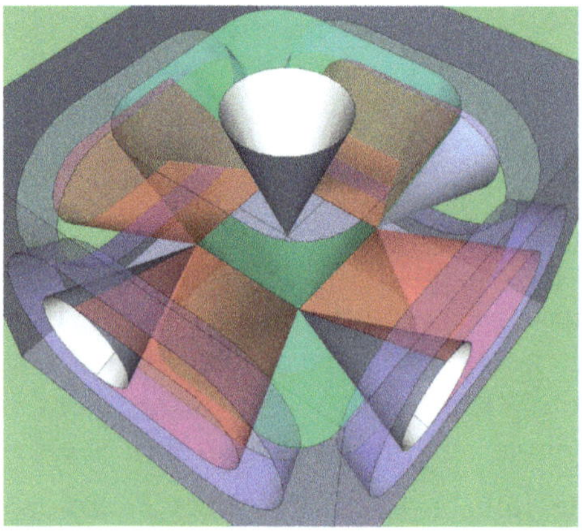

The light emission cones of a real LED wafer are far more complex than a single point-source light emission. The light emission zone is typically a two-dimensional plane between the wafers. Every atom across this plane has an individual set of emission cones. Drawing the billions of overlapping cones is impossible, so this is a simplified diagram showing the extents of all the emission cones combined. The larger side cones are clipped to show the interior features and reduce image complexity; they would extend to the opposite edges of the two-dimensional emission plane.

Bare uncoated semiconductors such as silicon exhibit a very high refractive index relative to open air, which prevents passage of photons arriving at sharp angles relative to the air-contacting surface of the semiconductor due to total internal reflection. This property affects both the light-emission efficiency of LEDs as well as the light-absorption efficiency of photovoltaic cells. The refractive index of silicon is 3.96 (at 590 nm), while air is 1.0002926.

In general, a flat-surface uncoated LED semiconductor chip will emit light only perpendicular to the semiconductor's surface, and a few degrees to the side, in a cone shape referred to as the light cone, cone of light, or the escape cone. The maximum angle of incidence is referred to as the critical angle. When this angle is exceeded, photons no longer escape the semiconductor but are instead reflected internally inside the semiconductor crystal as if it were a mirror.

Internal reflections can escape through other crystalline faces if the incidence angle is low enough and the crystal is sufficiently transparent to not re-absorb the photon emission. But for a simple square LED with 90-degree angled surfaces on all sides, the faces all act as equal angle mirrors. In this case, most of the light can not escape and is lost as waste heat in the crystal.

A convoluted chip surface with angled facets similar to a jewel or fresnel lens can increase light output by allowing light to be emitted perpendicular to the chip surface while far to the sides of the photon emission point.

The ideal shape of a semiconductor with maximum light output would be a microsphere with the photon emission occurring at the exact center, with electrodes penetrating to the center to contact at the emission point. All light rays emanating from the center would be perpendicular to the entire surface of the sphere, resulting in no internal reflections. A hemispherical semiconductor would also work, with the flat back-surface serving as a mirror to back-scattered photons.

Transition Coatings

After the doping of the wafer, it is cut apart into individual dies. Each die is commonly called a chip.

Many LED semiconductor chips are encapsulated or potted in clear or colored molded plastic shells. The plastic shell has three purposes:

- Mounting the semiconductor chip in devices is easier to accomplish.
- The tiny fragile electrical wiring is physically supported and protected from damage.
- The plastic acts as a refractive intermediary between the relatively high-index semiconductor and low-index open air.

The third feature helps to boost the light emission from the semiconductor by acting as a diffusing lens, allowing light to be emitted at a much higher angle of incidence from the light cone than the bare chip is able to emit alone.

Efficiency and Operational Parameters

Typical indicator LEDs are designed to operate with no more than 30–60 milliwatts (mW) of electrical power. Around 1999, Philips Lumileds introduced power LEDs capable of continuous use at one watt. These LEDs used much larger semiconductor die sizes to handle the large power inputs. Also, the semiconductor dies were mounted onto metal slugs to allow for heat removal from the LED die.

One of the key advantages of LED-based lighting sources is high luminous efficacy. White LEDs quickly matched and overtook the efficacy of standard incandescent lighting systems. In 2002, Lumileds made five-watt LEDs available with luminous efficacy of 18–22 lumens per watt (lm/W). For comparison, a conventional incandescent light bulb of 60–100 watts emits around 15 lm/W, and standard fluorescent lights emit up to 100 lm/W.

As of 2012, Philips had achieved the following efficacies for each color. The efficiency values show the physics – light power out per electrical power in. The lumen-per-watt efficacy value includes characteristics of the human eye and is derived using the luminosity function.

	Color	**Wavelength range (nm)**	**Typical efficiency coefficient**	**Typical efficacy (lm/W)**
	Red	$620 < \lambda < 645$	0.39	72
	Red-orange	$610 < \lambda < 620$	0.29	98
	Green	$520 < \lambda < 550$	0.15	93
	Cyan	$490 < \lambda < 520$	0.26	75
	Blue	$460 < \lambda < 490$	0.35	37

In September 2003, a new type of blue LED was demonstrated by Cree that consumes 24 mW at 20 milliamperes (mA). This produced a commercially packaged white light giving 65 lm/W at 20 mA, becoming the brightest white LED commercially available at the time, and more than four times as efficient as standard incandescents. In 2006, they demonstrated a prototype with a record white LED luminous efficacy of 131 lm/W at 20 mA. Nichia Corporation has developed a white LED with luminous efficacy of 150 lm/W at a forward current of 20 mA. Cree's XLamp XM-L LEDs, commercially available in 2011, produce 100 lm/W at their full power of 10 W, and up to 160 lm/W at around 2 W input power. In 2012, Cree announced a white LED giving 254 lm/W, and 303 lm/W in March 2014. Practical general lighting needs high-power LEDs, of one watt or more. Typical operating currents for such devices begin at 350 mA.

These efficiencies are for the light-emitting diode only, held at low temperature in a lab. Since LEDs installed in real fixtures operate at higher temperature and with driver losses, real-world efficiencies are much lower. United States Department of Energy (DOE) testing of commercial LED lamps designed to replace incandescent lamps or CFLs showed that average efficacy was still about 46 lm/W in 2009 (tested performance ranged from 17 lm/W to 79 lm/W).

Efficiency Droop

Efficiency droop is the decrease in luminous efficiency of LEDs as the electric current increases above tens of milliamperes.

This effect was initially theorized to be related to elevated temperatures. Scientists proved the opposite to be true: although the life of an LED would be shortened, the efficiency droop is less severe at elevated temperatures. The mechanism causing efficiency droop was identified in 2007 as Auger recombination, which was taken with mixed reaction. In 2013, a study confirmed Auger recombination as the cause of efficiency droop.

In addition to being less efficient, operating LEDs at higher electric currents creates higher heat levels which compromise the lifetime of the LED. Because of this increased heating at higher currents, high-brightness LEDs have an industry standard of operating at only 350 mA, which is a compromise between light output, efficiency, and longevity.

Possible Solutions

Instead of increasing current levels, luminance is usually increased by combining multiple LEDs in one bulb. Solving the problem of efficiency droop would mean that household LED light bulbs would need fewer LEDs, which would significantly reduce costs.

Researchers at the U.S. Naval Research Laboratory have found a way to lessen the efficiency droop. They found that the droop arises from non-radiative Auger recombination of the injected carriers. They created quantum wells with a soft confinement potential to lessen the non-radiative Auger processes.

Researchers at Taiwan National Central University and Epistar Corp are developing a way to lessen the efficiency droop by using ceramic aluminium nitride (AlN) substrates, which are more thermally conductive than the commercially used sapphire. The higher thermal conductivity reduces self-heating effects.

Lifetime and Failure

Solid-state devices such as LEDs are subject to very limited wear and tear if operated at low currents and at low temperatures. Typical lifetimes quoted are 25,000 to 100,000 hours, but heat and current settings can extend or shorten this time significantly.

The most common symptom of LED (and diode laser) failure is the gradual lowering of light output and loss of efficiency. Sudden failures, although rare, can also occur. Early red LEDs were notable for their short service life. With the development of high-power LEDs, the devices are subjected to higher junction temperatures and higher current densities than traditional devices. This causes stress on the material and may cause early light-output degradation. To quantitatively classify useful lifetime in a standardized manner it has been suggested to use L70 or L50, which are the runtimes (typically given in thousands of hours) at which a given LED reaches 70% and 50% of initial light output, respectively.

Whereas in most previous sources of light (incandescent lamps, discharge lamps, and those that burn combustible fuel, e.g. candles and oil lamps) the light results from heat, LEDs only operate if they are kept cool enough. The manufacturer commonly specifies a maximum junction temperature of 125 or 150 °C, and lower temperatures are advisable in the interests of long life. At these temperatures, relatively little heat is lost by radiation, which means that the light beam generated by an LED is cool.

The waste heat in a high-power LED (which as of 2015 can be less than half the power that it consumes) is conveyed by conduction through the substrate and package of the LED to a heat sink, which gives up the heat to the ambient air by convection. Careful thermal design is, therefore, essential, taking into account the thermal resistances of the LED's package, the heat sink and the interface between the two. Medium-power LEDs are often designed to be soldered directly to a printed circuit board that contains a thermally conductive metal layer. High-power LEDs are packaged in large-area ceramic packages designed to be attached to a metal heat sink, the interface being a material with high thermal conductivity (thermal grease, phase-change material, thermally conductive pad or thermal adhesive).

If an LED-based lamp is installed in an unventilated luminaire, or a luminaire is located in an environment that does not have free air circulation, the LED is likely to overheat, resulting in reduced life or early catastrophic failure. Thermal design is often based on an ambient temperature of 25 °C (77 °F). LEDs used in outdoor applications, such as traffic signals or in-pavement signal lights, and in climates where the temperature within the light fixture gets very high, could experience reduced output or even failure.

Since LED efficacy is higher at low temperatures, LED technology is well suited for supermarket freezer lighting.Because LEDs produce less waste heat than incandescent lamps, their use in freezers can save on refrigeration costs as well. However, they may be more susceptible to frost and snow buildup than incandescent lamps, so some LED lighting systems have been designed with an added heating circuit. Additionally, research has developed heat sink technologies that will transfer heat produced within the junction to appropriate areas of the light fixture.

Colors and Materials

Conventional LEDs are made from a variety of inorganic semiconductor materials. The following table shows the available colors with wavelength range, voltage drop, and material:

	Color	Wavelength [nm]	Voltage drop [ΔV]	Semiconductor material
	Infrared	$\lambda > 760$	$\Delta V < 1.63$	Gallium arsenide (GaAs) Aluminium gallium arsenide (AlGaAs)
	Red	$610 < \lambda < 760$	$1.63 < \Delta V < 2.03$	Aluminium gallium arsenide (AlGaAs) Gallium arsenide phosphide (GaAsP) Aluminium gallium indium phosphide (AlGaInP) Gallium(III) phosphide (GaP)
	Orange	$590 < \lambda < 610$	$2.03 < \Delta V < 2.10$	Gallium arsenide phosphide (GaAsP) Aluminium gallium indium phosphide (AlGaInP) Gallium(III) phosphide (GaP)
	Yellow	$570 < \lambda < 590$	$2.10 < \Delta V < 2.18$	Gallium arsenide phosphide (GaAsP) Aluminium gallium indium phosphide (AlGaInP) Gallium(III) phosphide (GaP)

	Color	Wavelength	Voltage	Semiconductor material
	Green	$500 < \lambda < 570$	$1.9 < \Delta V < 4.0$	**Traditional green:** Gallium(III) phosphide (GaP) Aluminium gallium indium phosphide (AlGaInP) Aluminium gallium phosphide (AlGaP) **Pure green:** Indium gallium nitride (InGaN) / Gallium(III) nitride (GaN)
	Blue	$450 < \lambda < 500$	$2.48 < \Delta V < 3.7$	Zinc selenide (ZnSe) Indium gallium nitride (InGaN) Silicon carbide (SiC) as substrate Silicon (Si) as substrate—under development
	Violet	$400 < \lambda < 450$	$2.76 < \Delta V < 4.0$	Indium gallium nitride (InGaN)
	Purple	Multiple types	$2.48 < \Delta V < 3.7$	Dual blue/red LEDs, blue with red phosphor, or white with purple plastic
	Ultraviolet	$\lambda < 400$	$3 < \Delta V < 4.1$	Indium gallium nitride (InGaN) (385-400 nm) Diamond (235 nm) Boron nitride (215 nm) Aluminium nitride (AlN) (210 nm) Aluminium gallium nitride (AlGaN) Aluminium gallium indium nitride (AlGaInN)—down to 210 nm
	Pink	Multiple types	$\Delta V \sim 3.3$	Blue with one or two phosphor layers, yellow with red, orange or pink phosphor added afterwards, white with pink plastic, or white phosphors with pink pigment or dye over top.
	White	Broad spectrum	$2.8 < \Delta V < 4.2$	**Cool / Pure White:** Blue/UV diode with yellow phosphor **Warm White:** Blue diode with orange phosphor

Blue and Ultraviolet

The first blue-violet LED using magnesium-doped gallium nitride was made at Stanford University in 1972 by Herb Maruska and Wally Rhines, doctoral students in materials science and engineering. At the time Maruska was on leave from RCA Laboratories, where he collaborated with Jacques Pankove on related work. In 1971, the year after Maruska left for Stanford, his RCA colleagues Pankove and Ed Miller demonstrated the first blue electroluminescence from zinc-doped gallium nitride, though the subsequent device Pankove and Miller built, the first actual gallium nitride light-emitting diode, emitted green light. In 1974 the U.S. Patent Office awarded Maruska, Rhines and Stanford professor David Stevenson a patent for their work in 1972 (U.S. Patent US3819974 A) and today magnesium-doping of gallium nitride continues to be the basis for all commercial blue LEDs and laser diodes. These devices built in the early 1970s had too little light output to be of practical use and research into gallium nitride devices slowed. In August 1989, Cree introduced the first commercially available blue LED based on the indirect bandgap semiconductor, silicon carbide (SiC). SiC LEDs had very low efficiency, no more than about 0.03%, but did emit in the blue portion of the visible light spectrum.

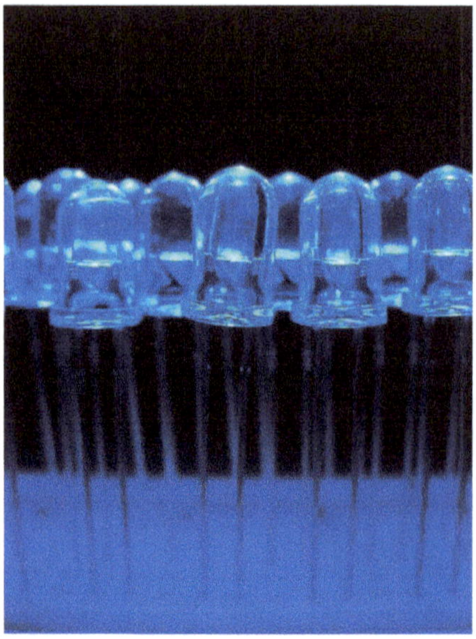

Blue LEDs.

In the late 1980s, key breakthroughs in GaN epitaxial growth and p-type dopingushered in the modern era of GaN-based optoelectronic devices. Building upon this foundation, Theodore Moustakas at Boston University patented a method for producing high-brightness blue LEDs using a new two-step process. Two years later, in 1993, high-brightness blue LEDs were demonstrated again by Shuji Nakamura of Nichia Corporation using a gallium nitride growth process similar to Moustakas's. Both Moustakas and Nakamura were issued separate patents, which confused the issue of who was the original inventor (partly because although Moustakas invented his first, Nakamura filed first). This new development revolutionized LED lighting, making high-power blue light sources practical, leading to the development of technologies like Blu-ray, as well as allowing the bright high-resolution screens of modern tablets and phones.

Nakamura was awarded the 2006 Millennium Technology Prize for his invention. Nakamura, Hiroshi Amano and Isamu Akasaki were awarded the Nobel Prize in Physics in 2014 for the invention of the blue LED. In 2015, a US court ruled that three companies (i.e. the litigants who had not previously settled out of court) that had licensed Nakamura's patents for production in the United States had infringed Moustakas's prior patent, and ordered them to pay licensing fees of not less than 13 million USD.

By the late 1990s, blue LEDs became widely available. They have an active region consisting of one or more InGaN quantum wells sandwiched between thicker layers of GaN, called cladding layers. By varying the relative In/Ga fraction in the InGaN quantum wells, the light emission can in theory be varied from violet to amber. Aluminium gallium nitride (AlGaN) of varying Al/Ga fraction can be used to manufacture the cladding and quantum well layers for ultraviolet LEDs, but these devices have not yet reached the level of efficiency and technological maturity of InGaN/GaN blue/green devices. If un-alloyed GaN is used in this case to form the active quantum well layers, the device will emit near-ultraviolet light with a peak wavelength centred around 365 nm. Green LEDs manufactured from the InGaN/GaN system are far more efficient and brighter than green LEDs produced with non-nitride material systems, but practical devices still exhibit efficiency too low for high-brightness applications.

With nitrides containing aluminium, most often AlGaN and AlGaInN, even shorter wavelengths are achievable. Ultraviolet LEDs in a range of wavelengths are becoming available on the market. Near-UV emitters at wavelengths around 375–395 nm are already cheap and often encountered, for example, as black light lamp replacements for inspection of anti-counterfeiting UV watermarks in some documents and paper currencies. Shorter-wavelength diodes, while substantially more expensive, are commercially available for wavelengths down to 240 nm. As the photosensitivity of microorganisms approximately matches the absorption spectrum of DNA, with a peak at about 260 nm, UV LED emitting at 250–270 nm are to be expected in prospective disinfection and sterilization devices. Recent research has shown that commercially available UVA LEDs (365 nm) are already effective disinfection and sterilization devices. UV-C wavelengths were obtained in laboratories using aluminium nitride (210 nm), boron nitride (215 nm) and diamond (235 nm).

RGB

RGB-SMD-LED.

RGB LEDs consist of one red, one green, and one blue LED. By independently adjusting each of the three, RGB LEDs are capable of producing a wide color gamut. Unlike dedicated-color LEDs, however, these obviously do not produce pure wavelengths. Moreover, such modules as commercially available are often not optimized for smooth color mixing.

White

There are two primary ways of producing white light-emitting diodes (WLEDs), LEDs that generate high-intensity white light. One is to use individual LEDs that emit three primary colors—red, green, and blue—and then mix all the colors to form white light. The other is to use a phosphor material to convert monochromatic light from a blue or UV LED to broad-spectrum white light, much in the same way a fluorescent light bulb works. It is important to note that the 'whiteness' of the light produced is essentially engineered to suit the human eye, and depending on the situation it may not always be appropriate to think of it as white light.

There are three main methods of mixing colors to produce white light from an LED:

- blue LED + green LED + red LED (color mixing; can be used as backlighting for displays)
- near-UV or UV LED + RGB phosphor (an LED producing light with a wavelength shorter than blue's is used to excite an RGB phosphor)
- blue LED + yellow phosphor (two complementary colors combine to form white light; more efficient than first two methods and more commonly used)

Because of metamerism, it is possible to have quite different spectra that appear white. However, the appearance of objects illuminated by that light may vary as the spectrum varies.

RGB Systems

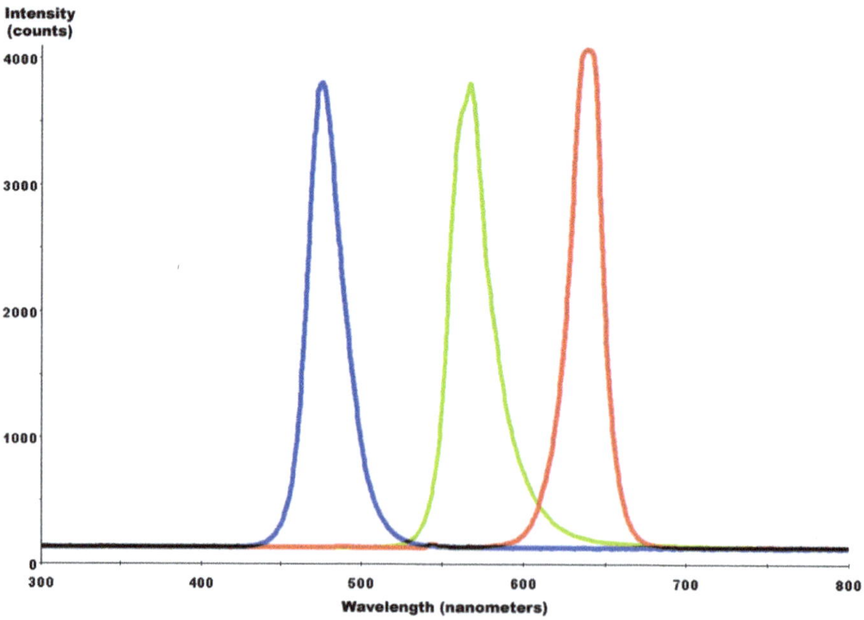

Combined spectral curves for blue, yellow-green, and high-brightness red solid-state semiconductor LEDs. FWHM spectral bandwidth is approximately 24–27 nm for all three colors.

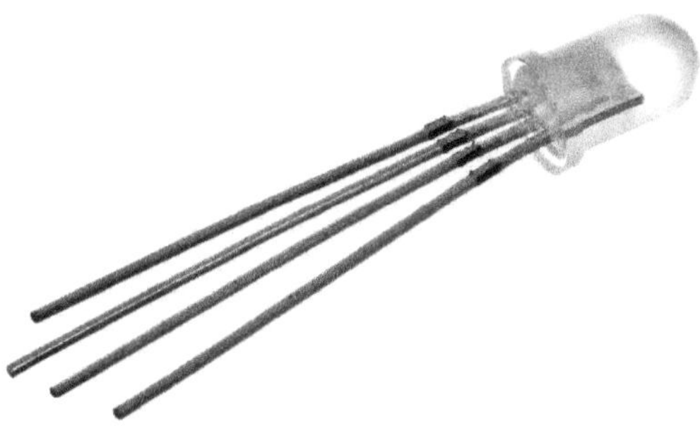

RGB LED.

White light can be formed by mixing differently colored lights; the most common method is to use red, green, and blue (RGB). Hence the method is called multi-color white LEDs (sometimes referred to as RGB LEDs). Because these need electronic circuits to control the blending and diffusion of different colors, and because the individual color LEDs typically have slightly different emission patterns (leading to variation of the color depending on direction) even if they are made as a single unit, these are seldom used to produce white lighting. Nonetheless, this method has many applications because of the flexibility of mixing different colors, and in principle, this mechanism also has higher quantum efficiency in producing white light.

There are several types of multi-color white LEDs: di-, tri-, and tetrachromatic white LEDs. Several key factors that play among these different methods include color stability, color rendering capability, and luminous efficacy. Often, higher efficiency will mean lower color rendering, presenting a trade-off between the luminous efficacy and color rendering. For example, the dichromatic white LEDs have the best luminous efficacy (120 lm/W), but the lowest color rendering capability. However, although tetrachromatic white LEDs have excellent color rendering capability, they often have poor luminous efficacy. Trichromatic white LEDs are in between, having both good luminous efficacy (>70 lm/W) and fair color rendering capability.

One of the challenges is the development of more efficient green LEDs. The theoretical maximum for green LEDs is 683 lumens per watt but as of 2010 few green LEDs exceed even 100 lumens per watt. The blue and red LEDs get closer to their theoretical limits.

Multi-color LEDs offer not merely another means to form white light but a new means to form light of different colors. Most perceivable colors can be formed by mixing different amounts of three primary colors. This allows precise dynamic color control. As more effort is devoted to investigating this method, multi-color LEDs should have profound influence on the fundamental method that we use to produce and control light color. However, before this type of LED can play a role on the market, several technical problems must be solved. These include that this type of LED's emission power decays exponentially with rising temperature, resulting in a substantial change in color stability.

Such problems inhibit and may preclude industrial use. Thus, many new package designs aimed at solving this problem have been proposed and their results are now being reproduced by researchers and scientists.

Correlated color temperature (CCT) dimming for LED technology is regarded as a difficult task since binning, age and temperature drift effects of LEDs change the actual color value output. Feedback loop systems are used for example with color sensors, to actively monitor and control the color output of multiple color mixing LEDs.

Phosphor-based LEDs

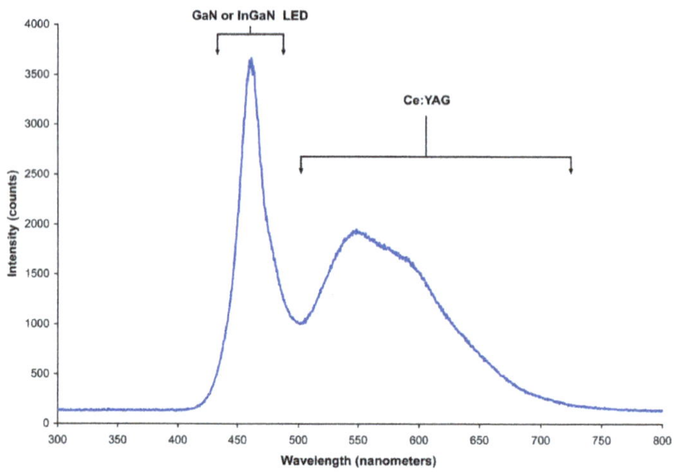

Spectrum of a white LED showing blue light directly emitted by the GaN-based LED (peak at about 465 nm) and the more broadband Stokes-shifted light emitted by the Ce^{3+}:YAG phosphor, which emits at roughly 500–700 nm.

This method involves coating LEDs of one color (mostly blue LEDs made of InGaN) with phosphors of different colors to form white light; the resultant LEDs are called phosphor-based or phosphor-converted white LEDs (pcLEDs). A fraction of the blue light undergoes the Stokes shift being transformed from shorter wavelengths to longer. Depending on the color of the original LED, phosphors of different colors can be employed. If several phosphor layers of distinct colors are applied, the emitted spectrum is broadened, effectively raising the color rendering index (CRI) value of a given LED.

Phosphor-based LED efficiency losses are due to the heat loss from the Stokes shift and also other phosphor-related degradation issues. Their luminous efficacies compared to normal LEDs depend on the spectral distribution of the resultant light output and the original wavelength of the LED itself. For example, the luminous efficacy of a typical YAG yellow phosphor based white LED ranges from 3 to 5 times the luminous efficacy of the original blue LED because of the human eye's greater sensitivity to yellow than to blue (as modeled in the luminosity function). Due to the simplicity of manufacturing, the phosphor method is still the most popular method for making high-intensity white

LEDs. The design and production of a light source or light fixture using a monochrome emitter with phosphor conversion is simpler and cheaper than a complex RGB system, and the majority of high-intensity white LEDs presently on the market are manufactured using phosphor light conversion.

Among the challenges being faced to improve the efficiency of LED-based white light sources is the development of more efficient phosphors. As of 2010, the most efficient yellow phosphor is still the YAG phosphor, with less than 10% Stoke shift loss. Losses attributable to internal optical losses due to re-absorption in the LED chip and in the LED packaging itself account typically for another 10% to 30% of efficiency loss. Currently, in the area of phosphor LED development, much effort is being spent on optimizing these devices to higher light output and higher operation temperatures. For instance, the efficiency can be raised by adapting better package design or by using a more suitable type of phosphor. Conformal coating process is frequently used to address the issue of varying phosphor thickness.

Some phosphor-based white LEDs encapsulate InGaN blue LEDs inside phosphor-coated epoxy. Alternatively, the LED might be paired with a remote phosphor, a preformed polycarbonate piece coated with the phosphor material. Remote phosphors provide more diffuse light, which is desirable for many applications. Remote phosphor designs are also more tolerant of variations in the LED emissions spectrum. A common yellow phosphor material is cerium-doped yttrium aluminium garnet (Ce^{3+}:YAG).

White LEDs can also be made by coating near-ultraviolet (NUV) LEDs with a mixture of high-efficiency europium-based phosphors that emit red and blue, plus copper and aluminium-doped zinc sulfide (ZnS:Cu, Al) that emits green. This is a method analogous to the way fluorescent lamps work. This method is less efficient than blue LEDs with YAG:Ce phosphor, as the Stokes shift is larger, so more energy is converted to heat, but yields light with better spectral characteristics, which render color better. Due to the higher radiative output of the ultraviolet LEDs than of the blue ones, both methods offer comparable brightness. A concern is that UV light may leak from a malfunctioning light source and cause harm to human eyes or skin.

Other white LEDs

Another method used to produce experimental white light LEDs used no phosphors at all and was based on homoepitaxially grown zinc selenide (ZnSe) on a ZnSe substrate that simultaneously emitted blue light from its active region and yellow light from the substrate.

A new style of wafers composed of gallium-nitride-on-silicon (GaN-on-Si) is being used to produce white LEDs using 200-mm silicon wafers. This avoids the typical costly sapphire substrate in relatively small 100- or 150-mm wafer sizes. The sapphire apparatus must be coupled with a mirror-like collector to reflect light that would otherwise be wasted. It is predicted that by 2020, 40% of all GaN LEDs will be made with GaN-on-Si. Manufacturing large sapphire material is difficult, while large silicon material is cheaper and more abundant. LED companies shifting from using sapphire to silicon should be a minimal investment.

Organic Light-Emitting Diodes (OLEDs)

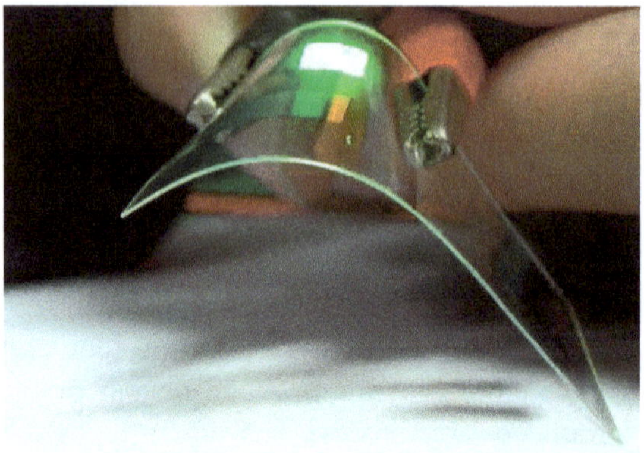

Demonstration of a flexible OLED device

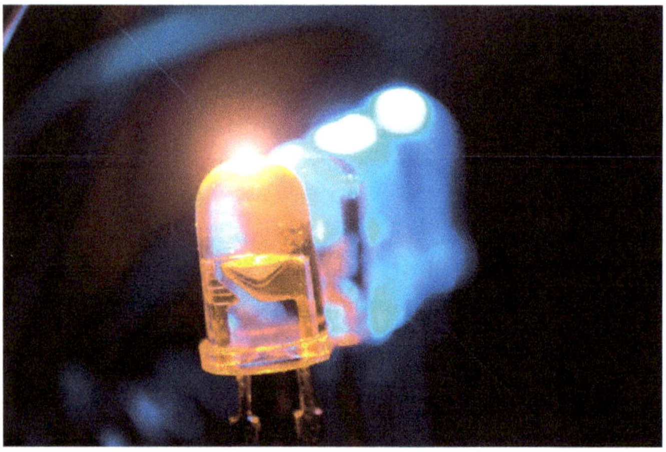

Orange light-emitting diode.

In an organic light-emitting diode (OLED), the electroluminescent material comprising the emissive layer of the diode is an organic compound. The organic material is electrically conductive due to the delocalization of pi electrons caused by conjugation over all or part of the molecule, and the material therefore functions as an organic semiconductor. The organic materials can be small organic molecules in a crystalline phase, or polymers.

The potential advantages of OLEDs include thin, low-cost displays with a low driving voltage, wide viewing angle, and high contrast and color gamut. Polymer LEDs have the added benefit of printable and flexible displays. OLEDs have been used to make visual displays for portable electronic devices such as cellphones, digital cameras, and MP3 players while possible future uses include lighting and televisions.

Quantum dot LEDs

Quantum dots (QD) are semiconductor nanocrystals that possess unique optical properties. Their emission color can be tuned from the visible throughout the infrared spectrum. This allows quantum dot LEDs to create almost any color on the CIE diagram. This provides more color options and better color rendering than white LEDs since the emission spectrum is much narrower, characteristic of quantum confined states. There are two types of schemes for QD excitation. One uses photo excitation with a primary light source LED (typically blue or UV LEDs are used). The other is direct electrical excitation first demonstrated by Alivisatos et al. One example of the photo-excitation scheme is a method developed by Michael Bowers, at Vanderbilt University in Nashville, involving coating a blue LED with quantum dots that glow white in response to the blue light from the LED. This method emits a warm, yellowish-white light similar to that made by incandescent light bulbs. Quantum dots are also being considered for use in white light-emitting diodes in liquid crystal display (LCD) televisions.

In February 2011 scientists at PlasmaChem GmbH were able to synthesize quantum dots for LED applications and build a light converter on their basis, which was able to efficiently convert light from blue to any other color for many hundred hours.Such QDs can be used to emit visible or near infrared light of any wavelength being excited by light with a shorter wavelength.

The structure of QD-LEDs used for the electrical-excitation scheme is similar to basic design of OLEDs. A layer of quantum dots is sandwiched between layers of electron-transporting and hole-transporting materials. An applied electric field causes electrons and holes to move into the quantum dot layer and recombine forming an exciton that excites a QD. This scheme is commonly studied for quantum dot display. The tunability of emission wavelengths and narrow bandwidth is also beneficial as excitation sources for fluorescence imaging. Fluorescence near-field scanning optical microscopy (NSOM) utilizing an integrated QD-LED has been demonstrated. In February 2008, a luminous efficacy of 300 lumens of visible light per watt of radiation (not per electrical watt) and warm-light emission was achieved by using nanocrystals.

Types

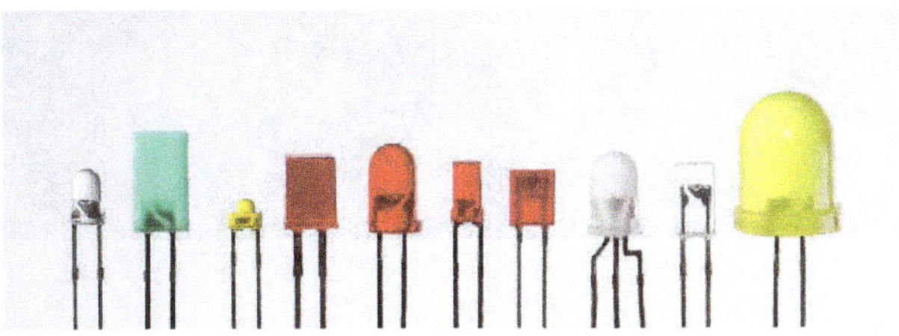

LEDs are produced in a variety of shapes and sizes. The color of the plastic lens is often the same as the actual color of light emitted, but not always. For instance, purple plastic is often used for infrared LEDs, and most blue devices have colorless housings. Modern high-power LEDs such as those used for lighting and backlighting are generally found in surface-mount technology (SMT) packages (not shown).

The main types of LEDs are miniature, high-power devices and custom designs such as alphanumeric or multi-color.

Miniature

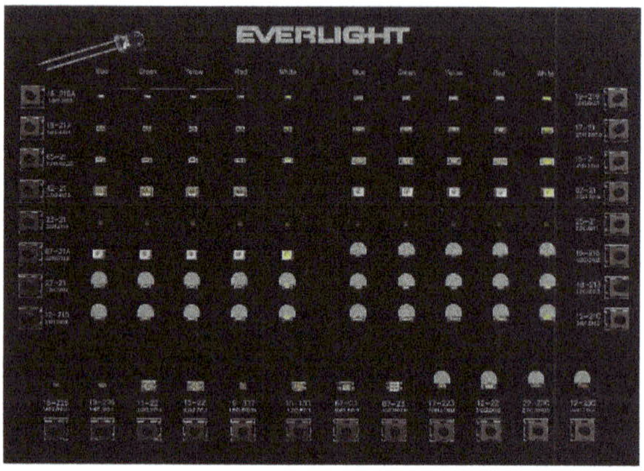

Photo of miniature surface mount LEDs in most common sizes. They can be much smaller than a traditional 5 mm lamp type LED which is shown on the upper left corner.

Very small (1.6x1.6x0.35 mm) red, green, and blue surface mount miniature LED package with gold wire bonding details.

These are mostly single-die LEDs used as indicators, and they come in various sizes from 2 mm to 8 mm, through-hole and surface mount packages. They usually do not use a separate heat sink. Typical current ratings range from around 1 mA to above 20 mA. The small size sets a natural upper boundary on power consumption due to heat caused by the high current density and need for a heat sink. Often daisy chained as used in LED tapes.

Common package shapes include round, with a domed or flat top, rectangular with a flat top (as used in bar-graph displays), and triangular or square with a flat top. The encapsulation may also be clear or tinted to improve contrast and viewing angle.

Researchers at the University of Washington have invented the thinnest LED. It is made of two-dimensional (2-D) flexible materials. It is three atoms thick, which is 10 to 20 times thinner than three-dimensional (3-D) LEDs and is also 10,000 times smaller than the thickness of a human hair. These 2-D LEDs are going to make it possible to create smaller, more energy-efficient lighting, optical communication and nano lasers.

There are three main categories of miniature single die LEDs:

Low-current

Typically rated for 2 mA at around 2 V (approximately 4 mW consumption)

Standard

20 mA LEDs (ranging from approximately 40 mW to 90 mW) at around:

- 1.9 to 2.1 V for red, orange, yellow, and traditional green
- 3.0 to 3.4 V for pure green and blue
- 2.9 to 4.2 V for violet, pink, purple and white

Ultra-high-output

20 mA at approximately 2 or 4–5 V, designed for viewing in direct sunlight

5 V and 12 V LEDs are ordinary miniature LEDs that incorporate a suitable series resistor for direct connection to a 5 V or 12 V supply.

High-power

High-power LEDs (HP-LEDs) or high-output LEDs (HO-LEDs) can be driven at currents from hundreds of mA to more than an ampere, compared with the tens of mA for other LEDs. Some can emit over a thousand lumens. LED power densities up to 300 W/cm^2 have been achieved. Since overheating is destructive, the HP-LEDs must be mounted on a heat sink to allow for heat dissipation. If the heat from an HP-LED is not removed, the device will fail in seconds. One HP-LED can often replace an incandescent bulb in a flashlight, or be set in an array to form a powerful LED lamp.

Some well-known HP-LEDs in this category are the Nichia 19 series, Lumileds Rebel Led, Osram Opto Semiconductors Golden Dragon, and Cree X-lamp. As of September 2009, some HP-LEDs manufactured by Cree now exceed 105 lm/W.

High-power light-emitting diodes attached to an LED star base (Luxeon, Lumileds).

Examples for Haitz's law, which predicts an exponential rise in light output and efficacy of LEDs over time, are the CREE XP-G series LED which achieved 105 lm/W in 2009 and the Nichia 19 series with a typical efficacy of 140 lm/W, released in 2010.

AC driven

LEDs have been developed by Seoul Semiconductor that can operate on AC power without the need for a DC converter. For each half-cycle, part of the LED emits light and part is dark, and this is reversed during the next half-cycle. The efficacy of this type of HP-LED is typically 40 lm/W. A large number of LED elements in series may be able to operate directly from line voltage. In 2009, Seoul Semiconductor released a high DC voltage LED, named as 'Acrich MJT', capable of being driven from AC power with a simple controlling circuit. The low-power dissipation of these LEDs affords them more flexibility than the original AC LED design.

Application-specific variations

Flashing

Flashing LEDs are used as attention seeking indicators without requiring external electronics. Flashing LEDs resemble standard LEDs but they contain an integrated multivibrator circuit that causes the LED to flash with a typical period of one second. In diffused lens LEDs, this circuit is visible as a small black dot. Most flashing LEDs emit light of one color, but more sophisticated devices can flash between multiple colors and even fade through a color sequence using RGB color mixing.

Bi-color

Bi-color LEDs contain two different LED emitters in one case. There are two types of these. One type consists of two dies connected to the same two leads antiparallel to each other. Current flow in one direction emits one color, and current in the opposite direction emits the other color. The other type consists of two dies with separate leads for both dies and another lead for common anode or cathode so that they can be controlled independently. The most common bi-color combination is red/traditional green, however, other available combinations include amber/traditional green, red/pure green, red/blue, and blue/pure green.

Tri-color

Tri-color LEDs contain three different LED emitters in one case. Each emitter is connected to a separate lead so they can be controlled independently. A four-lead arrangement is typical with one common lead (anode or cathode) and an additional lead for each color.

RGB

RGB LEDs are tri-color LEDs with red, green, and blue emitters, in general using a four-wire connection with one common lead (anode or cathode). These LEDs can have either common positive or common negative leads. Others, however, have only two leads (positive and negative) and have a built-in tiny electronic control unit.

Decorative-multicolor

Decorative-multicolor LEDs incorporate several emitters of different colors supplied by only two lead-out wires. Colors are switched internally by varying the supply voltage.

Alphanumeric

Alphanumeric LEDs are available in seven-segment, starburst, and dot-matrix format. Seven-segment displays handle all numbers and a limited set of letters. Starburst displays can display all letters. Dot-matrix displays typically use 5x7 pixels per character. Seven-segment LED displays were in widespread use in the 1970s and 1980s, but rising use of liquid crystal displays, with their lower power needs and greater display flexibility, has reduced the popularity of numeric and alphanumeric LED displays.

Digital-RGB

Digital-RGB LEDs are RGB LEDs that contain their own "smart" control electronics. In addition to power and ground, these provide connections for data-in, data-out, and sometimes a clock or strobe signal. These are connected in a daisy chain, with the data in of the first LED sourced by a microprocessor, which can control the brightness and color of each LED independently of the others. They are used where a combination of maximum control and minimum visible electronics are needed such as strings for

Christmas and LED matrices. Some even have refresh rates in the kHz range, allowing for basic video applications.

Filament

An LED filament consists of multiple LED chips connected in series on a common longitudinal substrate that forms a thin rod reminiscent of a traditional incandescent filament. These are being used as a low-cost decorative alternative for traditional light bulbs that are being phased out in many countries. The filaments require a rather high voltage to light to nominal brightness, allowing them to work efficiently and simply with mains voltages. Often a simple rectifier and capacitive current limiting are employed to create a low-cost replacement for a traditional light bulb without the complexity of creating a low voltage, high current converter which is required by single die LEDs. Usually, they are packaged in a sealed enclosure with a shape similar to lamps they were designed to replace (e.g. a bulb) and filled with inert nitrogen or carbon dioxide gas to remove heat efficiently.

Considerations for Use

Power Sources

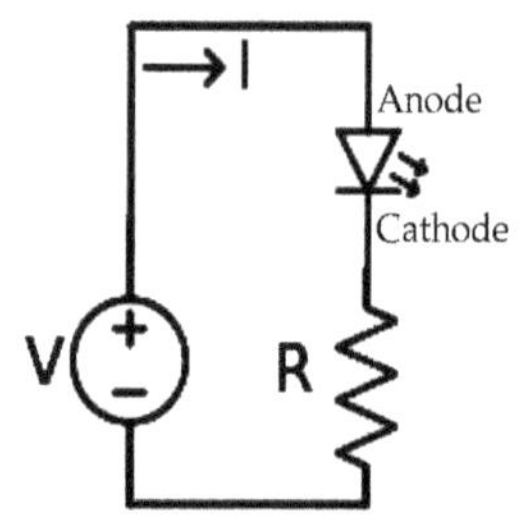

Simple LED circuit with resistor for current limiting.

The current–voltage characteristic of an LED is similar to other diodes, in that the current is dependent exponentially on the voltage (see Shockley diode equation). This means that a small change in voltage can cause a large change in current. If the applied voltage exceeds the LED's forward voltage drop by a small amount, the current rating may be exceeded by a large amount, potentially damaging or destroying the LED. The typical solution is to use constant-current power supplies to keep the current below the LED's maximum current rating. Since most common power sources (batteries, mains) are constant-voltage sources, most LED fixtures must include a power converter, at least a current-limiting resistor. However, the high resistance of three-volt coin cells combined with the high differential resistance of nitride-based LEDs makes it possible to power such an LED from such a coin cell without an external resistor.

Electrical Polarity

As with all diodes, current flows easily from p-type to n-type material. However, no current flows and no light is emitted if a small voltage is applied in the reverse direction. If the reverse voltage grows large enough to exceed the breakdown voltage, a large current flows and the LED may be damaged. If the reverse current is sufficiently limited to avoid damage, the reverse-conducting LED is a useful noise diode.

Safety and Health

The vast majority of devices containing LEDs are "safe under all conditions of normal use", and so are classified as "Class 1 LED product"/"LED Klasse 1". At present, only a few LEDs—extremely bright LEDs that also have a tightly focused viewing angle of 8° or less—could, in theory, cause temporary blindness, and so are classified as "Class 2". The opinion of the French Agency for Food, Environmental and Occupational Health & Safety (ANSES) of 2010, on the health issues concerning LEDs, suggested banning public use of lamps which were in the moderate Risk Group 2, especially those with a high blue component in places frequented by children. In general, laser safety regulations—and the "Class 1", "Class 2", etc. system—also apply to LEDs.

While LEDs have the advantage over fluorescent lamps that they do not contain mercury, they may contain other hazardous metals such as lead and arsenic. Regarding the toxicity of LEDs when treated as waste, a study published in 2011 stated: "According to federal standards, LEDs are not hazardous except for low-intensity red LEDs, which leached Pb [lead] at levels exceeding regulatory limits (186 mg/L; regulatory limit: 5). However, according to California regulations, excessive levels of copper (up to 3892 mg/kg; limit: 2500), lead (up to 8103 mg/kg; limit: 1000), nickel (up to 4797 mg/kg; limit: 2000), or silver (up to 721 mg/kg; limit: 500) render all except low-intensity yellow LEDs hazardous."

Advantages

- **Efficiency:** LEDs emit more lumens per watt than incandescent light bulbs. The efficiency of LED lighting fixtures is not affected by shape and size, unlike fluorescent light bulbs or tubes.
- **Color:** LEDs can emit light of an intended color without using any color filters as traditional lighting methods need. This is more efficient and can lower initial costs.
- **Size:** LEDs can be very small (smaller than 2 mm^2) and are easily attached to printed circuit boards.
- **Warmup time:** LEDs light up very quickly. A typical red indicator LED will achieve full brightness in under a microsecond. LEDs used in communications devices can have even faster response times.
- **Cycling:** LEDs are ideal for uses subject to frequent on-off cycling, unlike incandescent and fluorescent lamps that fail faster when cycled often, or high-intensity discharge lamps (HID lamps) that require a long time before restarting.

- **Dimming:** LEDs can very easily be dimmed either by pulse-width modulation or lowering the forward current. This pulse-width modulation is why LED lights, particularly headlights on cars, when viewed on camera or by some people, appear to be flashing or flickering. This is a type of stroboscopic effect.
- **Cool light:** In contrast to most light sources, LEDs radiate very little heat in the form of IR that can cause damage to sensitive objects or fabrics. Wasted energy is dispersed as heat through the base of the LED.
- **Slow failure:** LEDs mostly fail by dimming over time, rather than the abrupt failure of incandescent bulbs.
- **Lifetime:** LEDs can have a relatively long useful life. One report estimates 35,000 to 50,000 hours of useful life, though time to complete failure may be longer. Fluorescent tubes typically are rated at about 10,000 to 15,000 hours, depending partly on the conditions of use, and incandescent light bulbs at 1,000 to 2,000 hours. Several DOE demonstrations have shown that reduced maintenance costs from this extended lifetime, rather than energy savings, is the primary factor in determining the payback period for an LED product.
- **Shock resistance:** LEDs, being solid-state components, are difficult to damage with external shock, unlike fluorescent and incandescent bulbs, which are fragile.
- **Focus:** The solid package of the LED can be designed to focus its light. Incandescent and fluorescent sources often require an external reflector to collect light and direct it in a usable manner. For larger LED packages total internal reflection (TIR) lenses are often used to the same effect. However, when large quantities of light are needed many light sources are usually deployed, which are difficult to focus or collimate towards the same target.

Disadvantages

- **Initial price:** LEDs are currently slightly more expensive (price per lumen) on an initial capital cost basis, than other lighting technologies. As of March 2014, at least one manufacturer claims to have reached $1 per kilolumen. The additional expense partially stems from the relatively low lumen output and the drive circuitry and power supplies needed.
- **Temperature dependence:** LED performance largely depends on the ambient temperature of the operating environment – or thermal management properties. Overdriving an LED in high ambient temperatures may result in overheating the LED package, eventually leading to device failure. An adequate heat sink is needed to maintain long life. This is especially important in automotive, medical, and military uses where devices must operate over a wide range of temperatures, which require low failure rates. Toshiba has produced LEDs with an operating temperature range of −40 to 100 °C, which suits the LEDs for both indoor and outdoor use in applications such as lamps, ceiling lighting, street lights, and floodlights.
- **Voltage sensitivity:** LEDs must be supplied with a voltage above their threshold voltage and a current below their rating. Current and lifetime change greatly

with a small change in applied voltage. They thus require a current-regulated supply (usually just a series resistor for indicator LEDs).

- **Color rendition:** Most cool-white LEDs have spectra that differ significantly from a black body radiator like the sun or an incandescent light. The spike at 460 nm and dip at 500 nm can cause the color of objects to be perceived differently under cool-white LED illumination than sunlight or incandescent sources, due to metamerism, red surfaces being rendered particularly poorly by typical phosphor-based cool-white LEDs.
- **Area light source:** Single LEDs do not approximate a point source of light giving a spherical light distribution, but rather a lambertian distribution. So LEDs are difficult to apply to uses needing a spherical light field; however, different fields of light can be manipulated by the application of different optics or "lenses". LEDs cannot provide divergence below a few degrees. In contrast, lasers can emit beams with divergences of 0.2 degrees or less.
- **Electrical polarity:** Unlike incandescent light bulbs, which illuminate regardless of the electrical polarity, LEDs will only light with correct electrical polarity. To automatically match source polarity to LED devices, rectifiers can be used.
- **Blue hazard:** There is a concern that blue LEDs and cool-white LEDs are now capable of exceeding safe limits of the so-called blue-light hazard as defined in eye safety specifications such as ANSI/IESNA RP-27.1–05: Recommended Practice for Photobiological Safety for Lamp and Lamp Systems.
- **Light pollution:** Because white LEDs, especially those with high color temperature, emit much more short wavelength light than conventional outdoor light sources such as high-pressure sodium vapor lamps, the increased blue and green sensitivity of scotopic vision means that white LEDs used in outdoor lighting cause substantially more sky glow. The American Medical Association warned on the use of high blue content white LEDs in street lighting, due to their higher impact on human health and environment, compared to low blue content light sources (e.g. High-Pressure Sodium, PC amber LEDs, and low CCT LEDs).
- **Efficiency droop:** The efficiency of LEDs decreases as the electric current increases. Heating also increases with higher currents which compromises the lifetime of the LED. These effects put practical limits on the current through an LED in high power applications.
- **Impact on insects:** LEDs are much more attractive to insects than sodium-vapor lights, so much so that there has been speculative concern about the possibility of disruption to food webs.
- **Use in winter conditions:** Since they do not give off much heat in comparison to incandescent lights, LED lights used for traffic control can have snow obscuring them, leading to accidents.

Applications

LED uses fall into four major categories:

- Visual signals where light goes more or less directly from the source to the human eye, to convey a message or meaning
- Illumination where light is reflected from objects to give visual response of these objects
- Measuring and interacting with processes involving no human vision
- Narrow band light sensors where LEDs operate in a reverse-bias mode and respond to incident light, instead of emitting light

Indicators and Signs

The low energy consumption, low maintenance and small size of LEDs has led to uses as status indicators and displays on a variety of equipment and installations. Large-area LED displays are used as stadium displays, dynamic decorative displays, and dynamic message signs on freeways. Thin, lightweight message displays are used at airports and railway stations, and as destination displays for trains, buses, trams, and ferries.

Red and green LED traffic signals.

One-color light is well suited for traffic lights and signals, exit signs, emergency vehicle lighting, ships' navigation lights or lanterns (chromacity and luminance standards being set under the Convention on the International Regulations for Preventing Collisions at Sea 1972, Annex I and the CIE) and LED-based Christmas lights. In cold climates, LED traffic lights may remain snow-covered. Red or yellow LEDs are used in indicator and alphanumeric displays in environments where night vision must be retained: aircraft cockpits, submarine and ship bridges, astronomy observatories, and in the field, e.g. night time animal watching and military field use.

Automotive applications for LEDs continue to grow.

Because of their long life, fast switching times, and their ability to be seen in broad daylight due to their high output and focus, LEDs have been used in brake lights for cars' high-mounted brake lights, trucks, and buses, and in turn signals for some time, but many vehicles now use LEDs for their rear light clusters. The use in brakes improves safety, due to a great reduction in the time needed to light fully, or faster rise time, up to 0.5 second faster than an incandescent bulb. This gives drivers behind more time to react. In a dual intensity circuit (rear markers and brakes) if the LEDs are not pulsed at a fast enough frequency, they can create a phantom array, where ghost images of the LED will appear if the eyes quickly scan across the array. White LED headlamps are starting to be used. Using LEDs has styling advantages because LEDs can form much thinner lights than incandescent lamps with parabolic reflectors.

Due to the relative cheapness of low output LEDs, they are also used in many temporary uses such as glowsticks, throwies, and the photonic textile Lumalive. Artists have also used LEDs for LED art.

Weather and all-hazards radio receivers with Specific Area Message Encoding (SAME) have three LEDs: red for warnings, orange for watches, and yellow for advisories and statements whenever issued.

Lighting

With the development of high-efficiency and high-power LEDs, it has become possible to use LEDs in lighting and illumination. To encourage the shift to LED lamps and other high-efficiency lighting, the US Department of Energy has created the L Prize competition. The Philips Lighting North America LED bulb won the first competition on August 3, 2011, after successfully completing 18 months of intensive field, lab, and product testing.

LEDs are used as street lights and in other architectural lighting. The mechanical robustness and long lifetime are used in automotive lighting on cars, motorcycles, and bicycle lights. LED light emission may be efficiently controlled by using nonimaging optics principles.

LED street lights are employed on poles and in parking garages. In 2007, the Italian village of Torraca was the first place to convert its entire illumination system to LEDs.

LEDs are used in aviation lighting. Airbus has used LED lighting in its Airbus A320 Enhanced since 2007, and Boeing uses LED lighting in the 787. LEDs are also being used now in airport and heliport lighting. LED airport fixtures currently include medium-intensity runway lights, runway centerline lights, taxiway centerline and edge lights, guidance signs, and obstruction lighting.

LEDs are also used as a light source for DLP projectors, and to backlight LCD televisions (referred to as LED TVs) and laptop displays. RGB LEDs raise the color gamut by as much as 45%. Screens for TV and computer displays can be made thinner using LEDs for backlighting.

The lack of IR or heat radiation makes LEDs ideal for stage lights using banks of RGB LEDs that can easily change color and decrease heating from traditional stage lighting, as well as medical lighting where IR-radiation can be harmful. In energy conservation, the lower heat output of LEDs also means air conditioning (cooling) systems have less heat in need of disposal.

LEDs are small, durable and need little power, so they are used in handheld devices such as flashlights. LED strobe lights or camera flashes operate at a safe, low voltage, instead of the 250+ volts commonly found in xenon flashlamp-based lighting. This is especially useful in cameras on mobile phones, where space is at a premium and bulky voltage-raising circuitry is undesirable.

LEDs are used for infrared illumination in night vision uses including security cameras. A ring of LEDs around a video camera, aimed forward into a retroreflective background, allows chroma keying in video productions.

LEDs are used in mining operations, as cap lamps to provide light for miners. Research has been done to improve LEDs for mining, to reduce glare and to increase illumination, reducing risk of injury to the miners.

LEDs are now used commonly in all market areas from commercial to home use: standard lighting, AV, stage, theatrical, architectural, and public installations, and wherever artificial light is used.

LED to be used for miners, to increase visibility inside mines.

LEDs are increasingly finding uses in medical and educational applications, for example as mood enhancement, and new technologies such as AmBX, exploiting LED versatility. NASA has even sponsored research for the use of LEDs to promote health for astronauts.

Data Communication and other Signalling

Light can be used to transmit data and analog signals. For example, lighting white LEDs can be used in systems assisting people to navigate in closed spaces while searching necessary rooms or objects.

Assistive listening devices in many theaters and similar spaces use arrays of infrared LEDs to send sound to listeners' receivers. Light-emitting diodes (as well as semiconductor lasers) are used to send data over many types of fiber optic cable, from digital audio over TOSLINK cables to the very high bandwidth fiber links that form the Internet backbone. For some time, computers were commonly equipped with IrDA interfaces, which allowed them to send and receive data to nearby machines via infrared.

Because LEDs can cycle on and off millions of times per second, very high data bandwidth can be achieved.

Sustainable Lighting

Efficient lighting is needed for sustainable architecture. In 2009, US Department of Energy testing results on LED lamps showed an average efficacy of 35 lm/W, below that of typical CFLs, and as low as 9 lm/W, worse than standard incandescent bulbs. A typical 13-watt LED lamp emitted 450 to 650 lumens, which is equivalent to a standard 40-watt incandescent bulb.

However, as of 2011, there are LED bulbs available as efficient as 150 lm/W and even inexpensive low-end models typically exceed 50 lm/W, so that a 6-watt LED could achieve the same results as a standard 40-watt incandescent bulb. The latter has an expected lifespan of 1,000 hours, whereas an LED can continue to operate with reduced efficiency for more than 50,000 hours.

See the chart below for a comparison of common light types:

	LED	CFL	Incandescent
Lightbulb Projected Lifespan	50,000 hours	10,000 hours	1,200 hours
Watts Per Bulb (equiv. 60 watts)	10	14	60
Cost Per Bulb	$2.00	$7.00	$1.25
KWh of Electricity Used Over 50,000 Hours	500	700	3000
Cost of Electricity (@ 0.10 per KWh)	$50	$70	$300
Bulbs Needed for 50,000 Hours of Use	1	5	42
Equivalent 50,000 Hours Bulb Expense	$2.00	$35.00	$52.50
TOTAL Cost for 50,000 Hours	$52.00	$105.00	$352.50

Energy Consumption

In the US, one kilowatt-hour (3.6 MJ) of electricity currently causes an average 1.34 pounds (610 g) of CO_2 emission.Assuming the average light bulb is on for 10 hours a day, a 40-watt bulb will cause 196 pounds (89 kg) of CO_2 emission per year. The 6-watt LED equivalent will only cause 30 pounds (14 kg) of CO_2 over the same time span. A building's carbon footprint from lighting can, therefore, be reduced by 85% by exchanging all incandescent bulbs for new LEDs if a building previously used only incandescent bulbs.

In practice, most buildings that use a lot of lighting use fluorescent lighting, which has 22% luminous efficiency compared with 5% for filaments, so changing to LED lighting would still give a 34% reduction in electrical power use and carbon emissions.

The reduction in carbon emissions depends on the source of electricity. Nuclear power in the United States produced 19.2% of electricity in 2011, so reducing electricity consumption in the U.S. reduces carbon emissions more than in France (75% nuclear electricity) or Norway (almost entirely hydroelectric).

Replacing lights that spend the most time lit results in the most savings, so LED lights in infrequently used locations bring a smaller return on investment.

Light Sources for Machine Vision Systems

Machine vision systems often require bright and homogeneous illumination, so features of interest are easier to process. LEDs are often used for this purpose, and this is likely to remain one of their major uses until the price drops low enough to make signaling and illumination uses more widespread. Barcode scanners are the most common example of machine vision, and many low-cost products use red LEDs instead of lasers. Optical computer mice are an example of LEDs in machine vision, as it is used to provide an even light source on the surface for the miniature camera within the mouse. LEDs constitute a nearly ideal light source for machine vision systems for several reasons:

- The size of the illuminated field is usually comparatively small and machine vision systems are often quite expensive, so the cost of the light source is usually a minor concern. However, it might not be easy to replace a broken light source placed within complex machinery, and here the long service life of LEDs is a benefit.
- LED elements tend to be small and can be placed with high density over flat or even-shaped substrates (PCBs etc.) so that bright and homogeneous sources that direct light from tightly controlled directions on inspected parts can be designed. This can often be obtained with small, low-cost lenses and diffusers, helping to achieve high light densities with control over lighting levels and homogeneity. LED sources can be shaped in several configurations (spot lights for reflective illumination; ring lights for coaxial illumination; backlights for contour illumination; linear assemblies; flat, large format panels; dome sources for diffused, omnidirectional illumination).
- LEDs can be easily strobed (in the microsecond range and below) and synchronized with imaging. High-power LEDs are available allowing well-lit images even with very short light pulses. This is often used to obtain crisp and sharp "still" images of quickly moving parts.
- LEDs come in several different colors and wavelengths, allowing easy use of the best color for each need, where different color may provide better visibility of features of interest. Having a precisely known spectrum allows tightly matched filters to be used to separate informative bandwidth or to reduce disturbing effects of ambient light. LEDs usually operate at comparatively low working temperatures, simplifying heat management, and dissipation. This allows using plastic lenses, filters, and diffusers. Waterproof units can also easily be designed, allowing use in harsh or wet environments (food, beverage, oil industries).

A large LED display behind a disc jockey

LED digital display that can display four digits and points

Traffic light using LED

LED daytime running lights of Audi A4

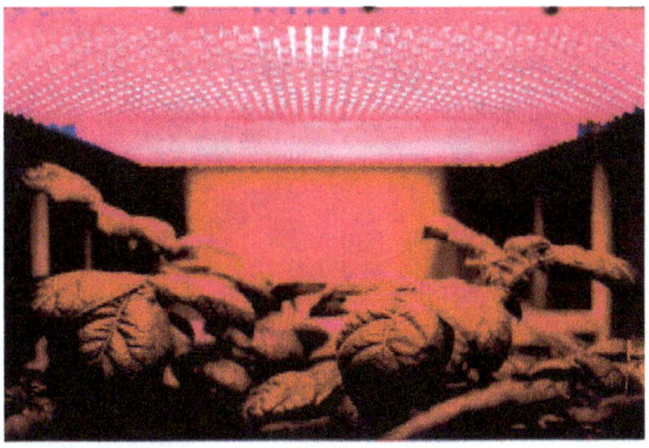

LED panel light source used in an experiment on plant growth. The findings of such experiments may be used to grow food in space on long duration missions.

LED lights reacting dynamically to video feed via AmBX

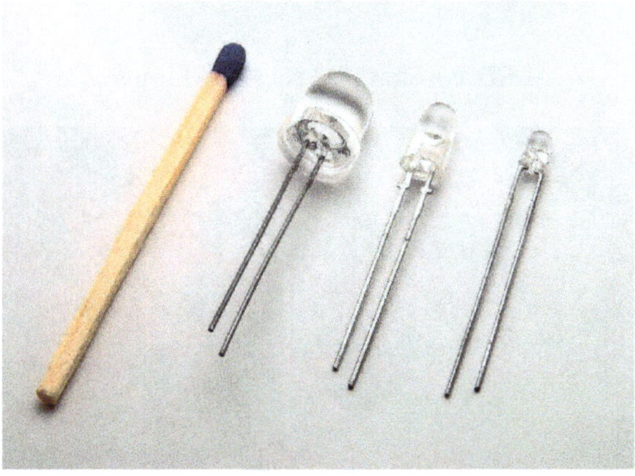

Different sized LEDs. 8 mm, 5 mm and 3 mm, with a wooden match-stick for scale.

A green surface-mount colored LED mounted on an Arduino circuit board

Other Applications

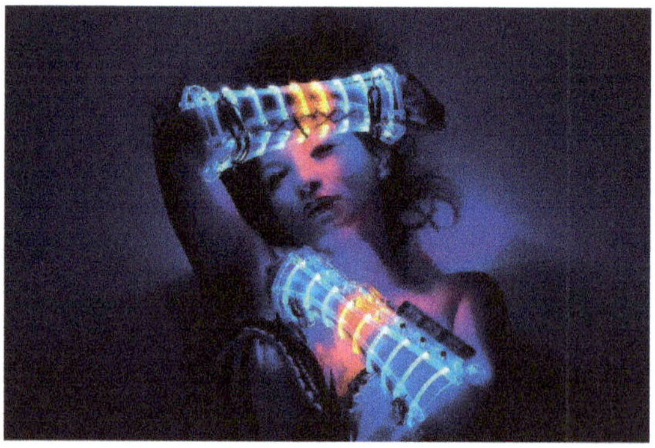

LED costume for stage performers

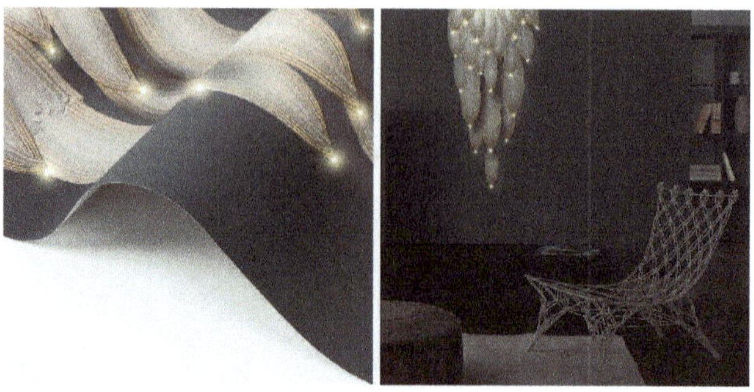

LED wallpaper by Meystyle

The light from LEDs can be modulated very quickly so they are used extensively in optical fiber and free space optics communications. This includes remote controls, such as for TVs, VCRs, and LED Computers, where infrared LEDs are often used. Opto-isolators use an LED combined with a photodiode or phototransistor to provide a signal path with electrical isolation between two circuits. This is especially useful in medical equipment where the signals from a low-voltage sensor circuit (usually battery-powered) in contact with a living organism must be electrically isolated from any possible electrical failure in a recording or monitoring device operating at potentially dangerous voltages. An optoisolator also allows information to be transferred between circuits not sharing a common ground potential.

Many sensor systems rely on light as the signal source. LEDs are often ideal as a light source due to the requirements of the sensors. LEDs are used as motion sensors, for example in optical computer mice. The Nintendo Wii's sensor bar uses infrared LEDs. Pulse oximeters use them for measuring oxygen saturation. Some flatbed scanners use arrays of RGB LEDs rather than the typical cold-cathode fluorescent lamp as the light source. Having independent control of three illuminated colors allows the scanner to calibrate itself for more accurate color balance, and there is no need for warm-up. Further, its sensors only need be monochromatic, since at any one time the page being scanned is only lit by one color of light. Since LEDs can also be used as photodiodes, they can be used for both photo emission and detection. This could be used, for example, in a touchscreen that registers reflected light from a finger or stylus. Many materials and biological systems are sensitive to, or dependent on, light. Grow lights use LEDs to increase photosynthesis in plants, and bacteria and viruses can be removed from water and other substances using UV LEDs for sterilization.

LEDs have also been used as a medium-quality voltage reference in electronic circuits. The forward voltage drop (e.g. about 1.7 V for a normal red LED) can be used instead of a Zener diode in low-voltage regulators. Red LEDs have the flattest I/V curve above the knee. Nitride-based LEDs have a fairly steep I/V curve and are useless for this purpose. Although LED forward voltage is far more current-dependent than a Zener diode, Zener diodes with breakdown voltages below 3 V are not widely available.

The progressive miniaturization of low-voltage lighting technology, such as LEDs and OLEDs, suitable to be incorporated into low-thickness materials has fostered in recent years the experimentation on combining light sources and wall covering surfaces to be applied onto interior walls. The new possibilities offered by these developments have prompted some designers and companies, such as Meystyle, Ingo Maurer, Lomox and Philips, to research and develop proprietary LED wallpaper technologies, some of which are currently available for commercial purchase. Other solutions mainly exist as prototypes or are in the process of being further refined.

A ‹Zener diode' allows current to flow from its anode to its cathode like a normal semiconductor diode, but it also permits current to flow in the reverse direction when its "Zener voltage" is reached. Zener diodes have a highly doped p-n junction. Normal diodes will also break down with a reverse voltage but the voltage and sharpness of the knee are not as well defined as for a Zener diode. Also normal diodes are not designed to operate in the breakdown region, but Zener diodes can reliably operate in this region.

The device was named after Clarence Melvin Zener, who discovered the Zener effect. Zener reverse breakdown is due to electron quantum tunnelling caused by a high strength electric field. However, many diodes described as "Zener" diodes rely instead on avalanche breakdown. Both breakdown types are used in Zener diodes with the Zener effect predominating under 5.6 V and avalanche breakdown above.

Zener diodes are widely used in electronic equipment of all kinds and are one of the basic building blocks of electronic circuits. They are used to generate low power stabilized supply rails from a higher voltage and to provide reference voltages for circuits, especially stabilized power supplies. They are also used to protect circuits from over-voltage, especially electrostatic discharge (ESD).

Operation

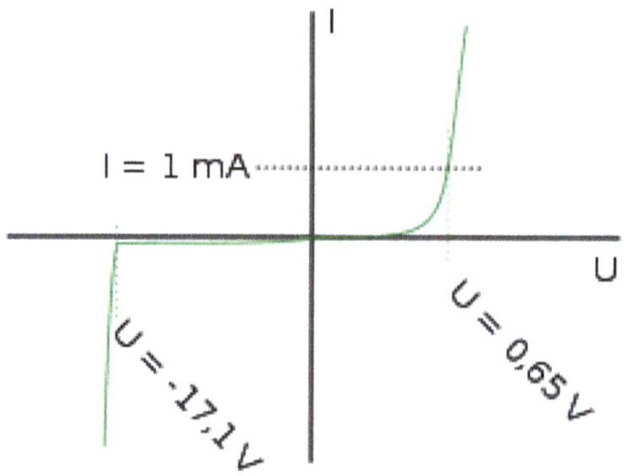

Current-voltage characteristic of a Zener diode with a breakdown voltage of 17 volts. Notice the change of voltage scale between the forward biased (positive) direction and the reverse biased (negative) direction.

A conventional solid-state diode allows significant current if it is reverse-biased above its reverse breakdown voltage. When the reverse bias breakdown voltage is exceeded, a conventional diode is subject to high current due to avalanche breakdown. Unless this current is limited by circuitry, the diode may be permanently damaged due to overheating. A Zener diode exhibits almost the same properties, except the device is specially designed so as to have a reduced breakdown voltage, the so-called Zener voltage. By contrast with the conventional device, a reverse-biased Zener diode exhibits a controlled breakdown and allows the current to keep the voltage across the Zener diode close to the Zener breakdown voltage. For example, a diode with a Zener breakdown voltage of 3.2 V exhibits a voltage drop of very nearly 3.2 V across a wide range of reverse currents. The Zener diode is therefore ideal for applications such as the generation of a reference voltage (e.g. for an amplifier stage), or as a voltage stabilizer for low-current applications.

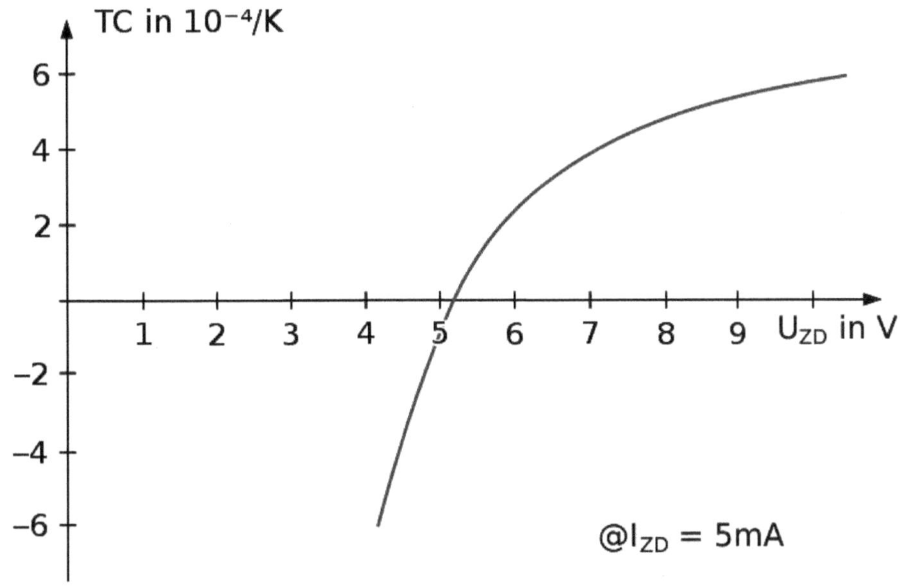

Temperature coefficient of Zener voltage against nominal Zener voltage.

Another mechanism that produces a similar effect is the avalanche effect as in the avalanche diode. The two types of diode are in fact constructed the same way and both effects are present in diodes of this type. In silicon diodes up to about 5.6 volts, the Zener effect is the predominant effect and shows a marked negative temperature coefficient. Above 5.6 volts, the avalanche effect becomes predominant and exhibits a positive temperature coefficient.

In a 5.6 V diode, the two effects occur together, and their temperature coefficients nearly cancel each other out, thus the 5.6 V diode is useful in temperature-critical applications. An alternative, which is used for voltage references that need to be highly stable over long periods of time, is to use a Zener diode with a temperature coefficient (TC) of +2 mV/°C (breakdown voltage 6.2–6.3 V) connected in series with a forward-biased silicon diode (or a transistor B-E junction) manufactured on the same chip. The forward-biased diode has a temperature coefficient of −2 mV/°C, causing the TCs to cancel out.

Modern manufacturing techniques have produced devices with voltages lower than 5.6 V with negligible temperature coefficients, but as higher-voltage devices are encountered, the temperature coefficient rises dramatically. A 75 V diode has 10 times the coefficient of a 12 V diode.

Zener and avalanche diodes, regardless of breakdown voltage, are usually marketed under the umbrella term of "Zener diode".

Under 5.6 V, where the Zener effect dominates, the IV curve near breakdown is much more rounded, which calls for more care in targeting its biasing conditions. The IV curve for Zeners above 5.6 V (being dominated by Avalanche), is much sharper at breakdown.

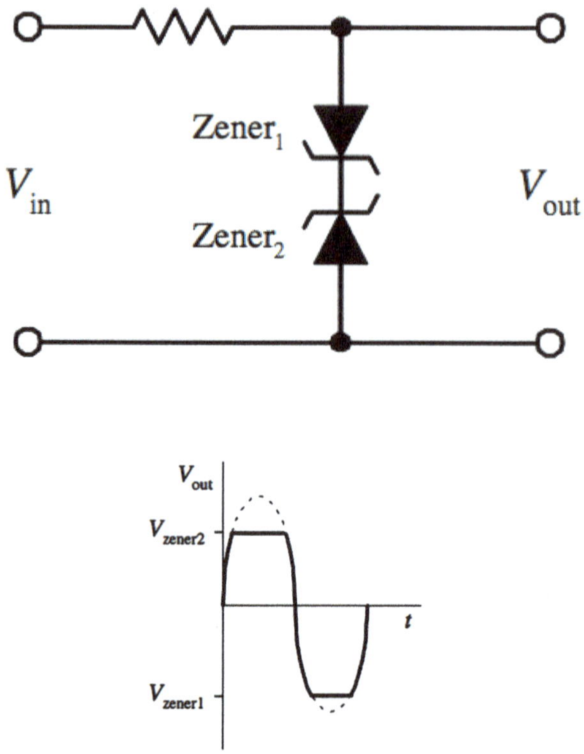

Examples of a Waveform Clipper.

Waveform Clipper

Two Zener diodes facing each other in series will act to clip both halves of an input signal. Waveform clippers can be used to not only reshape a signal, but also to prevent voltage spikes from affecting circuits that are connected to the power supply.

Voltage Shifter

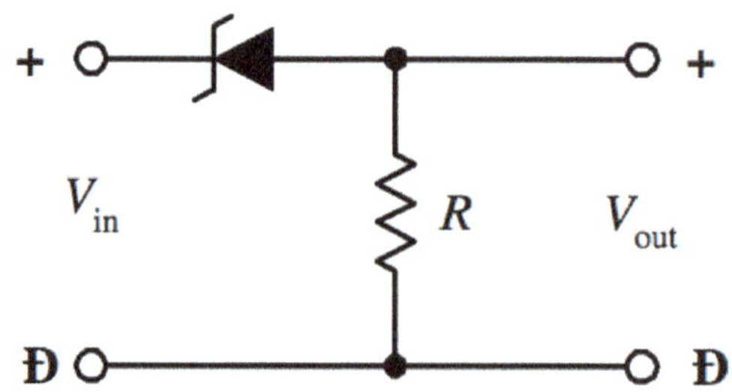

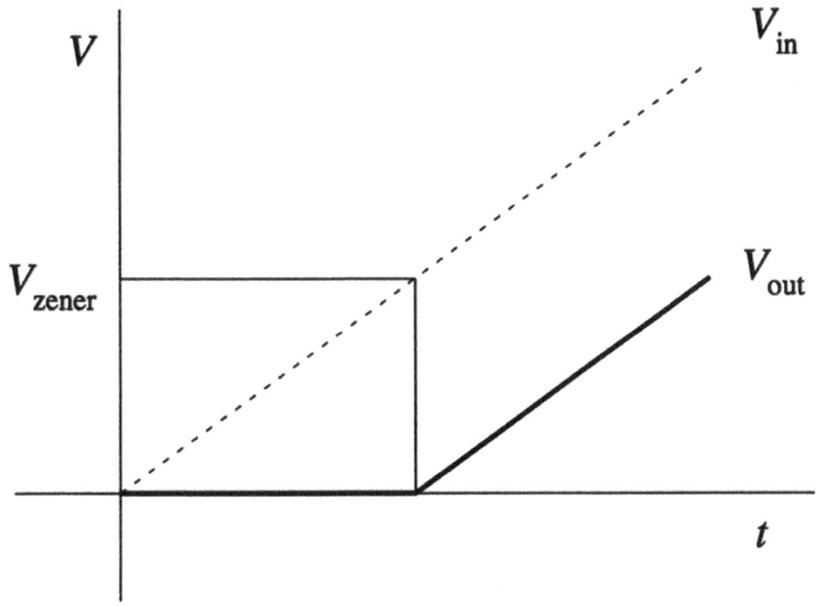

Examples of a Voltage Shifter.

A Zener diode can be applied to a circuit with a resistor to act as a voltage shifter. This circuit lowers the output voltage by a quantity that is equal to the Zener diode's breakdown voltage.

Voltage Regulator

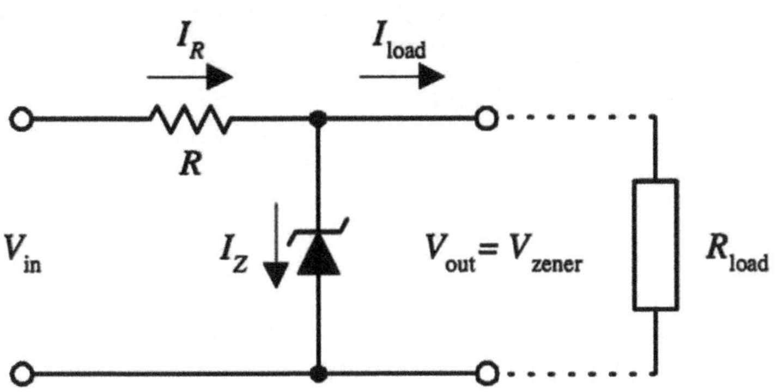

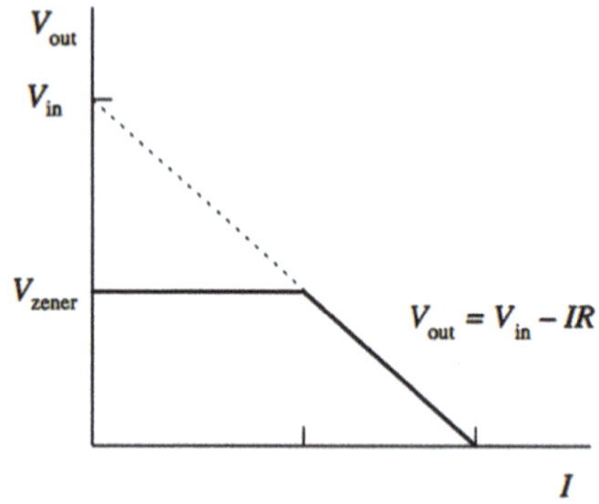

Examples of a Voltage Regulator.

A Zener diode can be applied in a voltage regulator circuit to regulate the voltage applied to a load, such as in a linear regulator.

Construction

The Zener diode's operation depends on the heavy doping of its p-n junction. The depletion region formed in the diode is very thin (<1 μm) and the electric field is consequently very high (about 500 kV/m) even for a small reverse bias voltage of about 5 V, allowing electrons to tunnel from the valence band of the p-type material to the conduction band of the n-type material.

At the atomic scale, this tunneling corresponds to the transport of valence band electrons into the empty conduction band states; as a result of the reduced barrier between these bands and high electric fields that are induced due to the relatively high levels of doping on both sides. The breakdown voltage can be controlled quite accurately in the doping process. While tolerances within 0.07% are available, the most widely used tolerances are 5% and 10%. Breakdown voltage for commonly available Zener diodes can vary widely from 1.2 volts to 200 volts.

For diodes that are lightly doped the breakdown is dominated by the avalanche effect rather than the Zener effect. Consequently, the breakdown voltage is higher (over 5.6 V) for these devices.

Surface Zeners

The emitter-base junction of a bipolar NPN transistor behaves as a Zener diode, with breakdown voltage at about 6.8 V for common bipolar processes and about 10 V for lightly doped base regions in BiCMOS processes. Older processes with poor control of doping characteristics had the variation of Zener voltage up to ±1 V, newer processes using ion implantation can achieve no more than ±0.25 V. The NPN transistor structure can be employed as a surface Zener diode, with collector and emitter connected together as its cathode and base region as anode. In this approach the base doping profile usually narrows towards the surface, creating a region with intensified electric field where the avalanche breakdown occurs. The hot carriers produced by acceleration in the intense field sometime shoot into the oxide layer above the junction and become trapped there. The accumulation of trapped charges can then cause 'Zener walkout', a corresponding change of the Zener voltage of the junction. The same effect can be achieved by radiation damage.

The emitter-base Zener diodes can handle only smaller currents as the energy is dissipated in the base depletion region which is very small. Higher amount of dissipated energy (higher current for longer time, or a short very high current spike) causes thermal damage to the junction and/or its contacts. Partial damage of the junction can shift its Zener voltage. Total destruction of the Zener junction by overheating it and causing migration of metallization across the junction ("spiking") can be used intentionally as a 'Zener zap' antifuse.

Subsurface Zeners

A subsurface Zener diode, also called 'buried Zener', is a device similar to the Surface Zener, but with the avalanche region located deeper in the structure, typically several micrometers below the oxide. The hot carriers then lose energy by collisions with the semiconductor lattice before reaching the oxide layer and cannot be trapped there. The Zener walkout phenomenon therefore does not occur here, and the buried Zeners have voltage constant over their entire lifetime. Most buried Zeners have breakdown voltage of 5–7 volts. Several different junction structures are used.

Uses

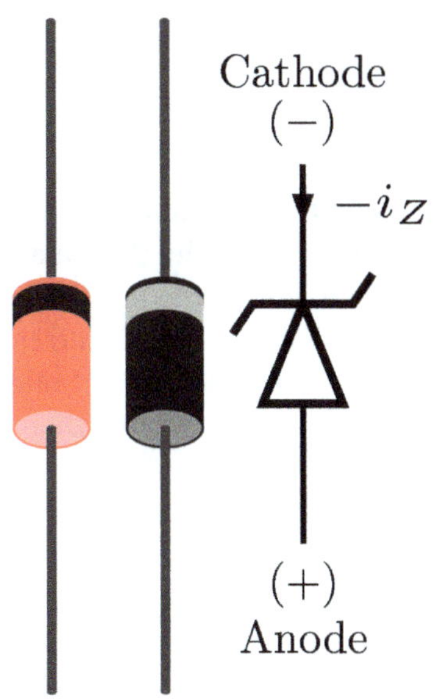

Zener diode shown with typical packages. Reverse current {\displaystyle -i_{Z}} is shown.

Zener diodes are widely used as voltage references and as shunt regulators to regulate the voltage across small circuits. When connected in parallel with a variable voltage source so that it is reverse biased, a Zener diode conducts when the voltage reaches the diode's reverse breakdown voltage. From that point on, the relatively low impedance of the diode keeps the voltage across the diode at that value.

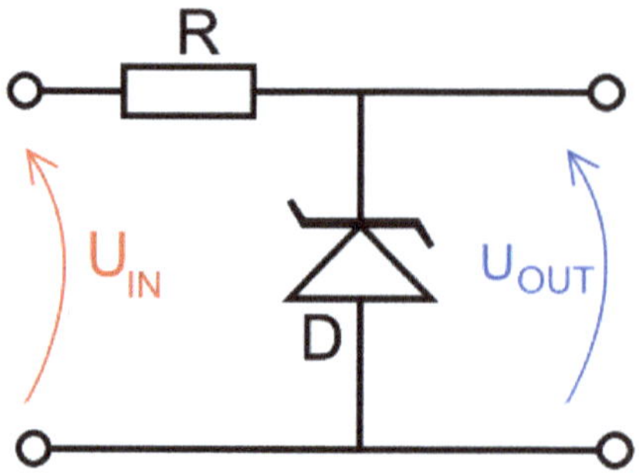

In this circuit, a typical voltage reference or regulator, an input voltage, U_{IN}, is regulated down to a stable output voltage U_{OUT}. The breakdown voltage of diode D is stable over a wide current range and holds U_{OUT} relatively constant even though the input voltage may fluctuate over a fairly wide range. Because of the low impedance of the diode when operated like this, resistor R is used to limit current through the circuit.

In the case of this simple reference, the current flowing in the diode is determined using Ohm's law and the known voltage drop across the resistor R;

$$I_{diode} = \frac{U_{IN} - U_{OUT}}{R_{\Omega}}$$

The value of R must satisfy two conditions :

- R must be small enough that the current through D keeps D in reverse breakdown. The value of this current is given in the data sheet for D. For example, the common BZX79C5V6 device, a 5.6 V 0.5 W Zener diode, has a recommended reverse current of 5 mA. If insufficient current exists through D, then U_{OUT} is unregulated and less than the nominal breakdown voltage (this differs to voltage-regulator tubes where the output voltage will be higher than nominal and could rise as high as U_{IN}). When calculating R, allowance must be made for any current through the external load, not shown in this diagram, connected across U_{OUT}.
- R must be large enough that the current through D does not destroy the device. If the current through D is I_D, its breakdown voltage V_B and its maximum power dissipation P_{MAX} correlate as such: $I_D V_B < P_{MAX}$.

A load may be placed across the diode in this reference circuit, and as long as the Zener stays in reverse breakdown, the diode provides a stable voltage source to the load. Zener diodes in this configuration are often used as stable references for more advanced voltage regulator circuits.

Shunt regulators are simple, but the requirements that the ballast resistor be small enough to avoid excessive voltage drop during worst-case operation (low input voltage concurrent with high load current) tends to leave a lot of current flowing in the diode much of the time, making for a fairly wasteful regulator with high quiescent power dissipation, only suitable for smaller loads.

These devices are also encountered, typically in series with a base-emitter junction, in transistor stages where selective choice of a device centered around the avalanche or Zener point can be used to introduce compensating temperature co-efficient balancing of the transistor p–n junction. An example of this kind of use would be a DC error amplifier used in a regulated power supply circuit feedback loop system.

Zener diodes are also used in surge protectors to limit transient voltage spikes.

Another application of the Zener diode is the use of noise caused by its avalanche breakdown in a random number generator.

A **tunnel diode** or **Esaki diode** is a type of semiconductor that is capable of very fast operation, well into the microwave frequency region, made possible by the use of the quantum mechanical effect called tunneling.

It was invented in August 1957 by Leo Esaki, Yuriko Kurose and Takashi Suzuki when they were working at Tokyo Tsushin Kogyo, now known as Sony.In 1973 Esaki received the Nobel Prize in Physics, jointly with Brian Josephson, for discovering the electron tunneling effect used in these diodes. Robert Noyce independently came up with the idea of a tunnel diode while working for William Shockley, but was discouraged from pursuing it.

These diodes have a heavily doped p–n junction that is about 10 nm (100 Å) wide. The heavy doping results in a broken band gap, where conduction bandelectron states on the n-side are more or less aligned with valence band hole states on the p-side.

Tunnel diodes were first manufactured by Sony in 1957 followed by General Electric and other companies from about 1960, and are still made in low volume today. Tunnel diodes are usually made from germanium, but can also be made from gallium arsenide and silicon materials. They are used in frequency converters and detectors. They have negative differential resistance in part of their operating range, and therefore are also used as oscillators, amplifiers, and in switching circuits using hysteresis.

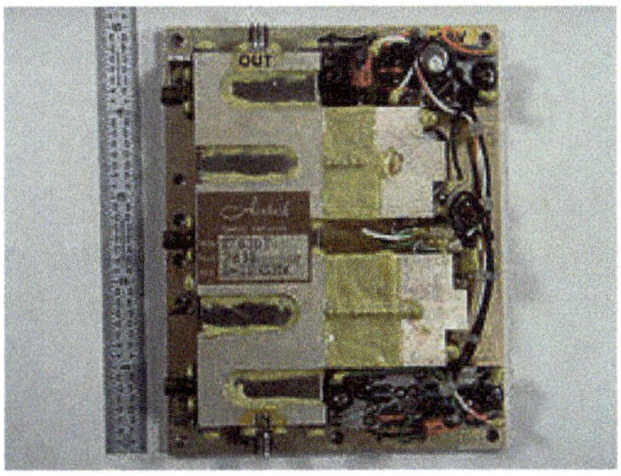

8–1 GHz tunnel diode amplifier, circa 1970.

In 1977, the Intelsat V satellite receiver used a microstrip tunnel diode amplifier (TDA) front-end in the 14 to 15.5 GHz frequency band. Such amplifiers were considered state-of-the-art, with better performance at high frequencies than any transistor-based front end.

The highest frequency room-temperature solid-state oscillators are based on the resonant-tunneling diode (RTD).

There is another type of tunnel diode called a metal–insulator–metal (MIM) diode, but its present application appears to be limited to research environments due to inherent sensitivities. There is also a metal–insulator–insulator–metal MIIM diode which has an additional insulator layer. The additional layer allows "step tunneling" for precise diode control.

Forward bias Operation

Under normal forward bias operation, as voltage begins to increase, electrons at first tunnel through the very narrow p–n junction barrier and fill electron states in the conduction band on the n-side which become aligned with empty valence band hole states on the p-side of the p-n junction. As voltage increases further, these states become increasingly misaligned and the current drops. This is called negative resistance because current decreases with increasing voltage. As voltage increases yet further, the diode begins to operate as a normal diode, where electrons travel by conduction across the p–n junction, and no longer by tunneling through the p–n junction barrier. The most important operating region for a tunnel diode is the negative resistance region. Its graph is different from normal p-n junction diode.

Reverse bias Operation

When used in the reverse direction, tunnel diodes are called back diodes (or backward diodes) and can act as fast rectifiers with zero offset voltage and extreme linearity for power signals (they have an accurate square law characteristic in the reverse direction). Under reverse bias, filled states on the p-side become increasingly aligned with empty states on the n-side and electrons now tunnel through the pn junction barrier in reverse direction.

Technical Comparisons

In a conventional semiconductor diode, conduction takes place while the p–n junction is forward biased and blocks current flow when the junction is reverse biased. This occurs up to a point known as the "reverse breakdown voltage" at which point conduction begins (often accompanied by destruction of the device). In the tunnel diode, the dopant concentrations in the p and n layers are increased to a level such that the **reverse breakdown voltage** becomes **zero** and the diode conducts in the reverse direction. However, when forward-biased, an effect occurs called quantum mechanical tunneling which gives rise to a region in its voltage-current behavior where an increase in forward voltage is accompanied by a decrease in forward current. This negative resistance region can be exploited in a solid state version of the dynatron oscillator which normally uses a tetrode thermionic valve (vacuum tube).

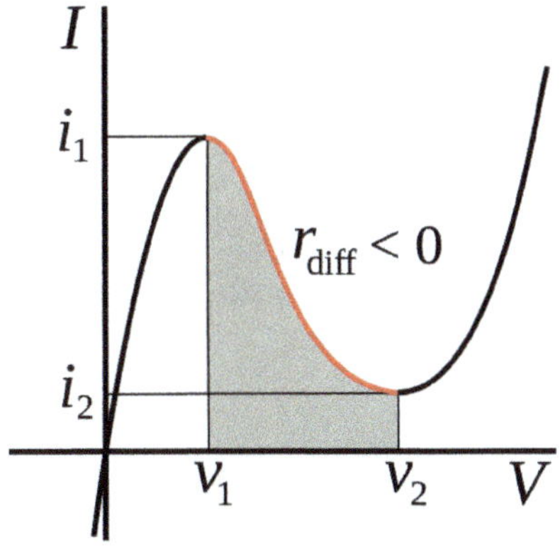

IV curve similar to a tunnel diode characteristic curve. It has negative differential resistance in the shaded voltage region, between v_1 and v_2.

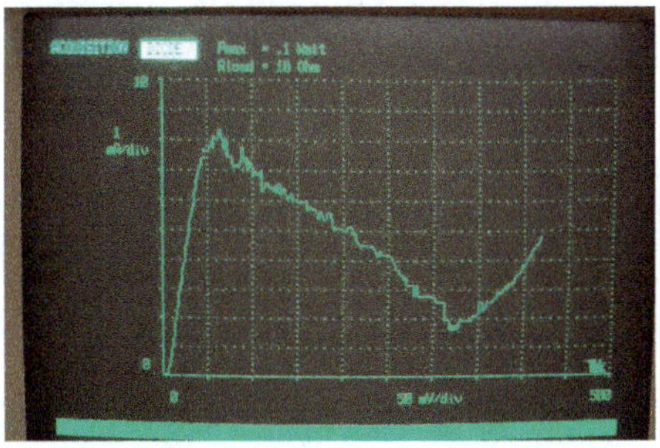

I-V curve of 10mA germanium tunnel diode, taken on a Tektronix model 571 curve tracer.

The tunnel diode showed great promise as an oscillator and high-frequency threshold (trigger) device since it operated at frequencies far greater than the tetrode could, well into the microwave bands. Applications for tunnel diodes included local oscillators for UHF television tuners, trigger circuits in oscilloscopes, high-speed counter circuits, and very fast-rise time pulse generator circuits. The tunnel diode can also be used as a low-noise microwave amplifier. However, since its discovery, more conventional semiconductor devices have surpassed its performance using conventional oscillator techniques. For many purposes, a three-terminal device, such as a field-effect transistor, is more flexible than a device with only two terminals. Practical tunnel diodes operate at a few milliamperes and a few tenths of a volt, making them low-power devices. The Gunn diode has similar high frequency capability and can handle more power.

Tunnel diodes are also more resistant to ionizing radiation than other diodes. This makes them well suited to higher radiation environments such as those found in space.

Longevity

Tunnel diodes are notable for their longevity, with devices made in the 1960s still functioning. Writing in Nature, Esaki and coauthors state that semiconductor devices in general are extremely stable, and suggest that their shelf life should be "infinite" if kept at room temperature. They go on to report that a small-scale test of 50-year-old devices revealed a "gratifying confirmation of the diode's longevity". As noticed on some samples of Esaki diodes, the gold plated iron pins can in fact corrode and short out to the case. This can usually be diagnosed, and the diode inside normally still works.

However, these components are susceptible to damage by overheating, and thus special care is needed when soldering them.

REFERENCES

1. Coldren, Larry; Corzine, Scott; Mashanovitch, Milan (2012). Diode Lasers and Photonic Integrated Circuits (Second ed.). John Wiley and Sons.
2. Arrigoni, M. et. al. (2009-09-28) "Optically Pumped Semiconductor Lasers: Green OPSLs poised to enter scientific pump-laser market", Laser Focus World
3. "Optically Pumped Semiconductor Laser (OPSL)", Sam's Laser FAQs.
4. Hecht, Jeff (1992). The Laser Guidebook (Second ed.). New York: McGraw-Hill, Inc. p. 317. ISBN 0-07-027738-9.
5. Voumard, C. (1977). "External-cavity-controlled 32-MHz narrow-band cw GaA1As-diode lasers". Optics Letters. 1 (2): 61–3. Bibcode:1977OptL....1...61V. doi:10.1364/OL.1.000061. PMID 19680331.
6. Fleming, M. W. and Mooradian, A. (1981). "Spectral characteristics of external-cavity controlled semiconductor lasers". IEEE J. Quantum Electron. 17: 44–59. Bibcode:1981IJQE...17...44F. doi:10.1109/JQE.1981.1070634.
7. Zorabedian, P. (1995). "8". In F. J. Duarte. Tunable Lasers Handbook. Academic Press. ISBN 0-12-222695-X.
8. Steele, Robert V. (2005). "Diode-laser market grows at a slower rate". Laser Focus World. 41 (2).
9. Kincade, Kathy; Anderson, Stephen (2005). "Laser Marketplace 2005: Consumer applications boost laser sales 10%". Laser Focus World. 41 (1). Archived from the original on June 28, 2006.
10. Yeh, S; Jain, K; Andreana, S (2005). "Using a diode laser to uncover dental implants in second-stage surgery". General Dentistry. 53 (6): 414–7. PMID 16366049.
11. Andreana, S (2005). "The use of diode lasers in periodontal therapy: literature review and suggested technique". Dentistry Today. 24 (11): 130, 132–5. PMID 16358809.
12. Deppe, Herbert; Horch, Hans-Henning (2007). "Laser applications in oral surgery and implant dentistry". Lasers in Medical Science. 22 (4): 217–221. doi:10.1007/s10103-007-0440-3. PMID 17268764.

13. Feuerstein, Paul. "Cuts Like A Knife". Dental Economics. Retrieved 2016-04-12.
14. Wright, V. Cecil; Fisher, John C. (1993-01-01). Laser Surgery in Gynecology: A Clinical Guide. Saunders. pp. 58–81. ISBN 9780721640075.
15. Shapshay, S. M. (1987-06-16). Endoscopic Laser Surgery Handbook. CRC Press. pp. 1–130. ISBN 9780824777111.
16. Romanos, Georgios E. (2013-12-01). "Diode laser soft-tissue surgery: advancements aimed at consistent cutting, improved clinical outcomes". Compendium of Continuing Education in Dentistry (Jamesburg, N.J.: 1995). 34 (10): 752–757; quiz 758. PMID 24571504.
17. Vitruk, PP (2015). "Oral Soft Tissue Laser Ablative and Coagulative Efficiencies Spectra". Implant Practice US. 7 (6): 19–27.
18. Hall, Robert N.; Fenner, G. E.; Kingsley, J. D.; Soltys, T. J.; Carlson, R. O. (November 1962). "Coherent Light Emission From GaAs Junctions". Physical Review Letters. 9 (9): 366–368. Bibcode:1962PhRvL...9..366H. doi:10.1103/PhysRevLett.9.366.
19. Nathan, Marshall I.; Dumke, William P.; Burns, Gerald; Dill, Frederick H.; Lasher, Gordon (1962). "Stimulated Emission of Radiation from GaAs p-n Junctions" (PDF). Applied Physics Letters. 1 (3): 62. Bibcode:1962ApPhL...1...62N. doi:10.1063/1.1777371.
20. Oral History Transcript — Dr. Marshall Nathan, American Institute of Physics
21. After Glow. Illinois Alumni Magazine. May–June 2007.
22. Nicolay G. Basov. Nobelprize.org. Retrieved 2009-06-06.
23. Chatak, Ajoy (2009). Optics. Tata McGraw-Hill Education. p. 1.14. ISBN 0-07-026215-2

CHAPTER 5

Microcontroller: Types and Design

Chapter Contents

- **Introduction**
- **History**
- **Embedded Design**
- **Higher Integration**
- **Programming Environments**
- **Types of Microcontrollers**
- **Interrupt Latency**
- **Microcontroller Embedded Memory Technology**

A microcontroller is an integrated chip that is often part of an embedded system. It includes a CPU, RAM, ROM, I/O ports, and timers like a standard computer, but because they are designed to execute only a single specific task to control a single system, they are much smaller and simplified so that they can include all the functions required on a single chip. Unlike a microprocessor, which is a general-purpose chip used to create a multi-function computer or device and requires multiple chips to handle various tasks, this device is meant to be more self-contained and independent, and functions as a tiny, dedicated computer.

INTRODUCTION

A **microcontroller** (or **MCU** for *microcontroller unit*) is a small computer on a single integrated circuit. In modern terminology, it is a System on a chip or SoC. A microcontroller contains one or more CPUs (processor cores) along with memory and programmable input/output peripherals. Program memory in the form of Ferroelectric RAM, NOR flash or OTP ROM is also often included on chip, as well as a small amount of RAM. Microcontrollers are designed for embedded applications, in contrast to the microprocessors used in personal computers or other general purpose applications consisting of various discrete chips.

The die from an Intel 8742, an 8-bit microcontroller that includes a CPU running at 12 MHz, 128 bytes of RAM, 2048 bytes of EPROM, and I/O in the same chip.

Two ATmega microcontrollers.

Microcontrollers are used in automatically controlled products and devices, such as automobile engine control systems, implantable medical devices, remote controls, office machines, appliances, power tools, toys and other embedded systems. By reducing the size and cost compared to a design that uses a separate microprocessor, memory, and input/output devices, microcontrollers make it economical to digitally control even more devices and processes. Mixed signal microcontrollers are common, integrating analog components needed to control non-digital electronic systems.

Some microcontrollers may use four-bit words and operate at frequencies as low as 4 kHz, for low power consumption (single-digit milliwatts or microwatts). They will generally have the ability to retain functionality while waiting for an event such as a button press or other interrupt; power consumption while sleeping (CPU clock and most peripherals off) may be just nanowatts, making many of them well suited for long lasting battery applications. Other microcontrollers may serve performance-critical roles, where they may need to act more like a digital signal processor (DSP), with higher clock speeds and power consumption.

HISTORY

The first microprocessor was the 4-bit Intel 4004 released in 1971, with the Intel 8008 and other more capable microprocessors becoming available over the next several years. However, both processors required external chips to implement a working system, raising total system cost, and making it impossible to economically computerize appliances.

One book credits TI engineers Gary Boone and Michael Cochran with the successful creation of the first microcontroller in 1971. The result of their work was the TMS 1000, which became commercially available in 1974. It combined read-only memory, read/write memory, processor and clock on one chip and was targeted at embedded systems.

Partly in response to the existence of the single-chip TMS 1000, Intel developed a computer system on a chip optimized for control applications, the Intel 8048, with commercial parts first shipping in 1977. It combined RAM and ROM on the same chip. This chip would find its way into over one billion PC keyboards, and other numerous applications. At that time Intel's President, Luke J. Valenter, stated that the microcontroller was one of the most successful in the company's history, and expanded the division's budget over 25%.

Most microcontrollers at this time had concurrent variants. One had an erasable EPROM program memory, with a transparent quartz window in the lid of the package to allow it to be erased by exposure to ultraviolet light, often used for prototyping. The other was either a mask programmed ROM from the manufacturer for large series, or a PROM variant which was only programmable once; sometimes this was signified with the designation OTP, standing for "one-time programmable". The PROM was of identical type of memory as the EPROM, but because there was no way to expose it to ultraviolet light, it could not be erased. The erasable versions required ceramic packages with quartz windows, making them significantly more expensive than the OTP versions, which could be made in lower-cost opaque plastic packages. For the erasable variants, quartz was required, instead of less expensive glass, for its transparency to ultraviolet—glass is largely opaque to UV—but the main cost differentiator was the ceramic package itself.

In 1993, the introduction of EEPROM memory allowed microcontrollers (beginning with the Microchip PIC16x84)to be electrically erased quickly without an expensive package as required for EPROM, allowing both rapid prototyping, and In System Programming. (EEPROM technology had been available prior to this time, but the earlier EEPROM was more expensive and less durable, making it unsuitable for low-cost mass-produced microcontrollers.) The same year, Atmel introduced the first microcontroller using Flash memory, a special type of EEPROM. Other companies rapidly followed suit, with both memory types.

Cost has plummeted over time, with the cheapest 8-bit microcontrollers being available for under 0.25 USD in quantity (thousands) in 2009, and some 32-bit microcontrollers around US$1 for similar quantities.

Nowadays microcontrollers are cheap and readily available for hobbyists, with large online communities around certain processors.

In the future, MRAM could potentially be used in microcontrollers as it has infinite endurance and its incremental semiconductor wafer process cost is relatively low.

Volumes

In 2002, about 55% of all CPUs sold in the world were 8-bit microcontrollers and microprocessors. Over two billion 8-bit microcontrollers were sold in 1997, and according to Semico, over four billion 8-bit microcontrollers were sold in 2006.More recently, Semico has claimed the MCU market grew 36.5% in 2010 and 12% in 2011.

A typical home in a developed country is likely to have only four general-purpose microprocessors but around three dozen microcontrollers. A typical mid-range automobile has as many as 30 or more microcontrollers. They can also be found in many electrical devices such as washing machines, microwave ovens, and telephones.

Historically, the 8-bit segment has dominated the MCU market [..] 16-bit microcontrollers became the largest volume MCU category in 2011, overtaking 8-bit devices for the first time that year [..] IC Insights believes the makeup of the MCU market will undergo substantial changes in the next five years with 32-bit devices steadily grabbing a greater share of sales and unit volumes. By 2017, 32-bit MCUs are expected to account for 55% of microcontroller sales [..] In terms of unit volumes, 32-bit MCUs are expected account for 38% of microcontroller shipments in 2017, while 16-bit devices will represent 34% of the total, and 4-/8-bit designs are forecast to be 28% of units sold that year. The 32-bit MCU market is expected to grow rapidly due to increasing demand for higher levels of precision in embedded-processing systems and the growth in connectivity using the Internet. [..] In the next few years, complex 32-bit MCUs are expected to account for over 25% of the processing power in vehicles.

—IC Insights, MCU Market on Migration Path to 32-bit and ARM-based Devices

In 2012, following a global crisis – a worst ever annual sales decline and recovery and average sales price year-over-year plunging 17% – the biggest reduction since the 1980s, the average price for a microcontroller was US$0.88 ($0.69 for 4-/8-bit, $0.59 for 16-bit, $1.76 for 32-bit).

In 2012, worldwide sales of 8-bit microcontrollers were around $4 billion, while 4-bit microcontrollers also saw significant sales.

In 2015, 8-bit microcontrollers can be bought for $0.311 (1,000 units), 16-bit for $0.385 (1,000 units), and 32-bit for $0.378 (1,000 units but at $0.35 for 5,000).

A PIC 18F8720 **microcontroller** in an 80-pin TQFP package.

EMBEDDED DESIGN

A microcontroller can be considered a self-contained system with a processor, memory and peripherals and can be used as an embedded system. The majority of microcontrollers in use today are embedded in other machinery, such as automobiles, telephones, appliances, and peripherals for computer systems.

While some embedded systems are very sophisticated, many have minimal requirements for memory and program length, with no operating system, and low software complexity. Typical input and output devices include switches, relays, solenoids, LED's, small or custom liquid-crystal displays, radio frequency devices, and sensors for data such as temperature, humidity, light level etc. Embedded systems usually have no keyboard, screen, disks, printers, or other recognizable I/O devices of a personal computer, and may lack human interaction devices of any kind.

Interrupts

Microcontrollers must provide real-time (predictable, though not necessarily fast) response to events in the embedded system they are controlling. When certain events occur, an interrupt system can signal the processor to suspend processing the current instruction sequence and to begin an interrupt service routine (ISR, or "interrupt handler") which will perform any processing required based on the source of the interrupt, before returning to the original instruction sequence. Possible interrupt sources are device dependent, and often include events such as an internal timer overflow, completing an analog to digital conversion, a logic level change on an input such as from a button being pressed, and data received on a communication link. Where power consumption is important as in battery devices, interrupts may also wake a microcontroller from a low-power sleep state where the processor is halted until required to do something by a peripheral event.

Programs

Typically micro-controller programs must fit in the available on-chip memory, since it would be costly to provide a system with external, expandable memory. Compilers and assemblers are used to convert both high-level and assembly language codes into a compact machine code for storage in the micro-controller's memory. Depending on the device, the program memory may be permanent, read-only memory that can only be programmed at the factory, or it may be field-alterable flash or erasable read-only memory.

Manufacturers have often produced special versions of their micro-controllers in order to help the hardware and software development of the target system. Originally these included EPROM versions that have a "window" on the top of the device through which program memory can be erased by ultraviolet light, ready for reprogramming after a programming ("burn") and test cycle. Since 1998, EPROM versions are rare and have been replaced by EEPROM and flash, which are easier to use (can be erased electronically) and cheaper to manufacture.

Other versions may be available where the ROM is accessed as an external device rather than as internal memory, however these are becoming rare due to the widespread availability of cheap microcontroller programmers.

The use of field-programmable devices on a micro controller may allow field update of the firmware or permit late factory revisions to products that have been assembled but not yet shipped. Programmable memory also reduces the lead time required for deployment of a new product.

Where hundreds of thousands of identical devices are required, using parts programmed at the time of manufacture can be economical. These "mask programmed" parts have the program laid down in the same way as the logic of the chip, at the same time.

A customized micro-controller incorporates a block of digital logic that can be personalized for additional processing capability, peripherals and interfaces that are adapted to the requirements of the application. One example is the AT91CAP from Atmel.

Other Microcontroller Features

Microcontrollers usually contain from several to dozens of general purpose input/output pins (GPIO). GPIO pins are software configurable to either an input or an output state. When GPIO pins are configured to an input state, they are often used to read sensors or external signals. Configured to the output state, GPIO pins can drive external devices such as LEDs or motors, often indirectly, through external power electronics.

Many embedded systems need to read sensors that produce analog signals. This is the purpose of the analog-to-digital converter (ADC). Since processors are built to interpret and process digital data, i.e. 1s and 0s, they are not able to do anything with the analog signals that may be sent to it by a device. So the analog to digital converter is used to convert the incoming data into a form that the processor can recognize. A less common feature on some microcontrollers is a digital-to-analog converter (DAC) that allows the processor to output analog signals or voltage levels.

In addition to the converters, many embedded microprocessors include a variety of timers as well. One of the most common types of timers is the Programmable Interval Timer (PIT). A PIT may either count down from some value to zero, or up to the capacity of the count register, overflowing to zero. Once it reaches zero, it sends an interrupt to the processor indicating that it has finished counting. This is useful for devices such as thermostats, which periodically test the temperature around them to see if they need to turn the air conditioner on, the heater on, etc.

A dedicated Pulse Width Modulation (PWM) block makes it possible for the CPU to control power converters, resistive loads, motors, etc., without using lots of CPU resources in tight timer loops.

Universal Asynchronous Receiver/Transmitter (UART) block makes it possible to receive and transmit data over a serial line with very little load on the CPU. Dedicated on-chip hardware also often includes capabilities to communicate with other devices (chips) in digital formats such as Inter-Integrated Circuit (I^2C), Serial Peripheral Interface (SPI), Universal Serial Bus (USB), and Ethernet.

HIGHER INTEGRATION

Die of a PIC12C508 8-bit, fully static, EEPROM/EPROM/ROM-based CMOS microcontroller manufactured by Microchip Technology using a 1200 nanometre process.

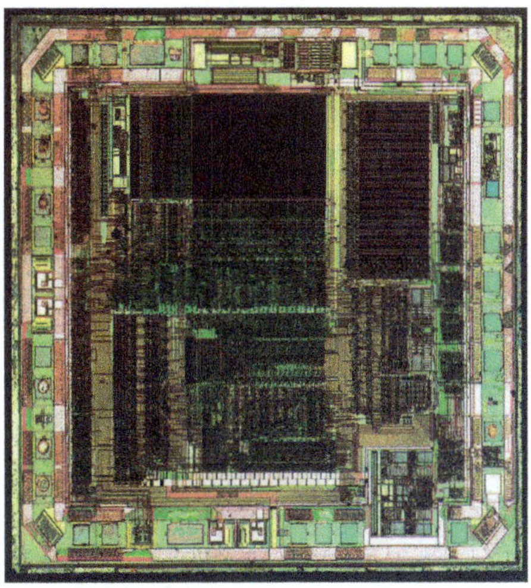

Die of a STM32F100C4T6B ARM Cortex-M3 microcontroller with 16 kilobytes flash memory, 24 MHzCentral Processing Unit (CPU), motor control and Consumer Electronics Control (CEC) functions. Manufactured by STMicroelectronics.

Micro-controllers may not implement an external address or data bus as they integrate RAM and non-volatile memory on the same chip as the CPU. Using fewer pins, the chip can be placed in a much smaller, cheaper package.

Integrating the memory and other peripherals on a single chip and testing them as a unit increases the cost of that chip, but often results in decreased net cost of the embedded system as a whole. Even if the cost of a CPU that has integrated peripherals is slightly more than the cost of a CPU and external peripherals, having fewer chips typically allows a smaller and cheaper circuit board, and reduces the labor required to assemble and test the circuit board, in addition to tending to decrease the defect rate for the finished assembly.

A micro-controller is a single integrated circuit, commonly with the following features:

- central processing unit - ranging from small and simple 4-bit processors to complex 32-bit or 64-bit processors
- volatile memory (RAM) for data storage
- ROM, EPROM, EEPROM or Flash memory for program and operating parameter storage
- discrete input and output bits, allowing control or detection of the logic state of an individual package pin
- serial input/output such as serial ports (UARTs)
- other serial communications interfaces like I²C, Serial Peripheral Interface and Controller Area Network for system interconnect
- peripherals such as timers, event counters, PWM generators, and watchdog
- clock generator - often an oscillator for a quartz timing crystal, resonator or RC circuit
- many include analog-to-digital converters, some include digital-to-analog converters
- in-circuit programming and in-circuit debugging support

This integration drastically reduces the number of chips and the amount of wiring and circuit board space that would be needed to produce equivalent systems using separate chips. Furthermore, on low pin count devices in particular, each pin may interface to several internal peripherals, with the pin function selected by software. This allows a part to be used in a wider variety of applications than if pins had dedicated functions.

Micro-controllers have proved to be highly popular in embedded systems since their introduction in the 1970s.

Some microcontrollers use a Harvard architecture: separate memory buses for instructions and data, allowing accesses to take place concurrently. Where a Harvard architecture is used, instruction words for the processor may be a different bit size than the length of internal memory and registers; for example: 12-bit instructions used with 8-bit data registers.

The decision of which peripheral to integrate is often difficult. The microcontroller vendors often trade operating frequencies and system design flexibility against time-to-market requirements from their customers and overall lower system cost. Manufacturers have to balance the need to minimize the chip size against additional functionality.

Microcontroller architectures vary widely. Some designs include general-purpose microprocessor cores, with one or more ROM, RAM, or I/O functions integrated onto the package. Other designs are purpose built for control applications. A micro-controller instruction set usually has many instructions intended for bit manipulation (bit-wise operations) to make control programs more compact. For example, a general purpose processor might require several instructions to test a bit in a register and branch if the bit is set, where a micro-controller could have a single instruction to provide that commonly required function.

Microcontrollers traditionally do not have a math coprocessor, so floating point arithmetic is performed by software. However, some recent designs do include an FPU and DSP optimized features. An example would be Microchip's PIC32 MIPS based line.

PROGRAMMING ENVIRONMENTS

Microcontrollers were originally programmed only in assembly language, but various high-level programming languages, such as C, Python and JavaScript, are now also in common use to target microcontrollers and embedded systems. These languages are either designed specially for the purpose, or versions of general purpose languages such as the C programming language. Compilers for general purpose languages will typically have some restrictions as well as enhancements to better support the unique characteristics of microcontrollers. Some microcontrollers have environments to aid developing certain types of applications. Microcontroller vendors often make tools freely available to make it easier to adopt their hardware.

Many microcontrollers are so quirky that they effectively require their own non-standard dialects of C, such as SDCC for the 8051, which prevent using standard tools (such as code libraries or static analysis tools) even for code unrelated to hardware features. Interpreters are often used to hide such low level quirks.

Interpreter firmware is also available for some microcontrollers. For example, BASIC on the early microcontrollers Intel8052; BASIC and FORTH on the Zilog Z8 as well as some modern devices. Typically these interpreters support interactive programming.

Simulators are available for some microcontrollers. These allow a developer to analyze what the behavior of the microcontroller and their program should be if they were using the actual part. A simulator will show the internal processor state and also that of the outputs, as well as allowing input signals to be generated. While on the one hand most simulators will be limited from being unable to simulate much other hardware in a system, they can exercise conditions that may otherwise be hard to reproduce at will in the physical implementation, and can be the quickest way to debug and analyze problems.

Recent microcontrollers are often integrated with on-chip debug circuitry that when accessed by an in-circuit emulator via JTAG, allow debugging of the firmware with a debugger. A real-time ICE may allow viewing and/or manipulating of internal states while running. A tracing ICE can record executed program and MCU states before/after a trigger point.

TYPES OF MICROCONTROLLERS

As of 2008, there are several dozen microcontroller architectures and vendors including:

- ARM core processors (many vendors)

- ARM Cortex-M cores are specifically targeted towards microcontroller applications
- Atmel AVR (8-bit), AVR32 (32-bit), and AT91SAM (32-bit)
- Cypress Semiconductor's M8C Core used in their PSoC (Programmable System-on-Chip)
- Freescale ColdFire (32-bit) and S08 (8-bit)
- Freescale 68HC11 (8-bit), and others based on the Motorola 6800 family
- Intel 8051, also manufactured by NXP Semiconductors, Infineon and many others
- Infineon: 8-bit XC800, 16-bit XE166, 32-bit XMC4000 (ARM based Cortex M4F), 32-bit TriCore and, 32-bit Aurix Tricore Bit microcontrollers
- MIPS
- Microchip Technology PIC, (8-bit PIC16, PIC18, 16-bit dsPIC33 / PIC24), (32-bit PIC32)
- NXP Semiconductors LPC1000, LPC2000, LPC3000, LPC4000 (32-bit), LPC900, LPC700 (8-bit)
- Parallax Propeller
- PowerPC ISE
- Rabbit 2000 (8-bit)
- Renesas Electronics: RL78 16-bit MCU; RX 32-bit MCU; SuperH; V850 32-bit MCU; H8; R8C 16-bit MCU
- Silicon Laboratories Pipelined 8-bit 8051 Microcontrollers and mixed-signal ARM-based 32-bit microcontrollers
- STMicroelectronics STM8 (8-bit), ST10 (16-bit) and STM32 (32-bit)
- Texas Instruments TI MSP430 (16-bit), MSP432 (32-bit), C2000 (32-bit)
- Toshiba TLCS-870 (8-bit/16-bit)

Many others exist, some of which are used in very narrow range of applications or are more like applications processors than microcontrollers. The microcontroller market is extremely fragmented, with numerous vendors, technologies, and markets. Note that many vendors sell or have sold multiple architectures.

INTERRUPT LATENCY

In contrast to general-purpose computers, microcontrollers used in embedded systems often seek to optimize interrupt latency over instruction throughput. Issues include both reducing the latency, and making it be more predictable (to support real-time control).

When an electronic device causes an interrupt, during the context switch the intermediate results (registers) have to be saved before the software responsible for handling the interrupt can run. They must also be restored after that interrupt handler is finished. If there are more processor registers, this saving and restoring process takes more time, increasing the latency. Ways to reduce such context/restore latency include having relatively few registers in their central processing units (undesirable because it slows down most non-interrupt processing substantially), or at least having the hardware not save them all (this fails if the software then needs to compensate by saving the rest "manually"). Another technique involves spending silicon gates on "shadow registers": One or more duplicate registers used only by the interrupt software, perhaps supporting a dedicated stack.

Other factors affecting interrupt latency include:

- Cycles needed to complete current CPU activities. To minimize those costs, microcontrollers tend to have short pipelines (often three instructions or less), small write buffers, and ensure that longer instructions are continuable or restartable. RISC design principles ensure that most instructions take the same number of cycles, helping avoid the need for most such continuation/restart logic.
- The length of any critical section that needs to be interrupted. Entry to a critical section restricts concurrent data structure access. When a data structure must be accessed by an interrupt handler, the critical section must block that interrupt. Accordingly, interrupt latency is increased by however long that interrupt is blocked. When there are hard external constraints on system latency, developers often need tools to measure interrupt latencies and track down which critical sections cause slowdowns.
    - One common technique just blocks all interrupts for the duration of the critical section. This is easy to implement, but sometimes critical sections get uncomfortably long.
    - A more complex technique just blocks the interrupts that may trigger access to that data structure. This is often based on interrupt priorities, which tend to not correspond well to the relevant system data structures. Accordingly, this technique is used mostly in very constrained environments.
    - Processors may have hardware support for some critical sections. Examples include supporting atomic access to bits or bytes within a word, or other atomic access primitives like the LDREX/STREX exclusive access primitives introduced in the ARMv6 architecture.
- Interrupt nesting. Some microcontrollers allow higher priority interrupts to interrupt lower priority ones. This allows software to manage latency by giving time-critical interrupts higher priority (and thus lower and more predictable latency) than less-critical ones.
- Trigger rate. When interrupts occur back-to-back, microcontrollers may avoid an extra context save/restore cycle by a form of tail call optimization.

Lower end microcontrollers tend to support fewer interrupt latency controls than higher end ones.

MICROCONTROLLER EMBEDDED MEMORY TECHNOLOGY

Since the emergence of microcontrollers, many different memory technologies have been used. Almost all microcontrollers have at least two different kinds of memory, a non-volatile memory for storing firmware and a read-write memory for temporary data.

Data

From the earliest microcontrollers to today, six-transistor SRAM is almost always used as the read/write working memory, with a few more transistors per bit used in the register file. FRAM or MRAM could potentially replace it as it is 4 to 10 times denser which would make it more cost effective.

In addition to the SRAM, some microcontrollers also have internal EEPROM for data storage; and even ones that do not have any (or not enough) are often connected to external serial EEPROM chip (such as the BASIC Stamp) or external serial flash memory chip.

A few recent microcontrollers beginning in 2003 have "self-programmable" flash memory.

Firmware

The earliest microcontrollers used mask ROM to store firmware. Later microcontrollers (such as the early versions of the Freescale 68HC11 and early PIC microcontrollers) had quartz windows that allowed ultraviolet light in to erase the EPROM.

The Microchip PIC16C84, introduced in 1993, was the first microcontroller to use EEPROM to store firmware. In the same year, Atmel introduced the first microcontroller using NOR Flash memory to store firmware.

REFERENCES

1. Augarten, Stan (1983). The Most Widely Used Computer on a Chip: The TMS 1000. State of the Art: A Photographic History of the Integrated Circuit. New Haven and New York: Ticknor & Fields. ISBN 0-89919-195-9. Retrieved 2009-12-23.
2. "Oral History Panel on the Development and Promotion of the Intel 8048 Microcontroller" (PDF). Computer History Museum Oral History, 2008. p. 4. Retrieved 2016-04-04.
3. "Atmel's Self-Programming Flash Microcontrollers" (PDF). 2012-01-24. Retrieved 2008-10-25. by Odd Jostein Svendsli 2003
4. Jim Turley. "The Two Percent Solution" 2002.
5. Tom Cantrell "Microchip on the March". Circuit Cellar. 1998.
6. http://www.semico.com
7. Momentum Carries MCUs Into 2011 http://semico.com/content/momentum-carries-mcus-2011
8. "MCU Market on Migration Path to 32-bit and ARM-based Devices". April 25, 2013. It typically takes a global economic recession to upset the diverse MCU marketplace, and that's exactly what occurred in 2009, when the microcontroller business suffered its worst-ever annual sales decline of 22% to $11.1 billion.

9. Bill Giovino. “Zilog Buys Microcontroller Product Lines from Samsung”. 2013.
10. http://www.mouser.com/ProductDetail/Silicon-Labs/EFM8BB10F2G-A-QFN20/?qs=sGAEpiMZZMu9ReDVvI6ax9sqO0qrXlDW4ZuhKcnb2c%252bQvyUXU1UbuQ%3d%3d
11. http://www.mouser.com/ProductDetail/Texas-Instruments/MSP430G2001IPW14R/?qs=sGAEpiMZZMvfhsTlJjecML5mLnp8Cec4esZ6%2f1aK7FQ%3d
12. http://www.mouser.com/ProductDetail/Cypress-Semiconductor/CY8C4013SXI-400/?qs=sGAEpiMZZMuI9neUTtPr752e7iT1qQqS4inl2jxeWgxWqjKLOdzceQ%3d%3d
13. Heath, Steve (2003). Embedded systems design. EDN series for design engineers (2 ed.). Newnes. pp. 11–12. ISBN 9780750655460.
14. David Harris & Sarah Harris (2012). Digital Design and Computer Architecture, Second Edition, p. 515. Morgan Kaufmann. ISBN 0123944244.
15. Easy Way to build a microcontroller project
16. Mazzei, Daniele; Montelisciani, Gabriele; Baldi, Giacomo; Fantoni, Gualtiero (2015). Changing the programming paradigm for the embedded in the IoT domain. Internet of Things (WF-IoT), 2015 IEEE 2nd World Forum on. Milan: IEEE. pp. 239–244. doi:10.1109/WF-IoT.2015.7389059.
17. “8052-Basic Microcontrollers” by Jan Axelson 1994
18. Edwards, Robert (1987). “Optimizing the Zilog Z8 Forth Microcontroller for Rapid Prototyping” (PDF). Martin Marietta: 3. Retrieved 9 December 2012.
19. www.infineon.com/mcu
20. Microchip unveils PIC16C84, a reprogrammable EEPROM-based 8-bit microcontroller 1993

CHAPTER 6

Amplifiers

Chapter Contents

- **Introduction**
- **History**
- **Figures of Merit**
- **Amplifier Categorisation**
- **Power Amplifier Classes**
- **Implementation**

A power amplifier is an electronic device that receives an electrical signal and reprocesses it to amplify, or increase, its power. The boost in power is achieved by significantly increasing the input signal's voltage. A power amplifier is used to power an output source, such as a stereo speaker, a relay or a motor.

One of the most common functions for a power amplifier is in audio applications. An audio amplifier is designed to deliver a signal to a speaker. In large systems, the electrical signal usually comes from an audio mixing board or a preamp component. Power amplifiers are also found in small systems, such as typical home stereo setups.

INTRODUCTION

An amplifier, electronic amplifier or (informally) amp is an electronic device that can increase the power of a signal (a time-varying voltage or current). An amplifier functions by taking power from a power supply and controlling the output to match the input signal shape but with a larger amplitude. In this sense, an amplifier modulates the output of the power supply based upon the properties of the input signal. An amplifier is effectively the opposite of an attenuator: while an amplifier provides gain, an attenuator provides loss.

An amplifier can either be a discrete piece of equipment or an electrical circuit contained within another device. Amplification is fundamental to modern electronics, and amplifiers are widely used in almost all electronic equipment. Amplifiers can be categorized in different ways. One is by the frequency of the electronic signal being amplified; audio amplifiers amplify signals in the audio (sound) range of less than 20 kHz, RF amplifiers amplify frequencies in the radio frequency range between 20 kHz and 300 GHz. Another is which quantity, voltage or current is being amplified; amplifiers can be divided into voltage amplifiers, current amplifiers, transconductance amplifiers, and transresistance amplifiers. A further distinction is whether the output is a linear or nonlinear representation of the input. Amplifiers can also be categorized by their physical placement in the signal chain.

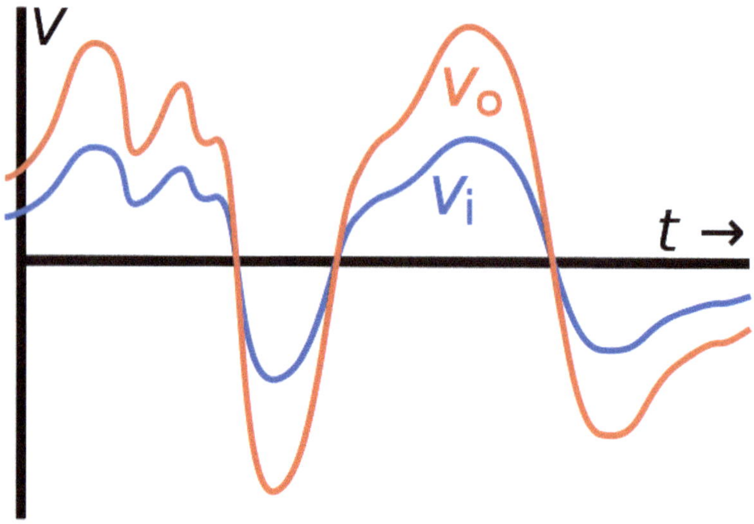

Graph of the input $v_i(t)$ *(blue)* and output voltage $v_o(t)$ *(red)* of an ideal linear amplifier with an arbitrary signal applied as input. *Amplification* means increasing the amplitude (voltage or current) of a time-varying signal by a given factor, as shown here. In this example the amplifier has a *voltage gain* of 2; that is at any instant $v_o = 2v_i$.

The first practical electronic device that could amplify was the triode vacuum tube, invented in 1906 by Lee De Forest, which led to the first amplifiers around 1912. Vacuum tubes were used in almost all amplifiers until the 1960s - 1970s when the transistor, invented in 1947, replaced them. Today most amplifiers use transistors, but vacuum tubes continue to be used in some applications.

HISTORY

The development of audio communication technology; the telephone and intercom around 1880 and the first AM radio transmitters and receivers around 1905 created a need to somehow make an electrical audio signal "louder". Before the invention of electronic amplifiers, mechanically coupled carbon microphones were used as crude amplifiers in telephone repeaters. After the turn of the century it was found that negative resistance mercury lamps could amplify, and were also tried in repeaters. The first practical electronic device that could amplify was the Audion (triode) vacuum tube, invented in 1906 by

Lee De Forest, which led to the first amplifiers around 1912. The terms "amplifier" and "amplification" (from the Latin *amplificare*, 'to enlarge or expand') were first used for this new capability around 1915 when triodes became widespread.

De Forest's prototype audio amplifier, 1914. His Audion (triode) vacuum tube had a voltage gain of about 5, so this 3 stage amplifier had a gain of about 125.

The amplifying vacuum tube revolutionized electrical technology, creating the new field of electronics, the technology of active electrical devices. It made possible long distance telephone lines, public address systems, radio broadcasting, talking motion pictures, practical audio recording, radar, television, and the first computers. For 50 years virtually all consumer electronic devices used vacuum tubes. Early tube amplifiers often had positive feedback (regeneration), which could increase gain but also make the amplifier unstable and prone to oscillation. Much of the mathematical theory of amplifiers was developed at Bell Telephone Laboratories during the 1920s to 1940s. Distortion levels in early amplifiers were high, usually around 5%, until 1934, when Harold Black developed negative feedback; this allowed the distortion levels to be greatly reduced, at the cost of lower gain. Other advances in the theory of amplification were made by Harry Nyquist and Hendrik Wade Bode.

The vacuum tube was the only amplifying device (besides specialized power devices such as the magnetic amplifier and amplidyne) for 40 years, and dominated electronics until 1947, when the first transistor, the BJT, was invented. The replacement of bulky, fragile vacuum tubes with transistors during the 1960s and 1970s created another revolution in electronics, making possible the first really portable electronic devices, such as the transistor radio developed in 1954. Today most amplifiers use transistors, but vacuum tubes are still used in some high power applications such as radio transmitters.

FIGURES OF MERIT

Amplifier quality is characterized by a list of specifications that include:

- Gain, the ratio between the magnitude of output and input signals
- Bandwidth, the width of the useful frequency range
- Efficiency, the ratio between the power of the output and total power consumption
- Linearity, the extent to which the proportion between input and output amplitude is the same for high amplitude and low amplitude input
- Noise, a measure of undesired noise mixed into the output
- Output dynamic range, the ratio of the largest and the smallest useful output levels
- Slew rate, the maximum rate of change of the output
- Rise time, settling time, ringing and overshoot that characterize the step response
- Stability, the ability to avoid self-oscillation

AMPLIFIER CATEGORISATION

Amplifiers are described according to the properties of their inputs, their outputs, and how they relate. All amplifiers have gain, a multiplication factor that relates the magnitude of some property of the output signal to a property of the input signal. The gain may be specified as the ratio of output voltage to input voltage (voltage gain), output power to input power (power gain), or some combination of current, voltage, and power. In many cases the property of the output that varies is dependent on the same property of the input, making the gain unitless (though often expressed in decibels (dB)).

Most amplifiers are designed to be linear. That is, they provide constant gain for any normal input level and output signal. If an amplifier's gain is not linear, the output signal can become distorted. There are, however, cases where variable gain is useful. Certain signal processing applications use exponential gain amplifiers.

Amplifiers are usually designed to function well in a specific application, for example: radio and television transmitters and receivers, high-fidelity ("hi-fi") stereo equipment, microcomputers and other digital equipment, and guitar and other instrument amplifiers. Every amplifier includes a least one active device, such as a vacuum tube or transistor.

Active Devices

All amplifiers include some form of active device: this is the device that does the actual amplification. The active device can be a vacuum tube, discrete solid state component, such as a single transistor, or part of an integrated circuit, as in an op-amp).

Transistor Amplifiers

Transistor amplifiers (or solid state amplifiers) are the most common type of amplifier in use today. A transistor is used as the active element. The gain of the amplifier is determined by the properties of the transistor itself as well as the circuit it is contained within.

Common active devices in transistor amplifiers include bipolar junction transistors (BJTs) and metal oxide semiconductor field-effect transistors (MOSFETs).

Applications are numerous, some common examples are audio amplifiers in a home stereo or public address system, RF high power generation for semiconductor equipment, to RF and microwave applications such as radio transmitters.

Transistor-based amplification can be realized using various configurations: for example a bipolar junction transistor can realize common base, common collector or common emitter amplification; a MOSFET can realize common gate, common source or common drain amplification. Each configuration has different characteristics.

Vacuum-tube Amplifiers

An ECC83 tube glowing inside a preamp

Vacuum-tube amplifiers (also known as tube amplifiers or valve amplifiers) use a vacuum tube as the active device. While semiconductor amplifiers have largely displaced valve amplifiers for low power applications, valve amplifiers can be much more cost effective in high power applications such as radar, countermeasures equipment, and communications equipment. Many microwave amplifiers are specially designed valve amplifiers, such as the klystron, gyrotron, traveling wave tube, and crossed-field amplifier, and these microwave valves provide much greater single-device power output at microwave frequencies than solid-state devices. Vacuum tubes remain in use in some high end audio equipment, as well as in musical instrument amplifiers, due to a preference for "tube sound".

Magnetic Amplifiers

Magnetic amplifiers are devices somewhat similar to a transformer where one winding is used to control the saturation of a magnetic core and hence alter the impedance of the other winding.

They have largely fallen out of use due to development in semiconductor amplifiers but are still useful in HVDC control, and in nuclear power control circuitry to their not being affected by radioactivity.

Negative Resistance Devices

Negative resistances can be used as amplifiers, such as the tunnel diode amplifier.

Amplifier Architectures

Amplifiers can also be categorised by the way they amplify the input signal.

Power Amplifier

A power amplifier is an amplifier designed primarily to increase the power available to a load. In practice, amplifier power gain depends on the source and load impedances, as well as the inherent voltage and current gain. A radio frequency (RF) amplifier design typically optimizes impedances for power transfer, while audio and instrumentation amplifier designs normally optimize input and output impedance for least loading and highest signal integrity. An amplifier that is said to have a gain of 20 dB might have a voltage gain of 20 dB and an available power gain of much more than 20 dB (power ratio of 100)—yet actually deliver a much lower power gain if, for example, the input is from a 600 ohm microphone and the output connects to a 47 kilohm input socket for a power amplifier. In general the power amplifier is the last 'amplifier' or actual circuit in a signal chain (the output stage) and is the amplifier stage that requires attention to power efficiency. Efficiency considerations lead to the various classes of power amplifier based on the biasing of the output transistors or tubes: see power amplifier classes below.

Power Amplifiers by Application

- Audio power amplifiers: typically used to drive loudspeakers, will often have two output channels and deliver equal power to each

- RF power amplifier—typical in transmitter final stages (see also: Linear amplifier)
- Servo motor controllers: amplify a control voltage where linearity is not important
- Piezoelectric audio amplifier—includes a DC-to-DC converter to generate the high voltage output required to drive piezoelectric speakers

Power Amplifier Circuits

Power amplifier circuits include the following types:

- Vacuum tube/valve, hybrid or transistor power amplifiers
- Push-pull output or single-ended output stages

Operational Amplifiers (op-amps)

An LM741 general purpose op-amp.

An operational amplifier is an amplifier circuit which typically has very high open loop gain and differential inputs. Op amps have become very widely used as standardized "gain blocks" in circuits due to their versatility; their gain, bandwidth and other characteristics can be controlled by feedback through an external circuit. Though the term today commonly applies to integrated circuits, the original operational amplifier design used valves, and later designs used discrete transistor circuits.

Differential Amplifiers

A fully differential amplifier is similar to the operational amplifier, but also has differential outputs. These are usually constructed using BJTs or FETs.

A differential amplifier is the first stage of an op-amp, a differential amplifier consists of two transistors which are emitter coupled. Types of differential amplifiers:

- Differential mode

Vd= V1-V2

- Common mode

It is the average between the input voltages V2 and V1 Vc=V1+V2/2

Distributed Amplifiers

These use balanced transmission lines to separate individual single stage amplifiers, the outputs of which are summed by the same transmission line. The transmission line is a balanced type with the input at one end and on one side only of the balanced transmission line and the output at the opposite end is also the opposite side of the balanced transmission line. The gain of each stage adds linearly to the output rather than multiplies one on the other as in a cascade configuration. This allows a higher bandwidth to be achieved than could otherwise be realised even with the same gain stage elements.

Switched Mode Amplifiers

These nonlinear amplifiers have much higher efficiencies than linear amps, and are used where the power saving justifies the extra complexity. Class-D amplifiers are the main example of this type of amplification—see below.

Applications

Video Amplifiers

Video amplifiers are designed to process video signals and have varying bandwidths depending on whether the video signal is for SDTV, EDTV, HDTV 720p or 1080i/p etc.. The specification of the bandwidth itself depends on what kind of filter is used—and at which point (-1 dB or -3 dB for example) the bandwidth is measured. Certain requirements for step response and overshoot are necessary for an acceptable TV image.

Oscilloscope Vertical Amplifiers

These deal with video signals that drive an oscilloscope display tube, and can have bandwidths of about 500 MHz. The specifications on step response, rise time, overshoot, and aberrations can make designing these amplifiers difficult. One of the pioneers in high bandwidth vertical amplifiers was the Tektronix company.

Microwave Amplifiers

Travelling Wave tube Amplifiers

Traveling wave tube amplifiers (TWTAs) are used for high power amplification at low microwave frequencies. They typically can amplify across a broad spectrum of frequencies; however, they are usually not as tunable as klystrons.

Klystrons

Klystrons are specialized linear-beam vacuum-devices, designed to provide high power, widely tunable amplification of millimetre and sub-millimetre waves. Klystrons are designed for large scale operations and despite having a narrower bandwidth than TWTAs, they have the advantage of coherently amplifying a reference signal so its output may be precisely controlled in amplitude, frequency and phase.

Musical Instrument Amplifiers

An audio power amplifier is usually used to amplify signals such as music or speech. In the mid 1960s, guitar and bass amplifiers began to gain popularity because of their relatively low price ($50) and guitars being the most popular instruments as well. Several factors are especially important in the selection of musical instrument amplifiers (such as guitar amplifiers) and other audio amplifiers (although the whole of the sound system – components such as microphones to loudspeakers – affect these parameters):

- Frequency response – not just the frequency range but the requirement that the signal level varies so little across the audible frequency range that the human ear notices no variation. A typical specification for audio amplifiers may be 20 Hz to 20 kHz +/- 0.5 dB.
- Power output – the power level obtainable with little distortion, to obtain a sufficiently loud sound pressure level from the loudspeakers.
- Low distortion – all amplifiers and transducers distort to some extent. They cannot be perfectly linear, but aim to pass signals without affecting the harmonic content of the sound more than the human ear can tolerate. That tolerance of distortion, and indeed the possibility that some "warmth" or second harmonic distortion (tube sound) improves the "musicality" of the sound, are subjects of great debate.

Before coming onto the music scene, amplifiers were heavily used in cinema. In the premiere of *Noah's Ark* in 1929, the movie's director (Michael Kurtiz) used the amplifier for a festival following the movie's premiere.

Classification of amplifier stages and systems

Many alternative classifications address different aspects of amplifier designs, and they all express some particular perspective relating the design parameters to the objectives of the circuit. Amplifier design is always a compromise of numerous factors, such as cost, power consumption, real-world device imperfections, and a multitude of performance specifications. Below are several different approaches to classification:

Input and Output Variables

Electronic amplifiers use one variable presented as either a current and voltage. Either current or voltage can be used as input and either as output, leading to four types of amplifiers. In idealized form they are represented by each of the four types of dependent source used in linear analysis, as shown in the figure, namely:

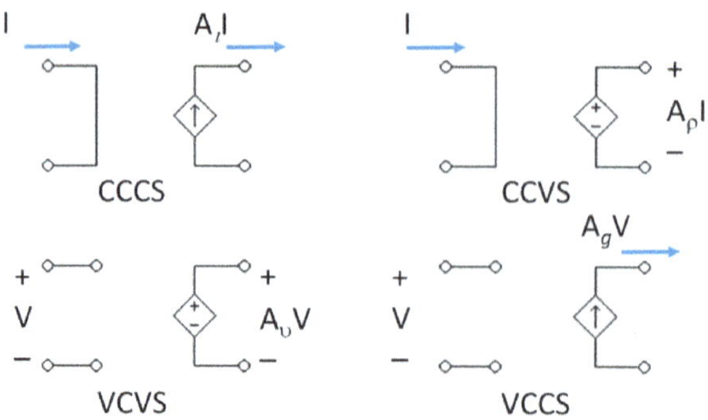

The four types of dependent source—control variable on left, output variable on right.

Input	Output	Dependent source	Amplifier type	Gain units
I	I	Current controlled current source CCCS	Current amplifier	Unitless
I	V	Current controlled voltage source CCVS	Transresistance amplifier	Ohm
V	I	Voltage controlled current source VCCS	Transconductance amplifier	Siemens
V	V	Voltage controlled voltage source VCVS	Voltage amplifier	Unitless

Each type of amplifier in its ideal form has an ideal input and output resistance that is the same as that of the corresponding dependent source:

Amplifier type	Dependent source	Input impedance	Output impedance
Current	CCCS	0	∞
Transresistance	CCVS	0	0
Transconductance	VCCS	∞	∞
Voltage	VCVS	∞	0

In practice the ideal impedances are not possible to achieve. For any particular circuit, a small-signal analysis is often used to find the actual impedance. A small-signal AC test current I_x is applied to the input or output node, all external sources are set to AC zero, and the corresponding alternating voltage V_x across the test current source determines the impedance seen at that node as $R = V_x / I_x$.

Amplifiers designed to attach to a transmission line at input and output, especially RF amplifiers, do not fit into this classification approach. Rather than dealing with voltage or current individually, they ideally couple with an input or output impedance matched to the transmission line impedance, that is, match *ratios* of voltage to current. Many real RF amplifiers come close to this ideal. Although, for a given appropriate source and load impedance, RF amplifiers can be characterized as amplifying voltage or current, they fundamentally are amplifying power.

Common Terminal

One set of classifications for amplifiers is based on which device terminal is common to both the input and the output circuit. In the case of bipolar junction transistors, the three classes are common emitter, common base, and common collector. For field-effect transistors, the corresponding configurations are common source, common gate, and common drain; for vacuum tubes, common cathode, common grid, and common plate. The common emitter (or common source, common cathode, etc.) is most often configured to provide amplification of a voltage applied between base and emitter, and the output signal taken between collector and emitter is inverted, relative to the input. The common collector arrangement applies the input voltage between base and collector, and to take the output voltage between emitter and collector. This causes negative feedback, and the output voltage tends to follow the input voltage. This arrangement is also used as the input presents a high impedance and does not load the signal source, though the voltage amplification is less than one. The common-collector circuit is, therefore, better known as an emitter follower, source follower, or cathode follower.

Unilateral or Bilateral

An amplifier whose output exhibits no feedback to its input side is described as 'unilateral'. The input impedance of a unilateral amplifier is independent of load, and output impedance is independent of signal source impedance.

An amplifier that uses feedback to connect part of the output back to the input is a *bilateral* amplifier. Bilateral amplifier input impedance depends on the load, and output impedance on the signal source impedance. All amplifiers are bilateral to some degree; however they may often be modeled as unilateral under operating conditions where feedback is small enough to neglect for most purposes, simplifying analysis (see the common base article for an example).

An amplifier design often deliberately applies negative feedback to tailor amplifier behavior. Some feedback, positive or negative, is unavoidable and often undesirable—introduced, for example, by parasitic elements, such as inherent capacitance between input and output of devices such as transistors, and capacitive coupling of external wiring. Excessive frequency-dependent positive feedback can turn an amplifier into an oscillator.

Linear unilateral and bilateral amplifiers can be represented as two-port networks.

Inverting or Non-Inverting

Another way to classify amplifiers is by the phase relationship of the input signal to the output signal. An 'inverting' amplifier produces an output 180 degrees out of phase with the input signal (that is, a polarity inversion or mirror image of the input as seen on an oscilloscope). A 'non-inverting' amplifier maintains the phase of the input signal waveforms. An emitter follower is a type of non-inverting amplifier, indicating that the signal at the emitter of a transistor is following (that is, matching with unity gain but perhaps an offset) the input signal. Voltage follower is also non inverting type of amplifier having unity gain.

This description can apply to a single stage of an amplifier, or to a complete amplifier system.

Function

Other amplifiers may be classified by their function or output characteristics. These functional descriptions usually apply to complete amplifier systems or sub-systems and rarely to individual stages.

- A **servo amplifier** indicates an integrated feedback loop to actively control the output at some desired level. A DC servo indicates use at frequencies down to DC levels, where the rapid fluctuations of an audio or RF signal do not occur. These are often used in mechanical actuators, or devices such as DC motors that must maintain a constant speed or torque. An **AC servo** amp can do this for some ac motors.
- A **linear** amplifier responds to different frequency components independently, and does not generate harmonic distortion or Intermodulation distortion. No amplifier can provide *perfect* linearity (even the most linear amplifier has some nonlinearities, since the amplifying devices—transistors or vacuum tubes—follow nonlinear power laws such as square-laws and rely on circuitry techniques to reduce those effects).
- A **nonlinear** amplifier generates significant distortion and so changes the harmonic content; there are situations where this is useful. Amplifier circuits intentionally providing a non-linear transfer function include:
  - a device like a Silicon Controlled Rectifier or a transistor used as a switch may be employed to turn either fully ON or OFF a load such as a lamp based on a threshold in a continuously variable input.
  - a non-linear amplifier in an analog computer or true RMS converter for example can provide a special transfer function, such as logarithmic or square-law.
  - a Class C RF amplifier may be chosen because it can be very efficient—but is non-linear. Following such an amplifier with a "tank" tuned circuit can reduce unwanted harmonics (distortion) sufficiently to make it useful in transmitters, or some desired harmonic may be selected by setting the resonant frequency of the tuned circuit to a higher frequency rather than fundamental frequency in frequency multiplier circuits.
  - Automatic gain control circuits require an amplifier's gain be controlled by the time-averaged amplitude so that the output amplitude varies little when weak stations are being received. The non-linearities are assumed arranged so the relatively small signal amplitude suffers from little distortion (cross-channel interference or intermodulation) yet is still modulated by the relatively large gain-control DC voltage.
  - AM detector circuits that use amplification such as Anode-bend detectors, Precision rectifiers and Infinite impedance detectors (so excluding

unamplified detectors such as Cat's-whisker detectors), as well as peak detector circuits, rely on changes in amplification based on the signal's instantaneous amplitude to derive a direct current from an alternating current input.

- Operational amplifier comparator and detector circuits.

■ A wideband amplifier has a precise amplification factor over a wide frequency range, and is often used to boost signals for relay in communications systems. A narrowband amp amplifies a specific narrow range of frequencies, to the exclusion of other frequencies.

■ An RF amplifier amplifies signals in the radio frequency range of the electromagnetic spectrum, and is often used to increase the sensitivity of a receiver or the output power of a transmitter.

■ An audio amplifier amplifies audio frequencies. This category subdivides into small signal amplification, and power amps that are optimised to driving speakers, sometimes with multiple amps grouped together as separate or bridgeable channels to accommodate different audio reproduction requirements. Frequently used terms within audio amplifiers include:

  - Preamplifier (preamp), which may include a phono preamp with RIAA equalization, or tape head preamps with CCIR equalisation filters. They may include filters or tone control circuitry.
  - Power amplifier (normally drives loudspeakers), headphone amplifiers, and public address amplifiers.
  - Stereo amplifiers imply two channels of output (left and right), though the term simply means "solid" sound (referring to three-dimensional)—so quadraphonic stereo was used for amplifiers with four channels. 5.1 and 7.1 systems refer to Home theatre systems with 5 or 7 normal spatial channels, plus a subwoofer channel.

■ Buffer amplifiers, which may include emitter followers, provide a high impedance input for a device (perhaps another amplifier, or perhaps an energy-hungry load such as lights) that would otherwise draw too much current from the source. Line drivers are a type of buffer that feeds long or interference-prone interconnect cables, possibly with differential outputs through twisted pair cables.

■ A special type of amplifier—originally used in analog computers—is widely used in measuring instruments for signal processing, and many other uses. These are called **operational amplifiers** or **op-amps**. The "operational" name is because this type of amplifier can be used in circuits that perform mathematical algorithmic functions, or "operations" on input signals to obtain specific types of output signals. Modern op-amps are usually provided as integrated circuits, rather than constructed from discrete components. A typical modern op-amp has differential inputs (one "inverting", one "non-inverting") and one output. An *idealised* op-amp has the following characteristics:

  - Infinite input impedance (so it does not load the circuitry at its input)
  - Zero output impedance

- Infinite gain
- Zero propagation delay

The performance of an op-amp with these characteristics is entirely defined by the (usually passive) components that form a negative feedback loop around it. The amplifier itself does not affect the output. All real-world op-amps fall short of the idealised specification above—but some modern components have remarkable performance and come close in some respects.

Interstage Coupling Method

Amplifiers are sometimes classified by the coupling method of the signal at the input, output, or between stages. Different types of these include:

Resistive-capacitive (RC) coupled Amplifier, using a Network of Resistors and Capacitors

By design these amplifiers cannot amplify DC signals as the capacitors block the DC component of the input signal. RC-coupled amplifiers were used very often in circuits with vacuum tubes or discrete transistors. In the days of the integrated circuit a few more transistors on a chip are much cheaper and smaller than a capacitor.

Inductive-capacitive (LC) Coupled Amplifier, using a Network of Inductors and Capacitors

This kind of amplifier is most often used in selective radio-frequency circuits.

Transformer Coupled Amplifier, using a Transformer to match Impedances or to Decouple Parts of the Circuits

Quite often LC-coupled and transformer-coupled amplifiers cannot be distinguished as a transformer is some kind of inductor.

Direct Coupled Amplifier, Using no Impedance and Bias Matching Components

This class of amplifier was very uncommon in the vacuum tube days when the anode (output) voltage was at greater than several hundred volts and the grid (input) voltage at a few volts minus. So they were only used if the gain was specified down to DC (e.g., in an oscilloscope). In the context of modern electronics developers are encouraged to use directly coupled amplifiers whenever possible. In FET and CMOS technologies direct coupling is dominant since gates of MOSFETs theoretically pass no current through themselves. Therefore, DC component of the input signals is automatically filtered.

Frequency Range

Depending on the frequency range and other properties amplifiers are designed according to different principles.

- Frequency ranges down to DC are only used when this property is needed. DC amplification leads to specific complications that are avoided if possible; DC-blocking capacitors can be added to remove DC and sub-sonic frequencies from audio amplifiers.
- Depending on the frequency range specified different design principles must be used. Up to the MHz range only "discrete" properties need be considered; e.g., a terminal has an input impedance.
- As soon as any connection within the circuit gets longer than perhaps 1% of the wavelength of the highest specified frequency (e.g., at 100 MHz the wavelength is 3 m, so the critical connection length is approx. 3 cm) design properties radically change. For example, a specified length and width of a PCB trace can be used as a selective or impedance-matching entity.
- Above a few hundred MHz, it gets difficult to use discrete elements, especially inductors. In most cases, PCB traces of very closely defined shapes are used instead.

The frequency range handled by an amplifier might be specified in terms of bandwidth (normally implying a response that is 3 dB down when the frequency reaches the specified bandwidth), or by specifying a **frequency response** that is within a certain number of decibels between a lower and an upper frequency (e.g. "20 Hz to 20 kHz plus or minus 1 dB").

POWER AMPLIFIER CLASSES

Power amplifier circuits (output stages) are classified as A, B, AB and C for analog designs—and class D and E for switching designs—based on the proportion of each input cycle (conduction angle) during which an amplifying device passes current.The image of the conduction angle derives from amplifying a sinusoidal signal. If the device is always on, the conducting angle is 360°. If it is on for only half of each cycle, the angle is 180°. The angle of flow is closely related to the amplifier power efficiency. The various classes are introduced below, followed by a more detailed discussion under their individual headings further down.

In the illustrations below, a bipolar junction transistor is shown as the amplifying device. However the same attributes are found with MOSFETs or vacuum tubes.

Conduction Angle Classes

Class A

100% of the input signal is used (conduction angle $\Theta = 360°$). The active element remains conducting all of the time.

Class B

50% of the input signal is used (Θ = 180°); the active element carries current half of each cycle, and is turned off for the other half.

Class AB

Class AB is intermediate between class A and B, the two active elements conduct more than half of the time.

Class C

Less than 50% of the input signal is used (conduction angle Θ < 180°).

A "Class D" amplifier uses some form of pulse-width modulation to control the output devices; the conduction angle of each device is no longer related directly to the input signal but instead varies in pulse width. These are sometimes called "digital" amplifiers because the output device is switched fully on or off, and not carrying current proportional to the signal amplitude.

Additional classes

There are several other amplifier classes, although they are mainly variations of the previous classes. For example, class-G and class-H amplifiers are marked by variation of the supply rails (in discrete steps or in a continuous fashion, respectively) following the input signal. Wasted heat on the output devices can be reduced as excess voltage is kept to a minimum. The amplifier that is fed with these rails itself can be of any class. These kinds of amplifiers are more complex, and are mainly used for specialized applications, such as very high-power units. Also, class-E and class-F amplifiers are commonly described in literature for radio-frequency applications where efficiency of the traditional classes is important, yet several aspects deviate substantially from their ideal values. These classes use harmonic tuning of their output networks to achieve higher efficiency and can be considered a subset of class C due to their conduction-angle characteristics.

Class A

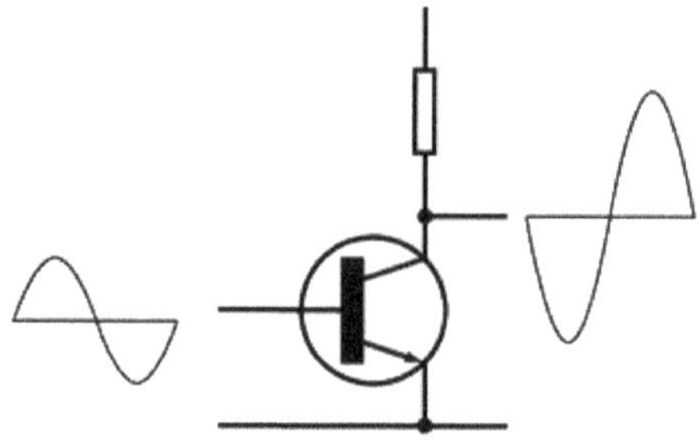

Class-A amplifier.

Amplifying devices operating in class A conduct over the entire range of the input cycle. A *class-A amplifier* is distinguished by the output stage devices being biased for class A operation. Subclass A2 is sometimes used to refer to vacuum-tube class-A stages that drive the grid slightly positive on signal peaks for slightly more power than normal class A (A1; where the grid is always negative). This, however, incurs higher signal distortion.

Advantages of class-A amplifiers

- Class-A designs are simpler than other classes; for example class -AB and -B designs require two connected devices in the circuit (push–pull output), each to handle one half of the waveform; class A can use a single device (single-ended).
- The amplifying element is biased so the device is always conducting, the quiescent (small-signal) collector current (for transistors; drain current for FETs or anode/plate current for vacuum tubes) is close to the most linear portion of its transconductance curve.
- Because the device is never ‹off' there is no "turn on" time, no problems with charge storage, and generally better high frequency performance and feedback loop stability (and usually fewer high-order harmonics).
- The point where the device comes closest to being ‹off' is not at ‹zero signal', so the problems of crossover distortion associated with class-AB and -B designs is avoided.
- Best for low signal levels of radio receivers due to low distortion.

Disadvantage of class-A amplifiers

- Class-A amplifiers are inefficient. A theoretical efficiency of 50% is obtainable in a push-pull topology, and only 25% in a single-ended topology, unless deliberate use of nonlinearities is made (such as in **square-law output stages**). In a power amplifier, this not only wastes power and limits operation with batteries, but increases operating costs and requires higher-rated output devices. Inefficiency comes from the standing current that must be roughly half the maximum output current, and a large part of the power supply voltage is present across the output device at low signal levels. If high output power is needed from a class-A circuit, the power supply and accompanying heat becomes significant. For every watt delivered to the load, the amplifier itself, at best, uses an extra watt. For high power amplifiers this means very large and expensive power supplies and heat sinks.

Class-A power amplifier designs have largely been superseded by more efficient designs, though their simplicity makes them popular with some hobbyists. There is a market for expensive **high fidelity** class-A amps considered a "cult item" among audiophiles mainly for their absence of crossover distortion and reduced odd-harmonic and high-order harmonic distortion. They also fill a niche market for recreations of vintage guitar amplifiers, due to their unique tone.

Single-Ended and Triode class-A amplifiers

Some hobbyists who prefer class-A amplifiers also prefer the use of thermionic valve (tube) designs instead of transistors, for several reasons:

- Single-ended output stages have an asymmetrical transfer function, meaning that even-order harmonics in the created distortion tend to not cancel out (as they do in push–pull output stages). For tubes, or FETs, most distortion is second-order harmonics, from the square law transfer characteristic, which to some produces a "warmer" and more pleasant sound.
- For those who prefer low distortion figures, the use of tubes with class A (generating little odd-harmonic distortion, as mentioned above) together with symmetrical circuits (such as push–pull output stages, or balanced low-level stages) results in the cancellation of most of the even distortion harmonics, hence the removal of most of the distortion.
- Historically, valve amplifiers often used a class-A power amplifier simply because valves are large and expensive; many class-A designs use only a single device.

Transistors are much cheaper, and so more elaborate designs that give greater efficiency but use more parts are still cost-effective. A classic application for a pair of class-A devices is the long-tailed pair, which is exceptionally linear, and forms the basis of many more complex circuits, including many audio amplifiers and almost all op-amps.

Class-A amplifiers may be used in output stages of op-amps (although the accuracy of the bias in low cost op-amps such as the **741** may result in class A or class AB or class B performance, varying from device to device or with temperature). They are sometimes used as medium-power, low-efficiency, and high-cost audio power amplifiers. The power consumption is unrelated to the output power. At idle (no input), the power consumption is essentially the same as at high output volume. The result is low efficiency and high heat dissipation.

Class B

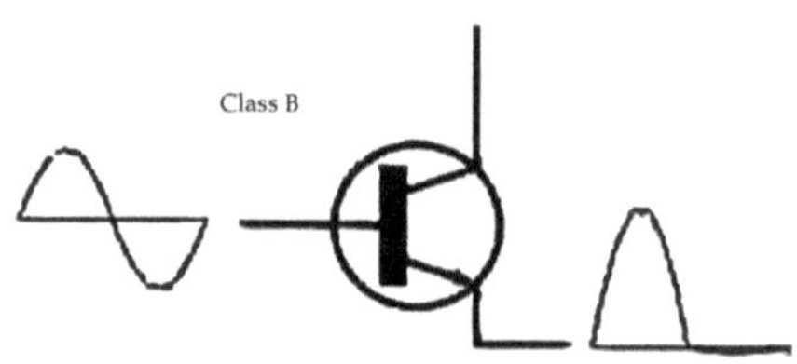

Class-B amplifier.

Class-B amplifiers only amplify half of the input wave cycle, thus creating a large amount of distortion, but their efficiency is greatly improved and is much better than class A. Class-B amplifiers are also favoured in battery-operated devices, such as transistor radios. Class B has a maximum theoretical efficiency of $\pi/4$ ($\approx$ 78.5%).This is because the amplifying element is switched off altogether half of the time, and so cannot dissipate power. A single class-B element is rarely found in practice, though it has been used for driving the loudspeaker in the early IBM Personal Computers with beeps, and it can be used in RF power amplifier where the distortion levels are less important. However, class C is more commonly used for this.

A practical circuit using class-B elements is the push–pull stage, such as the very simplified complementary pair arrangement shown below. Here, complementary or quasi-complementary devices are each used for amplifying the opposite halves of the input signal, which is then recombined at the output. This arrangement gives excellent efficiency, but can suffer from the drawback that there is a small mismatch in the cross-over region – at the "joins" between the two halves of the signal, as one output device has to take over supplying power exactly as the other finishes. This is called crossover distortion. An improvement is to bias the devices so they are not completely off when they are not in use. This approach is called *class AB* operation.

Class B amplifiers offer higher efficiency than class A amplifier using a single active device.

Class AB

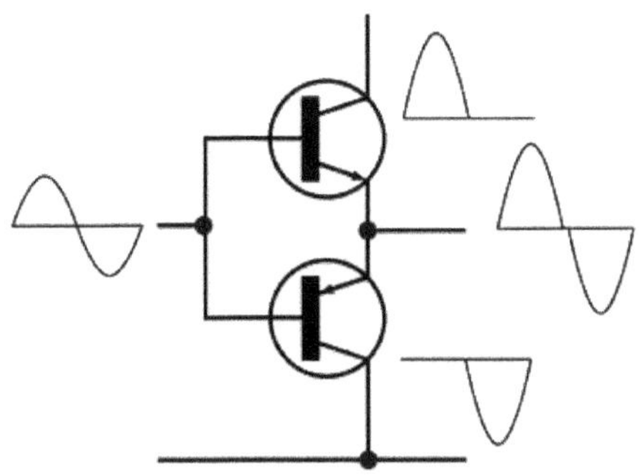

Class-AB push–pull amplifier.

Class AB is widely considered a good compromise for amplifiers, since much of the time the music signal is quiet enough that the signal stays in the "class A" region, where it is amplified with good fidelity, and by definition if passing out of this region, is large enough that the distortion products typical of class B are relatively small. The crossover distortion can be reduced further by using negative feedback.

In class-AB operation, each device operates the same way as in class B over half the waveform, but also conducts a small amount on the other half. As a result, the region where both devices simultaneously are nearly off (the "dead zone") is reduced. The result is that when the waveforms from the two devices are combined, the crossover is greatly minimised or eliminated altogether. The exact choice of **quiescent current** (the standing current through both devices when there is no signal) makes a large difference to the level of distortion (and to the risk of thermal runaway, that may damage the devices). Often, bias voltage applied to set this quiescent current must be adjusted with the temperature of the output transistors. (For example, in the circuit at the beginning of the article, the diodes would be mounted physically close to the output transistors, and specified to have a matched temperature coefficient.) Another approach (often used with thermally tracking bias voltages) is to include small value resistors in series with the emitters.

Class AB sacrifices some efficiency over class B in favor of linearity, thus is less efficient (below 78.5% for full-amplitude sine waves in transistor amplifiers, typically; much less is common in class-AB vacuum-tube amplifiers). It is typically much more efficient than class A.

Sometimes a numeral is added for vacuum-tube stages. If grid current is not permitted to flow, the class is AB_1. If grid current is allowed to flow (adding more distortion, but giving slightly higher output power) the class is AB_2.

Class C

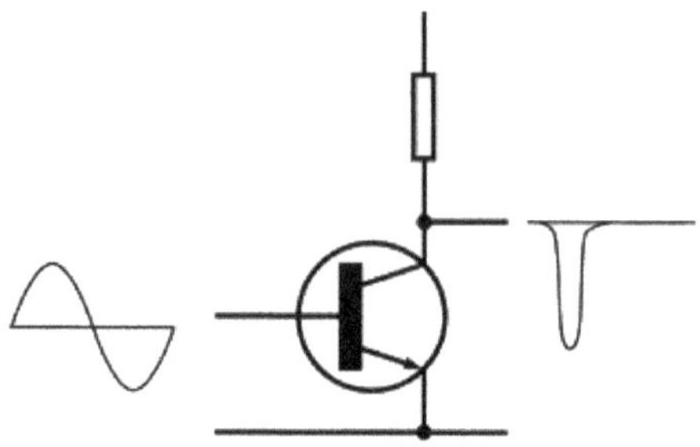

Class-C amplifier.

Class-C amplifiers conduct less than 50% of the input signal and the distortion at the output is high, but high efficiencies (up to 90%) are possible. The usual application for class-C amplifiers is in RF transmitters operating at a single fixed carrier frequency, where the distortion is controlled by a tuned load on the amplifier. The input signal is used to switch the active device causing pulses of current to flow through a tuned circuit forming part of the load.

The class-C amplifier has two modes of operation: tuned and untuned. The diagram shows a waveform from a simple class-C circuit without the tuned load. This is called untuned operation, and the analysis of the waveforms shows the massive distortion that appears in the signal. When the proper load (e.g., an inductive-capacitive filter plus a load resistor) is used, two things happen. The first is that the output's bias level is clamped with the average output voltage equal to the supply voltage. This is why tuned operation is sometimes called a *clamper*. This restores the waveform to its proper shape, despite the amplifier having only a one-polarity supply. This is directly related to the second phenomenon: the waveform on the center frequency becomes less distorted. The residual distortion is dependent upon the bandwidth of the tuned load, with the center frequency seeing very little distortion, but greater attenuation the farther from the tuned frequency that the signal gets.

The tuned circuit resonates at one frequency, the fixed carrier frequency, and so the unwanted frequencies are suppressed, and the wanted full signal (sine wave) is extracted by the tuned load. The signal bandwidth of the amplifier is limited by the Q-factor of the tuned circuit but this is not a serious limitation. Any residual harmonics can be removed using a further filter.

In practical class-C amplifiers a tuned load is invariably used. In one common arrangement the resistor shown in the circuit above is replaced with a parallel-tuned circuit consisting of an inductor and capacitor in parallel, whose components are chosen to resonate the frequency of the input signal. Power can be coupled to a load by transformer action with a secondary coil wound on the inductor. The average voltage at the collector is then equal to the supply voltage, and the signal voltage appearing across the tuned circuit varies from near zero to near twice the supply voltage during the RF cycle. The input circuit is biased so that the active element (e.g., transistor) conducts for only a fraction of the RF cycle, usually one third (120 degrees) or less.

The active element conducts only while the collector voltage is passing through its minimum. By this means, power dissipation in the active device is minimised, and efficiency increased. Ideally, the active element would pass only an instantaneous current pulse while the voltage across it is zero: it then dissipates no power and 100% efficiency is achieved. However practical devices have a limit to the peak current they can pass, and the pulse must therefore be widened, to around 120 degrees, to obtain a reasonable amount of power, and the efficiency is then 60–70%.

Class D

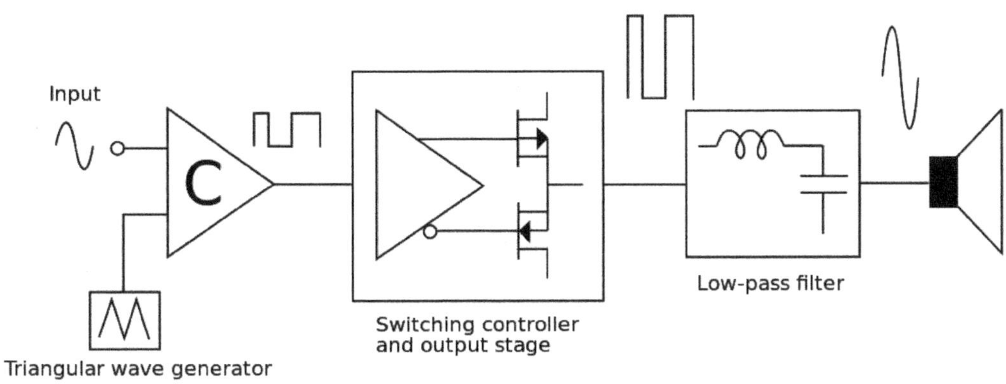

Block diagram of a basic switching or PWM (class-D) amplifier.

Boss Audio class-D mono amplifier with a low-pass filter for powering subwoofers.

In the class-D amplifier the active devices (transistors) function as electronic switches instead of linear gain devices; they are either on or off. The analog signal is converted to a stream of pulses that represents the signal by pulse-width modulation, pulse-density modulation, delta-sigma modulation or a related modulation technique before being applied to the amplifier. The time average power value of the pulses is directly proportional to the analog signal, so after amplification the signal can be converted back to an analog signal by a passive low-pass filter. The purpose of the output filter is to smooth the pulse stream to an analog signal, removing the high frequency spectral components of the pulses. The frequency of the output pulses is typically ten or more times the highest frequency in the input signal to amplify, so that the filter can adequately reduce the unwanted harmonics and accurately reproduce the input.

The main advantage of a class-D amplifier is power efficiency. Because the output pulses have a fixed amplitude, the switching elements (usually MOSFETs, but vacuum tubes, and at one time bipolar transistors, were used) are switched either completely on or completely off, rather than operated in linear mode. A MOSFET operates with the lowest resistance when fully on and thus (excluding when fully off) has the lowest power dissipation when in that condition. Compared to an equivalent class-AB device, a class-D amplifier's lower losses permit the use of a smaller heat sink for the MOSFETs while also reducing the amount of input power required, allowing for a lower-capacity power supply design. Therefore, class-D amplifiers are typically smaller than an equivalent class-AB amplifier.

Another advantage of the class-D amplifier is that it can operate from a digital signal source without requiring a digital-to-analog converter (DAC) to convert the signal to analog form first. If the signal source is in digital form, such as in a digital media player or computer sound card, the digital circuitry can convert the binary digital signal directly to a pulse-width modulation signal that is applied to the amplifier, simplifying the circuitry considerably.

Class-D amplifiers are widely used to control motors—but are now also used as power amplifiers, with extra circuitry that converts analogue to a much higher frequency pulse width modulated signal. Switching power supplies have even been modified into crude class-D amplifiers (though typically these only reproduce low-frequencies with acceptable accuracy). High quality class-D audio power amplifiers have now appeared on the market. These designs have been said to rival traditional AB amplifiers in terms of quality. An early use of class-D amplifiers was high-power subwoofer amplifiers in cars. Because subwoofers are generally limited to a bandwidth of no higher than 150 Hz, switching speed for the amplifier does not have to be as high as for a full range amplifier, allowing simpler designs. Class-D amplifiers for driving subwoofers are relatively inexpensive in comparison to class-AB amplifiers.

The letter *D* used to designate this amplifier class is simply the next letter after *C* and, although occasionally used as such, does not stand for *digital*. Class-D and class-E amplifiers are sometimes mistakenly described as "digital" because the output waveform superficially resembles a pulse-train of digital symbols, but a class-D amplifier merely converts an input waveform into a continuously pulse-width modulated analog signal. (A digital waveform would be pulse-code modulated.)

Additional Classes

Class E

The class-E/F amplifier is a highly efficient switching power amplifier, typically used at such high frequencies that the switching time becomes comparable to the duty time. As said in the class-D amplifier, the transistor is connected via a serial LC circuit to the load, and connected via a large L (inductor) to the supply voltage. The supply voltage is connected to ground via a large capacitor to prevent any RF signals leaking into the supply. The class-E amplifier adds a C (capacitor) between the transistor and ground and uses a defined L_1 to connect to the supply voltage.

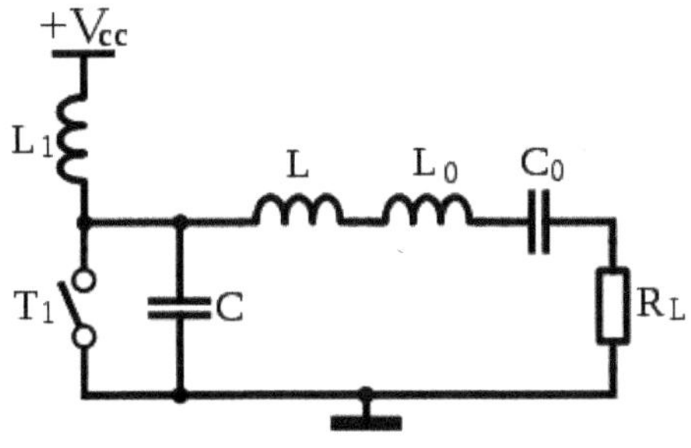

Class-E amplifier.

The following description ignores DC, which can be added easily afterwards. The above-mentioned C and L are in effect a parallel LC circuit to ground. When the transistor is on, it pushes through the serial LC circuit into the load and some current begins to flow to the parallel LC circuit to ground. Then the serial LC circuit swings back and compensates the current into the parallel LC circuit. At this point the current through the transistor is zero and it is switched off. Both LC circuits are now filled with energy in C and L_0. The whole circuit performs a damped oscillation. The damping by the load has been adjusted so that some time later the energy from the Ls is gone into the load, but the energy in both C_0 peaks at the original value to in turn restore the original voltage so that the voltage across the transistor is zero again and it can be switched on.

With load, frequency, and duty cycle (0.5) as given parameters and the constraint that the voltage is not only restored, but peaks at the original voltage, the four parameters (L, L_0, C and C_0) are determined. The class-E amplifier takes the finite on resistance into account and tries to make the current touch the bottom at zero. This means that the voltage and the current at the transistor are symmetric with respect to time. The Fourier transform allows an elegant formulation to generate the complicated LC networks and says that the first harmonic is passed into the load, all even harmonics are shorted and all higher odd harmonics are open.

Class E uses a significant amount of second-harmonic voltage. The second harmonic can be used to reduce the overlap with edges with finite sharpness. For this to work, energy on the second harmonic has to flow from the load into the transistor, and no source for this is visible in the circuit diagram. In reality, the impedance is mostly reactive and the only reason for it is that class E is a class F (see below) amplifier with a much simplified load network and thus has to deal with imperfections.

In many amateur simulations of class-E amplifiers, sharp current edges are assumed nullifying the very motivation for class E and measurements near the transit frequency of the transistors show very symmetric curves, which look much similar to class-F simulations.

The class-E amplifier was invented in 1972 by Nathan O. Sokal and Alan D. Sokal, and details were first published in 1975.Some earlier reports on this operating class have been published in Russian and Polish.

Class F

In push–pull amplifiers and in CMOS, the even harmonics of both transistors just cancel. Experiment shows that a square wave can be generated by those amplifiers. Theoretically square waves consist of odd harmonics only. In a class-D amplifier, the output filter blocks all harmonics; i.e., the harmonics see an open load. So even small currents in the harmonics suffice to generate a voltage square wave. The current is in phase with the voltage applied to the filter, but the voltage across the transistors is out of phase. Therefore, there is a minimal overlap between current through the transistors and voltage across the transistors. The sharper the edges, the lower the overlap.

While in class D, transistors and the load exist as two separate modules, class F admits imperfections like the parasitics of the transistor and tries to optimise the global system to have a high impedance at the harmonics. Of course there must be a finite voltage across the transistor to push the current across the on-state resistance. Because the combined current through both transistors is mostly in the first harmonic, it looks like a sine. That means that in the middle of the square the maximum of current has to flow, so it may make sense to have a dip in the square or in other words to allow some overswing of the voltage square wave. A class-F load network by definition has to transmit below a cutoff frequency and reflect above.

Any frequency lying below the cutoff and having its second harmonic above the cutoff can be amplified, that is an octave bandwidth. On the other hand, an inductive-capacitive series circuit with a large inductance and a tunable capacitance may be simpler to implement. By reducing the duty cycle below 0.5, the output amplitude can be modulated. The voltage square waveform degrades, but any overheating is compensated by the lower overall power flowing. Any load mismatch behind the filter can only act on the first harmonic current waveform, clearly only a purely resistive load makes sense, then the lower the resistance, the higher the current.

Class F can be driven by sine or by a square wave, for a sine the input can be tuned by an inductor to increase gain. If class F is implemented with a single transistor, the filter is complicated to short the even harmonics. All previous designs use sharp edges to minimise the overlap.

Classes G and H

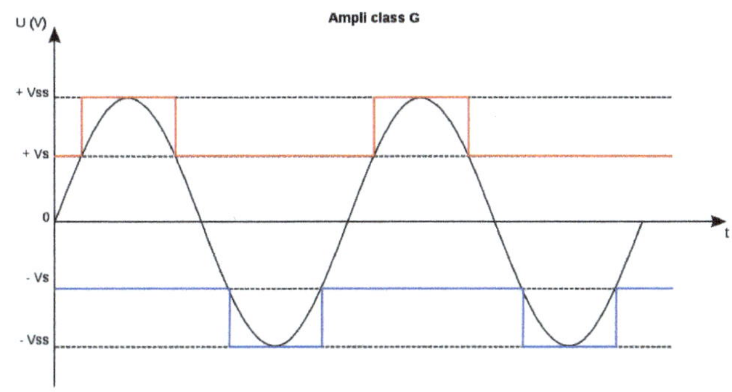

Idealized class-G rail voltage modulation.

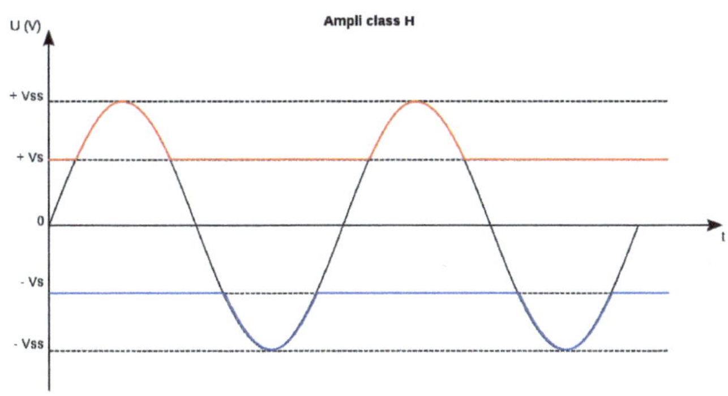

Idealized class-H rail voltage modulation.

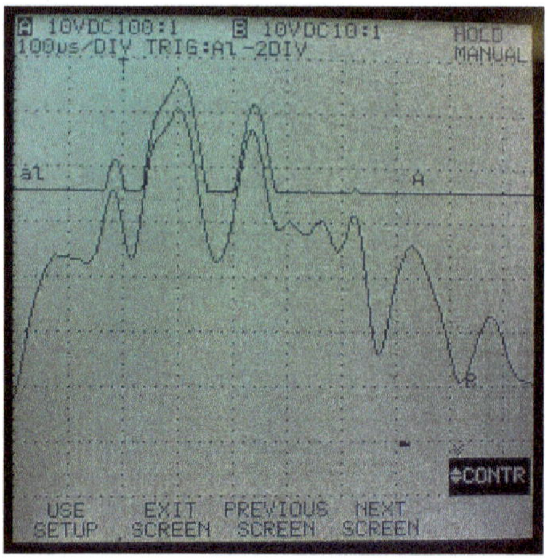

Rail voltage modulation.

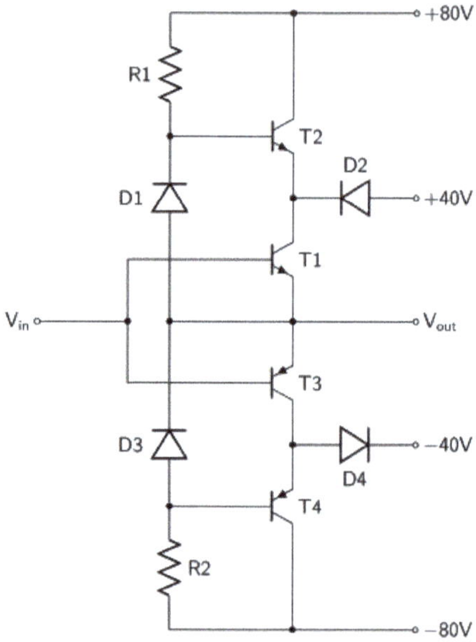

Basic schematic of a class-H configuration.

There is a variety of amplifier designs that enhance class-AB output stages with more efficient techniques to achieve greater efficiency with low distortion. These designs are common in large audio amplifiers since the heatsinks and power transformers would be prohibitively large (and costly) without the efficiency increases. The terms "class G" and "class H" are used interchangeably to refer to different designs, varying in definition from one manufacturer or paper to another.

Class-G amplifiers (which use "rail switching" to decrease power consumption and increase efficiency) are more efficient than class-AB amplifiers. These amplifiers provide several power rails at different voltages and switch between them as the signal output approaches each level. Thus, the amplifier increases efficiency by reducing the wasted power at the output transistors. Class-G amplifiers are more efficient than class AB but less efficient when compared to class D, however, they do not have the electromagnetic interference effects of class D.

Class-H amplifiers take the idea of class G one step further creating an infinitely variable supply rail. This is done by modulating the supply rails so that the rails are only a few volts larger than the output signal at any given time. The output stage operates at its maximum efficiency all the time. Switched-mode power supplies can be used to create the tracking rails. Significant efficiency gains can be achieved but with the drawback of more complicated supply design and reduced THD performance. In common designs, a voltage drop of about 10V is maintained over the output transistors in Class H circuits. The picture above shows positive supply voltage of the output stage and the voltage at the speaker output. The boost of the supply voltage is shown for a real music signal.

The voltage signal shown is thus a larger version of the input, but has been changed in sign (inverted) by the amplification. Other arrangements of amplifying device are possible, but that given (that is, common emitter, common source or common cathode) is the easiest to understand and employ in practice. If the amplifying element is linear, the output is a faithful copy of the input, only larger and inverted. In practice, transistors are not linear, and the output only approximates the input. nonlinearity from any of several sources is the origin of distortion within an amplifier. The class of amplifier (A, B, AB or C) depends on how the amplifying device is biased. The diagrams omit the bias circuits for clarity.

Any real amplifier is an imperfect realization of an ideal amplifier. An important limitation of a real amplifier is that the output it generates is ultimately limited by the power available from the power supply. An amplifier saturates and clips the output if the input signal becomes too large for the amplifier to reproduce or exceeds operational limits for the device.

Doherty Amplifiers

The Doherty amplifier is a hybrid configuration. It was invented in 1934 by William H. Doherty for Bell Laboratories—whose sister company, Western Electric, manufactured radio transmitters. The Doherty amplifier consists of a class-B *primary* or *carrier* stages in parallel with a class-C *auxiliary* or *peak* stage. The input signal splits to drive the two amplifiers, and a combining network sums the two output signals. Phase shifting networks are used in inputs and outputs. During periods of low signal level, the class-B amplifier efficiently operates on the signal and the class-C amplifier is cutoff and consumes little power. During periods of high signal level, the class-B amplifier delivers its maximum power and the class-C amplifier delivers up to its maximum power. The efficiency of previous AM transmitter designs was proportional to modulation but, with average modulation typically around 20%, transmitters were limited to less than 50% efficiency. In Doherty's design, even with zero modulation, a transmitter could achieve at least 60% efficiency.

As a successor to Western Electric for broadcast transmitters, the Doherty concept was considerably refined by Continental Electronics Manufacturing Company of Dallas, TX. Perhaps, the ultimate refinement was the screen-grid modulation scheme invented by Joseph B. Sainton. The Sainton amplifier consists of a class-C primary or carrier stage in parallel with a class-C auxiliary or peak stage. The stages are split and combined through 90-degree phase shifting networks as in the Doherty amplifier. The unmodulated radio frequency carrier is applied to the control grids of both tubes. Carrier modulation is applied to the screen grids of both tubes. The bias point of the carrier and peak tubes is different, and is established such that the peak tube is cutoff when modulation is absent (and the amplifier is producing rated unmodulated carrier power) whereas both tubes contribute twice the rated carrier power during 100% modulation (as four times the carrier power is required to achieve 100% modulation). As both tubes operate in class C, a significant improvement in efficiency is thereby achieved in the final stage. In addition, as the tetrode carrier and peak tubes require very little drive power, a significant improvement in efficiency within the driver stage is achieved as well (317C, et al.). The released version of the Sainton amplifier employs a cathode-follower modulator, not a push–pull modulator. Previous Continental Electronics designs, by James O. Weldon and others, retained most of the characteristics of the Doherty amplifier but added screen-grid modulation of the driver (317B, et al.).

The Doherty amplifier remains in use in very-high-power AM transmitters, but for lower-power AM transmitters, vacuum-tube amplifiers in general were eclipsed in the 1980s by arrays of solid-state amplifiers, which could be switched on and off with much finer granularity in response to the requirements of the input audio. However, interest in the Doherty configuration has been revived by cellular-telephone and wireless-Internet applications where the sum of several constant envelope users creates an aggregate AM result. The main challenge of the Doherty amplifier for digital transmission modes is in aligning the two stages and getting the class-C amplifier to turn on and off very quickly.

Recently, Doherty amplifiers have found widespread use in cellular base station transmitters for GHz frequencies. Implementations for transmitters in mobile devices have also been demonstrated.

IMPLEMENTATION

Amplifiers are implemented using active elements of different kinds:

The first active elements were relays. They were for example used in transcontinental telegraph lines: a weak current was used to switch the voltage of a battery to the outgoing line.

- For transmitting audio, carbon microphones were used as the active element. This was used to modulate a radio-frequency source in one of the first AM audio transmissions, by Reginald Fessenden on Dec. 24, 1906.
- Power control circuitry used magnetic amplifiers until the latter half of the twentieth century when high power FETs, and their easy interfacing to the newly developed digital circuitry, took over.

- Audio and most low power amplifiers used vacuum tubes exclusively until the 1960s. Today, tubes are used for specialist audio applications such as guitar amplifiers and audiophile amplifiers. Many broadcast transmitters still use vacuum tubes.
- In the 1960s, the transistor started to take over. These days, discrete transistors are still used in high-power amplifiers and in specialist audio devices.
- Beginning in the 1970s, more and more transistors were connected on a single chip therefore creating the integrated circuit. A large number of amplifiers commercially available today are based on integrated circuits.

For special purposes, other active elements have been used. For example, in the early days of the satellite communication, parametric amplifiers were used. The core circuit was a diode whose capacitance was changed by an RF signal created locally. Under certain conditions, this RF signal provided energy that was modulated by the extremely weak satellite signal received at the earth station.

Amplifier Circuit

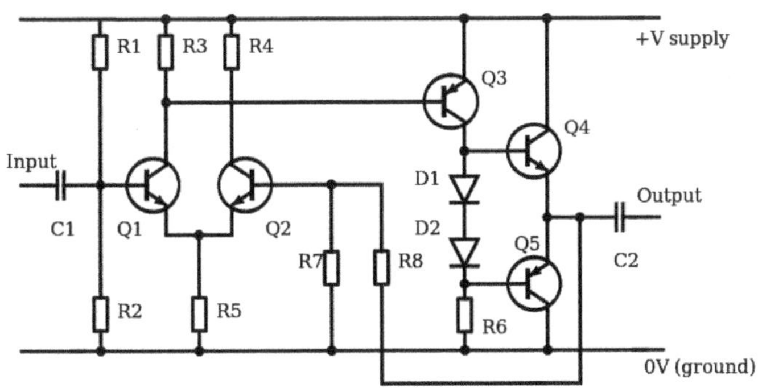

A practical amplifier circuit.

The practical amplifier circuit to the right could be the basis for a moderate-power audio amplifier. It features a typical (though substantially simplified) design as found in modern amplifiers, with a class-AB push–pull output stage, and uses some overall negative feedback. Bipolar transistors are shown, but this design would also be realizable with FETs or valves.

The input signal is coupled through capacitor C1 to the base of transistor Q1. The capacitor allows the AC signal to pass, but blocks the DC bias voltage established by resistors R1 and R2 so that any preceding circuit is not affected by it. Q1 and Q2 form a differential amplifier (an amplifier that multiplies the difference between two inputs by some constant), in an arrangement known as a long-tailed pair. This arrangement is used to conveniently allow the use of negative feedback, which is fed from the output to Q2 via R7 and R8.

The negative feedback into the difference amplifier allows the amplifier to compare the input to the actual output. The amplified signal from Q1 is directly fed to the second stage, Q3, which is a common emitter stage that provides further amplification of the signal and the DC bias for the output stages, Q4 and Q5. R6 provides the load for Q3 (a better design would probably use some form of active load here, such as a constant-current sink). So far, all of the amplifier is operating in class A. The output pair are arranged in class-AB push–pull, also called a complementary pair. They provide the majority of the current amplification (while consuming low quiescent current) and directly drive the load, connected via DC-blocking capacitor C2. The diodes D1 and D2 provide a small amount of constant voltage bias for the output pair, just biasing them into the conducting state so that crossover distortion is minimized. That is, the diodes push the output stage firmly into class-AB mode (assuming that the base-emitter drop of the output transistors is reduced by heat dissipation).

This design is simple, but a good basis for a practical design because it automatically stabilises its operating point, since feedback internally operates from DC up through the audio range and beyond. Further circuit elements would probably be found in a real design that would roll-off the frequency response above the needed range to prevent the possibility of unwanted oscillation. Also, the use of fixed diode bias as shown here can cause problems if the diodes are not both electrically and thermally matched to the output transistors – if the output transistors turn on too much, they can easily overheat and destroy themselves, as the full current from the power supply is not limited at this stage.

A common solution to help stabilise the output devices is to include some emitter resistors, typically one ohm or so. Calculating the values of the circuit's resistors and capacitors is done based on the components employed and the intended use of the amp.

Two most Common Circuits

- A Cascode amplifier is a two-stage circuit consisting of a transconductance amplifier followed by a buffer amplifier.
- A Log amplifier is a linear circuit in which output voltage is a constant times the natural logarithm of input.

For the basics of radio frequency amplifiers using valves, see Valved RF amplifiers.

Notes on Implementation

Real world amplifiers are imperfect.

- The power supply may influence the output, so must be considered in the design.
- A power amplifier is effectively an input signal controlled power regulator. It regulates the power sourced from the power supply or mains to the amplifier's load. The power output from a power amplifier cannot exceed the power input to it.
- The amplifier circuit has an "open loop" performance. This is described by various parameters (gain, slew rate, output impedance, distortion, bandwidth, signal to noise ratio, etc.).

- Many modern amplifiers use negative feedback techniques to hold the gain at the desired value and reduce distortion. Negative loop feedback has the intended effect of electrically damping loudspeaker motion, thereby damping the mechanical dynamic performance of the loudspeaker.
- When assessing rated amplifier power output, it is useful to consider the applied load, the signal type (e.g., speech or music), required power output duration (i.e., short-time or continuous), and required dynamic range (e.g., recorded or live audio).
- In high-powered audio applications that require long cables to the load (e.g., cinemas and shopping centres) it may be more efficient to connect to the load at line output voltage, with matching transformers at source and loads. This avoids long runs of heavy speaker cables.
- Prevent instability or overheating requires care to ensure solid state amplifiers are adequately loaded. Most have a rated minimum load impedance.
- A summing circuit is typical in applications that must combine many inputs or channels to form a composite output. It is best to combine multiple channels for this.
- All amplifiers generate heat through electrical losses. The amplifier must dissipate this heat via convection or forced air cooling. Heat can damage or reduce electronic component service life. Designers and installers must also consider heating effects on adjacent equipment.

Different power supply types result in many different methods of bias. Bias is a technique by which active devices are set to operate in a particular region, or by which the DC component of the output signal is set to the midpoint between the maximum voltages available from the power supply. Most amplifiers use several devices at each stage; they are typically matched in specifications except for polarity. Matched inverted polarity devices are called complementary pairs. Class-A amplifiers generally use only one device, unless the power supply is set to provide both positive and negative voltages, in which case a dual device symmetrical design may be used. Class-C amplifiers, by definition, use a single polarity supply.

Amplifiers often have multiple stages in cascade to increase gain. Each stage of these designs may be a different type of amp to suit the needs of that stage. For instance, the first stage might be a class-A stage, feeding a class-AB push–pull second stage, which then drives a class-G final output stage, taking advantage of the strengths of each type, while minimizing their weaknesses.

REFERENCES

1. Patronis, Gene (1987). “Amplifiers”. In Glen Ballou. Handbook for Sound Engineers: The New Audio Cyclopedia. Howard W. Sams & Co. p. 493. ISBN 0-672-21983-2.
2. Sungook, Hong (2001). Wireless: From Marconi’s Black-Box to the Audion. MIT Press. p. 165. ISBN 0262082985.

3. Harper, Douglas (2001). "Amplify". Online Etymology Dictionary. Etymonline.com. Retrieved July 10, 2015.
4. Robert Boylestad and Louis Nashelsky (1996). Electronic Devices and Circuit Theory, 7th Edition. Prentice Hall College Division. ISBN 978-0-13-375734-7.
5. Robert S. Symons (1998). "Tubes: Still vital after all these years". IEEE Spectrum. **35** *(4): 52–63. doi:10.1109/6.666962.*
6. Mammano, Bob (2001). "Magnetic Amplifier Control for Simple, Low-Cost, Secondary Regulation" (PDF). Texas Instruments.
7. "Negative Resistance Revived". users.tpg.com.au. Retrieved 2016-06-20.
8. Munsterman, G.T. (June 1965). "Tunnel-Diode Microwave Amplifiers" (PDF). APL Technical Digest. **4**: 2–10.
9. Mark Cherry, Maxim Engineering journal, volume 62, Amplifier Considerations in Ceramic Speaker Applications, p.3, accessed 2012-10-01
10. "What is a video amplifier, video booster amplifiers - Future Electronics". www.futureelectronics.com. Retrieved 2016-06-20.
11. Orwiler, Bob (December 1969). "Vertical Amplifier Circuits" (PDF). Tektronix, Inc.
12. "Travelling Wave Tube Amplifiers". www.r-type.org. Retrieved 2016-06-20.
13. Rood, George. "Music Concerns Seek New Volume With Amplifier". New York Times. Retrieved 23 February 2015.
14. "Amplifier Fills Need in Picture: Loud Speaker Only Method Found to Carry Directions During Turmoil". Los Angeles Times.
15. This table is a "Zwicky box"; in particular, it encompasses all possibilities. See Fritz Zwicky.
16. "Small signal analysis of Complex amplifier circuits". www.eeherald.com. Retrieved 2016-06-20.
17. John Everett (1992). Vsats: Very Small Aperture Terminals. IET. ISBN 0-86341-200-9.
18. Administrator. "Microwaves101 | Active Directivity of Amplifiers". www.microwaves101.com. Retrieved 2016-06-20.
19. Roy, Apratim; Rashid, S. M. S. (5 June 2012). "A power efficient bandwidth regulation technique for a low-noise high-gain RF wideband amplifier". Central European Journal of Engineering. **2** *(3): 383–391. Bibcode:*2012CEJE....2..383R. doi:10.2478/s13531-012-0009-1.
20. "Understanding Amplifier Operating "Classes"". electronicdesign.com. Retrieved 2016-06-20.
21. RCA Receiving Tube Manual, RC-14 (1940) p 12
22. ARRL Handbook, 1968; page 65
23. "Amplifier classes". www.duncanamps.com. Retrieved 2016-06-20.
24. "Amplifier Efficiency". sound.whsites.net. Retrieved 2016-06-20.

25. Jerry Del Colliano (20 February 2012), Pass Labs XA30.5 Class-A Stereo Amp Reviewed, Home Theater Review, Luxury Publishing Group Inc.
26. Ask the Doctors: Tube vs. Solid-State Harmonics
27. Volume cranked up in amp debate
28. "Biasing Op-Amps into Class A". tangentsoft.net. Retrieved 2016-06-20.
29. "Class B Amplifier - Class-B Transistor Amplifier Tutorial". Basic Electronics Tutorials. 2013-07-25. Retrieved 2016-06-20.
30. "Class AB Power Amplifiers". www.learnabout-electronics.org. Retrieved 2016-06-20.
31. "Class C power amplifier circuit diagram and theory. Output characteristics DC load line". www.circuitstoday.com. Retrieved 2016-06-20.
32. A.P. Malvino, Electronic Principles (2nd Ed.1979. ISBN 0-07-039867-4) p.299.
33. Electronic and Radio Engineering, R.P.Terman, McGraw Hill, 1964
34. "Class D Amplifiers: Fundamentals of Operation and Recent Developments - Application Note - Maxim". www.maximintegrated.com. Retrieved 2016-06-20.
35. N. O. Sokal and A. D. Sokal, "Class E – A New Class of High-Efficiency Tuned Single-Ended Switching Power Amplifiers", IEEE Journal of Solid-State Circuits, vol. SC-10, pp. 168–176, June 1975. HVK
36. US patent 2210028, William H. Doherty, "Amplifier", issued 1940-08-06, assigned to Bell Telephone Laboratories
37. US patent 3314034, Joseph B. Sainton, "High Efficiency Amplifier and Push–Pull Modulator", issued 1967-04-11, assigned to Continental Electronics Manufacturing Company
38. Lee, Thomas (2004). The Design of CMOS Radio-Frequency Integrated Circuits. New York, NY: Cambridge University Press. p. 8. ISBN 978-0-521-83539-8.
39. Malina, Roger. "Visual Art, Sound, Music and Technology".
40. Shortess, George. "Interactive Sound Installations Using Microcomputers". JSTOR 1578331.

CHAPTER 7

Bipolar Junction Transistor Theory and Modeling

Chapter Contents

- **Introduction**
- **Note on Current Direction**
- **Structure**
- **Regions of Operation**
- **History**
- **Theory and Modeling**
- **Applications**
- **Vulnerabilities**

A bipolar junction transistor, abbreviated BJT, is a type of transistor with three terminals that is made from a doped semiconductor material. Its primary purpose is for use in electrical devices to amplify or change electrical signals. To understand the bipolar junction transistor, it's best to understand the basic function of transistors and what the terms "doped" and "semiconductor" mean.

Transistors simply take an electrical signal that passes through them and either amplify it or alter it. The electrical current enters into the transistor and comes out stronger, thus amplified, when it leaves. Transistors are one of the basic units that make up all electrical devices and help control and manipulate the electric currents that pass through them. A bipolar junction transistor is simply a specific kind of transistor used to send currents in two different directions.

INTRODUCTION

A **bipolar junction transistor** (**bipolar transistor** or **BJT**) is a type of transistor that uses both electron and hole charge carriers. In contrast, unipolar transistors, such as field-effect transistors, only use one kind of charge carrier. For their operation, BJTs use two junctions between two semiconductor types, n-type and p-type.

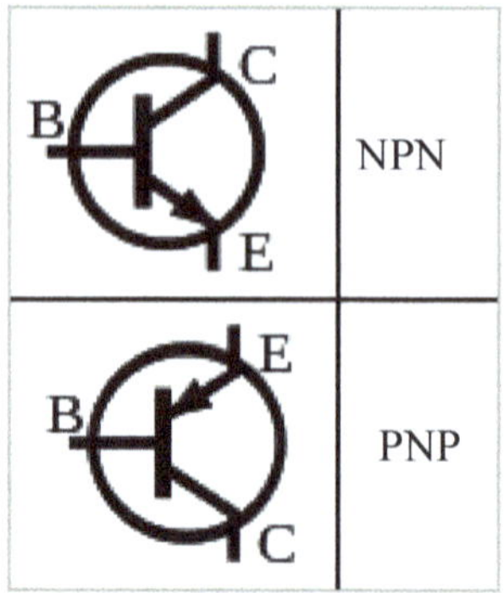

BJT schematic symbols

Typical individual BJT packages. From top to bottom: TO-3, TO-126, TO-92, SOT-23

BJTs are manufactured in two types, NPN and PNP, and are available as individual components, or fabricated in integrated circuits, often in large numbers. The basic function of a BJT is to amplify current. This allows BJTs to be used as amplifiers or switches, giving them wide applicability in electronic equipment, including computers, televisions, mobile phones, audio amplifiers, industrial control, and radio transmitters.

NOTE ON CURRENT DIRECTION

By convention, the direction of current on diagrams is shown as the direction that a positive charge would move. This is called *conventional current*. However, current in metal conductors is due to the flow of electrons which, because they carry a negative charge, move in the opposite direction to conventional current. On the other hand, inside a bipolar transistor, currents can be composed of both positively charged holes and negatively charged electrons. In this article, current arrows are shown in the conventional direction, but labels for the movement of holes and electrons show their actual direction inside the transistor.

Function

BJTs come in two types, or polarities, known as PNP and NPN based on the doping types of the three main terminal regions. An NPN transistor comprises two semiconductor junctions that share a thin p-doped anode region, and a PNP transistor comprises two semiconductor junctions that share a thin n-doped cathode region.

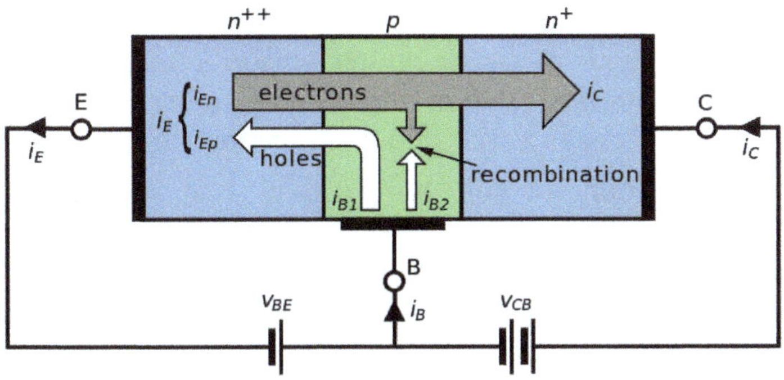

NPN BJT with forward-biased E–B junction and reverse-biased B–C junction.

Charge flow in a BJT is due to diffusion of charge carriers across a junction between two regions of different charge concentrations. The regions of a BJT are called *emitter*, *collector*, and *base*. A discrete transistor has three leads for connection to these regions. Typically, the emitter region is heavily doped compared to the other two layers, whereas the majority charge carrier concentrations in base and collector layers are about the same. By design, most of the BJT collector current is due to the flow of charges injected from a high-concentration emitter into the base where they are minority carriers that diffuse toward the collector, and so BJTs are classified as minority-carrier devices.

In typical operation, the base–emitter junction is forward biased, which means that the p-doped side of the junction is at a more positive potential than the n-doped side, and the base–collector junction is reverse biased. In an NPN transistor, when positive bias is applied to the base–emitter junction, the equilibrium is disturbed between the thermally generated carriers and the repelling electric field of the n-doped emitter depletion region. This allows thermally excited electrons to inject from the emitter into the base region. These electrons diffuse through the base from the region of high concentration near the emitter towards the region of low concentration near the collector. The electrons in the base are called *minority carriers* because the base is doped p-type, which makes holes the *majority carrier* in the base.

To minimize the percentage of carriers that recombine before reaching the collector–base junction, the transistor's base region must be thin enough that carriers can diffuse across it in much less time than the semiconductor's minority carrier lifetime. In particular, the thickness of the base must be much less than the diffusion length of the electrons. The collector–base junction is reverse-biased, and so little electron injection occurs from the collector to the base, but electrons that diffuse through the base towards the collector are swept into the collector by the electric field in the depletion region of the collector–base junction. The thin *shared* base and asymmetric collector–emitter doping are what differentiates a bipolar transistor from two *separate* and oppositely biased diodes connected in series.

Voltage, Current, and Charge Control

The collector–emitter current can be viewed as being controlled by the base–emitter current (current control), or by the base–emitter voltage (voltage control). These views are related by the current–voltage relation of the base–emitter junction, which is just the usual exponential current–voltage curve of a p-n junction (diode).

The physical explanation for collector current is the concentration of minority carriers in the base region. Due to low level injection (in which there are much fewer excess carriers than normal majority carriers) the ambipolar transport rates (in which the excess majority and minority carriers flow at the same rate) is in effect determined by the excess minority carriers.

Detailed transistor models of transistor action, such as the Gummel–Poon model, account for the distribution of this charge explicitly to explain transistor behaviour more exactly. The charge-control view easily handles phototransistors, where minority carriers in the base region are created by the absorption of photons, and handles the dynamics of turn-off, or recovery time, which depends on charge in the base region recombining. However, because base charge is not a signal that is visible at the terminals, the current- and voltage-control views are generally used in circuit design and analysis.

In analog circuit design, the current-control view is sometimes used because it is approximately linear. That is, the collector current is approximately {\displaystyle \beta _{F}} times the base current. Some basic circuits can be designed by assuming that the emitter–base voltage is approximately constant, and that collector current is beta times the base current. However, to accurately and reliably design production BJT circuits, the voltage-control (for example, Ebers–Moll) model is required. The voltage-control model requires an exponential function to be taken into account, but when it is linearized such that the transistor can be modeled as a transconductance, as in the Ebers–Moll model, design for circuits such as differential amplifiers again becomes a mostly linear problem, so the voltage-control view is often preferred. For translinear circuits, in which the exponential I–V curve is key to the operation, the transistors are usually modeled as voltage-controlled current sources whose transconductance is proportional to their collector current. In general, transistor-level circuit design is performed using SPICE or a comparable analog circuit simulator, so model complexity is usually not of much concern to the designer.

Turn-on, Turn-off, and Storage Delay

The bipolar transistor exhibits a few delay characteristics when turning on and off. Most transistors, and especially power transistors, exhibit long base-storage times that limit maximum frequency of operation in switching applications. One method for reducing this storage time is by using a Baker clamp.

Transistor Parameters: Alpha (α) and beta (β)

The proportion of electrons able to cross the base and reach the collector is a measure of the BJT efficiency. The heavy doping of the emitter region and light doping of the base region causes many more electrons to be injected from the emitter into the base than holes to be injected from the base into the emitter.

The *common-emitter current gain* is represented by β_F or the h-parameter h_{FE}; it is approximately the ratio of the DC collector current to the DC base current in forward-active region. It is typically greater than 50 for small-signal transistors but can be smaller in transistors designed for high-power applications.

Another important parameter is the *common-base current gain*, α_F. The common-base current gain is approximately the gain of current from emitter to collector in the forward-active region. This ratio usually has a value close to unity; between 0.980 and 0.998. It is less than unity due to recombination of charge carriers as they cross the base region.

Alpha and beta are more precisely related by the following identities (NPN transistor):

$$\alpha_F = \frac{I_C}{I_E}, \qquad \beta_F = \frac{I_C}{I_B}$$

$$\alpha_F = \frac{\beta_F}{1+\beta_F} \iff \beta_F = \frac{\alpha_F}{1-\alpha_F}$$

STRUCTURE

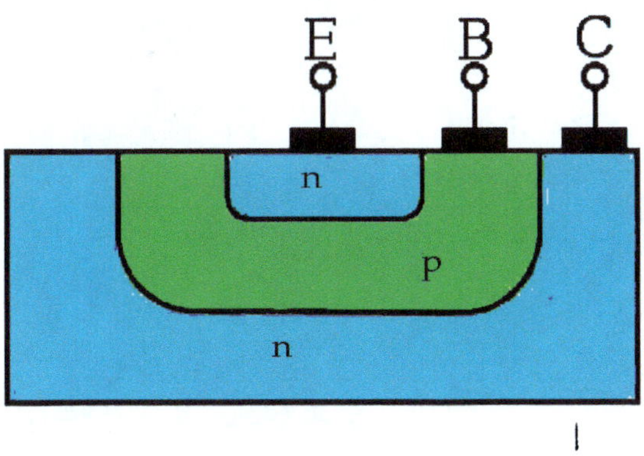

Simplified cross section of a planar *NPN* bipolar junction transistor.

A BJT consists of three differently doped semiconductor regions: the *emitter* region, the *base* region and the *collector* region. These regions are, respectively, *p* type, *n* type and *p* type in a PNP transistor, and *n* type, *p* type and *n* type in an NPN transistor. Each semiconductor region is connected to a terminal, appropriately labeled: *emitter* (E), *base* (B) and *collector* (C).

The *base* is physically located between the *emitter* and the *collector* and is made from lightly doped, high-resistivity material. The collector surrounds the emitter region, making it almost impossible for the electrons injected into the base region to escape without being collected, thus making the resulting value of α very close to unity, and so, giving the transistor a large β. A cross-section view of a BJT indicates that the collector–base junction has a much larger area than the emitter–base junction.

The bipolar junction transistor, unlike other transistors, is usually not a symmetrical device. This means that interchanging the collector and the emitter makes the transistor leave the forward active mode and start to operate in reverse mode. Because the transistor's internal structure is usually optimized for forward-mode operation, interchanging the collector and the emitter makes the values of α and β in reverse operation much smaller than those in forward operation; often the α of the reverse mode is lower than 0.5. The lack of symmetry is primarily due to the doping ratios of the emitter and the collector. The emitter is heavily doped, while the collector is lightly doped, allowing a large reverse bias voltage to be applied before the collector–base junction breaks down. The collector–base junction is reverse biased in normal operation. The reason the emitter is heavily doped is to increase the emitter injection efficiency: the ratio of carriers injected by the emitter to those injected by the base. For high current gain, most of the carriers injected into the emitter–base junction must come from the emitter.

Die of a KSY34 high-frequency NPN transistor. Bond wires connect to the base and emitter.

The low-performance "lateral" bipolar transistors sometimes used in CMOS processes are sometimes designed symmetrically, that is, with no difference between forward and backward operation.

Small changes in the voltage applied across the base–emitter terminals cause the current between the *emitter* and the *collector* to change significantly. This effect can be used to amplify the input voltage or current. BJTs can be thought of as voltage-controlled current sources, but are more simply characterized as current-controlled current sources, or current amplifiers, due to the low impedance at the base.

Early transistors were made from germanium but most modern BJTs are made from silicon. A significant minority are also now made from gallium arsenide, especially for very high speed applications (see HBT, below).

NPN

The symbol of an NPN BJT. A mnemonic for the symbol is "**n**ot **p**ointing **in**".

NPN is one of the two types of bipolar transistors, consisting of a layer of P-doped semiconductor (the "base") between two N-doped layers. A small current entering the base is amplified to produce a large collector and emitter current. That is, when there is a positive potential difference measured from the base of an NPN transistor to its emitter (that is, when the base is **high** relative to the emitter), as well as a positive potential difference measured from the collector to the emitter, the transistor becomes active. In this "on" state, charge flows from the collector to the emitter of the transistor. Most of the current is carried by electrons moving from emitter to collector as minority carriers in the P-type base region. To allow for greater current and faster operation, most bipolar transistors used today are NPN because electron mobility is higher than hole mobility.

A mnemonic device for the NPN transistor symbol is "*n*ot *p*ointing i*n*", based on the arrows in the symbol and the letters in the name.

PNP

The symbol of a PNP BJT. A mnemonic for the symbol is "**p**oints **in** **p**roudly".

The other type of BJT is the PNP, consisting of a layer of N-doped semiconductor between two layers of P-doped material. A small current leaving the base is amplified in the collector output. That is, a PNP transistor is "on" when its base is pulled **low** relative to the emitter. In a PNP transistor, the emitter–base region is forward biased, so holes are injected into the base as minority carriers. The base is very thin, and most of the holes cross the reverse-biased base–collector junction to the collector.

The arrows in the NPN and PNP transistor symbols are on the emitter legs and point in the direction of the conventional current when the device is in forward active or forward saturated mode.

A mnemonic device for the PNP transistor symbol is*p*ointing i*n* (*p*roudly/*p*ermanently)", based on the arrows in the symbol and the letters in the name.

Heterojunction Bipolar Transistor

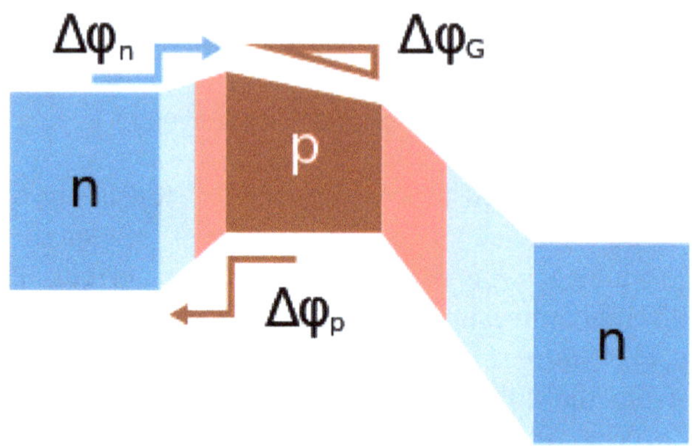

Bands in graded heterojunction NPN bipolar transistor. Barriers indicated for electrons to move from emitter to base, and for holes to be injected backward from base to emitter; also, grading of bandgap in base assists electron transport in base region; light colors indicate depleted regions.

The heterojunction bipolar transistor (HBT) is an improvement of the BJT that can handle signals of very high frequencies up to several hundred GHz. It is common in modern ultrafast circuits, mostly RF systems.

Heterojunction transistors have different semiconductors for the elements of the transistor. Usually the emitter is composed of a larger bandgap material than the base. The figure shows that this difference in bandgap allows the barrier for holes to inject backward from the base into the emitter, denoted in the figure as $\Delta\varphi_p$, to be made large, while the barrier for electrons to inject into the base $\Delta\varphi_n$ is made low. This barrier arrangement helps reduce minority carrier injection from the base when the emitter-base junction is under forward bias, and thus reduces base current and increases emitter injection efficiency.

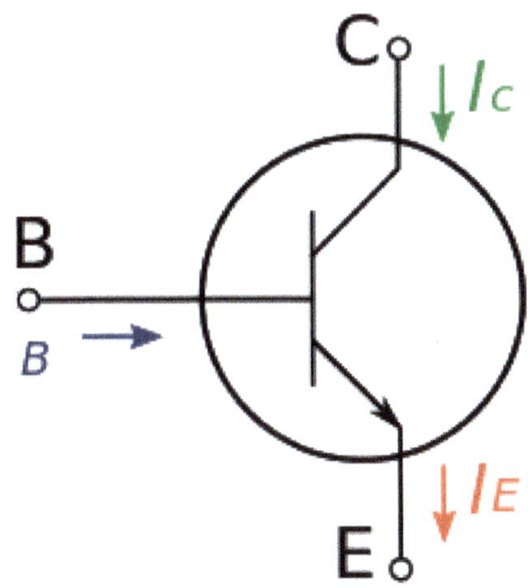

Symbol for NPN Bipolar Transistor with current flow direction.

The improved injection of carriers into the base allows the base to have a higher doping level, resulting in lower resistance to access the base electrode. In the more traditional BJT, also referred to as homojunction BJT, the efficiency of carrier injection from the emitter to the base is primarily determined by the doping ratio between the emitter and base, which means the base must be lightly doped to obtain high injection efficiency, making its resistance relatively high. In addition, higher doping in the base can improve figures of merit like the Early voltage by lessening base narrowing.

The grading of composition in the base, for example, by progressively increasing the amount of germanium in a SiGe transistor, causes a gradient in bandgap in the neutral base, denoted in the figure by $\Delta\varphi_G$, providing a "built-in" field that assists electron transport across the base. That drift component of transport aids the normal diffusive transport, increasing the frequency response of the transistor by shortening the transit time across the base.

Two commonly used HBTs are silicon–germanium and aluminum gallium arsenide, though a wide variety of semiconductors may be used for the HBT structure. HBT structures are usually grown by epitaxy techniques like MOCVD and MBE.

REGIONS OF OPERATION

Applied voltages	B-E junction bias (NPN)	B-C junction bias (NPN)	Mode (NPN)
E < B < C	Forward	Reverse	Forward-active
E < B > C	Forward	Forward	Saturation
E > B < C	Reverse	Reverse	Cut-off

E > B > C	Reverse	Forward	Reverse-active

Applied voltages	**B-E junction bias (PNP)**	**B-C junction bias (PNP)**	**Mode (PNP)**
E < B < C	Reverse	Forward	Reverse-active
E < B > C	Reverse	Reverse	Cut-off
E > B < C	Forward	Forward	Saturation
E > B > C	Forward	Reverse	Forward-active

Bipolar transistors have four distinct regions of operation, defined by BJT junction biases.

Forward-Active (or SImply *Active*)

The base–emitter junction is forward biased and the base–collector junction is reverse biased. Most bipolar transistors are designed to afford the greatest common-emitter current gain, β_F, in forward-active mode. If this is the case, the collector–emitter current is approximately proportional to the base current, but many times larger, for small base current variations.

Reverse-active (or *Inverse-active* or *Inverted*)

By reversing the biasing conditions of the forward-active region, a bipolar transistor goes into reverse-active mode. In this mode, the emitter and collector regions switch roles. Because most BJTs are designed to maximize current gain in forward-active mode, the β_F in inverted mode is several times smaller (2–3 times for the ordinary germanium transistor). This transistor mode is seldom used, usually being considered only for failsafe conditions and some types of bipolar logic. The reverse bias breakdown voltage to the base may be an order of magnitude lower in this region.

Saturation

With both junctions forward-biased, a BJT is in saturation mode and facilitates high current conduction from the emitter to the collector (or the other direction in the case of NPN, with negatively charged carriers flowing from emitter to collector). This mode corresponds to a logical "on", or a closed switch.

Cut-off

In cut-off, biasing conditions opposite of saturation (both junctions reverse biased) are present. There is very little current, which corresponds to a logical "off", or an open switch.

Avalanche Breakdown Region

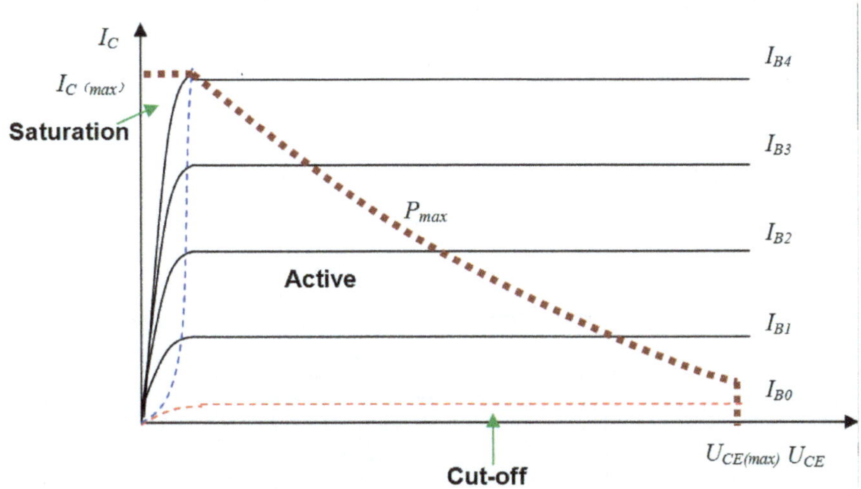

The relationship between I_C, UC_E and I_B

The modes of operation can be described in terms of the applied voltages (this description applies to NPN transistors; polarities are reversed for PNP transistors):

Forward-Active

Base higher than emitter, collector higher than base (in this mode the collector current is proportional to base current by β_F).

Saturation

Base higher than emitter, but collector is not higher than base.

Cut-off

Base lower than emitter, but collector is higher than base. It means the transistor is not letting conventional current go through from collector to emitter.

Reverse-active

Base lower than emitter, collector lower than base: reverse conventional current goes through transistor.

In terms of junction biasing: (*reverse biased base–collector junction* means $V_{bc} < 0$ for NPN, opposite for PNP)

Although these regions are well defined for sufficiently large applied voltage, they overlap somewhat for small (less than a few hundred millivolts) biases. For example, in the typical grounded-emitter configuration of an NPN BJT used as a pulldown switch in digital logic, theoff state never involves a reverse-biased junction because the base voltage never goes below ground; nevertheless the forward bias is close enough to zero that essentially no current flows, so this end of the forward active region can be regarded as the cutoff region.

Active-mode NPN Transistors in Circuits

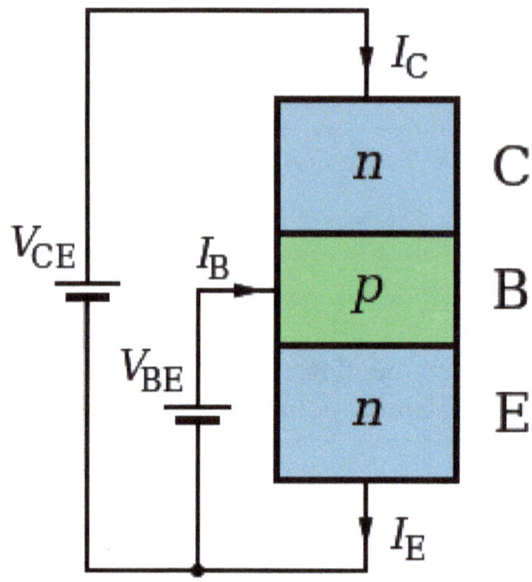

Structure and use of NPN transistor. Arrow according to schematic.

The diagram shows a schematic representation of an NPN transistor connected to two voltage sources. To make the transistor conduct appreciable current (on the order of 1 mA) from C to E, V_{BE} must be above a minimum value sometimes referred to as the **cut-in voltage**. The cut-in voltage is usually about 650 mV for silicon BJTs at room temperature but can be different depending on the type of transistor and its biasing. This applied voltage causes the lower P-N junction to 'turn on', allowing a flow of electrons from the emitter into the base. In active mode, the electric field existing between base and collector (caused by V_{CE}) will cause the majority of these electrons to cross the upper P-N junction into the collector to form the collector current I_C. The remainder of the electrons recombine with holes, the majority carriers in the base, making a current through the base connection to form the base current, I_B. As shown in the diagram, the emitter current, I_E, is the total transistor current, which is the sum of the other terminal currents, (i.e., $I_E = I_B + I_C$).

In the diagram, the arrows representing current point in the direction of conventional current – the flow of electrons is in the opposite direction of the arrows because electrons carry negative electric charge. In active mode, the ratio of the collector current to the base current is called the *DC current gain*. This gain is usually 100 or more, but robust circuit designs do not depend on the exact value (for example see op-amp). The value of this gain for DC signals is referred to as h_{FE}, and the value of this gain for small signals is referred to as h_{fe}. That is, when a small change in the currents occurs, and sufficient time has passed for the new condition to reach a steady state h_{fe} is the ratio of the change in collector current to the change in base current. The symbol β is used for both h_{FE} and h_{fe}.

The emitter current is related to V_{BE} exponentially. At room temperature, an increase V_{BE} by approximately 60 mV increases the emitter current by a factor of 10. Because the base current is approximately proportional to the collector and emitter currents, they vary in the same way.

Active-mode PNP Transistors in Circuits

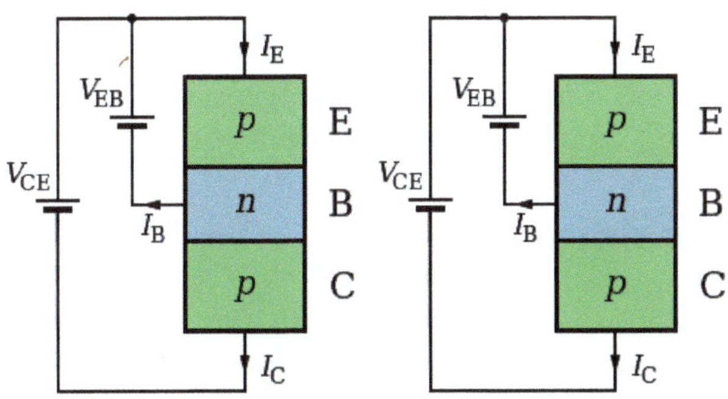

Structure and use of PNP transistor.

The diagram shows a schematic representation of a PNP transistor connected to two voltage sources. To make the transistor conduct appreciable current (on the order of 1 mA) from E to C, V_{EB} must be above a minimum value sometimes referred to as the **cut-in voltage**. The cut-in voltage is usually about 650 mV for silicon BJTs at room temperature but can be different depending on the type of transistor and its biasing. This applied voltage causes the upper P-N junction to 'turn-on' allowing a flow of holes from the emitter into the base. In active mode, the electric field existing between the emitter and the collector (caused by V_{CE}) causes the majority of these holes to cross the lower p-n junction into the collector to form the collector current I_C. The remainder of the holes recombine with electrons, the majority carriers in the base, making a current through the base connection to form the base current, I_B. As shown in the diagram, the emitter current, I_E, is the total transistor current, which is the sum of the other terminal currents (i.e., $I_E = I_B + I_C$).

In the diagram, the arrows representing current point in the direction of conventional current – the flow of holes is in the same direction of the arrows because holes carry positive electric charge. In active mode, the ratio of the collector current to the base current is called the *DC current gain*. This gain is usually 100 or more, but robust circuit designs do not depend on the exact value. The value of this gain for DC signals is referred to as h_{fe}, and the value of this gain for AC signals is referred to as h_{fe}. However, when there is no particular frequency range of interest, the symbol β is used.

The emitter current is related to V_{EB} exponentially. At room temperature, an increase in V_{EB} by approximately 60 mV increases the emitter current by a factor of 10. Because the base current is approximately proportional to the collector and emitter currents, they vary in the same way.

HISTORY

The bipolar point-contact transistor was invented in December 1947 at the Bell Telephone Laboratories by John Bardeen and Walter Brattain under the direction of William Shockley. The junction version known as the bipolar junction transistor (BJT), invented by Shockley in 1948, was for three decades the device of choice in the design of discrete and integrated circuits. Nowadays, the use of the BJT has declined in favor of CMOS technology in the design of digital integrated circuits. The incidental low performance BJTs inherent in CMOS ICs, however, are often utilized as bandgap voltage reference, silicon bandgap temperature sensor and to handle electrostatic discharge.

Germanium Transistors

The germanium transistor was more common in the 1950s and 1960s, and while it exhibits a lower “cut-off” voltage, typically around 0.2 V, making it more suitable for some applications, it also has a greater tendency to exhibit thermal runaway.

Early Manufacturing Techniques

Various methods of manufacturing bipolar transistors were developed.

Bipolar Transistors

- Point-contact transistor – first transistor ever constructed (December 1947), a bipolar transistor, limited commercial use due to high cost and noise.
    - Tetrode point-contact transistor – Point-contact transistor having two emitters. It became obsolete in the middle 1950s.
- Junction transistors
    - Grown-junction transistor – first bipolar *junction* transistor made. Invented by William Shockley at Bell Labs on June 23, 1948. Patent filed on June 26, 1948.

- Alloy-junction transistor – emitter and collector alloy beads fused to base. Developed at General Electric and RCA in 1951.
- Micro-alloy transistor (MAT) – high speed type of alloy junction transistor. Developed at Philco.
- Micro-alloy diffused transistor (MADT) – high speed type of alloy junction transistor, speedier than MAT, a diffused-base transistor. Developed at Philco.
- Post-alloy diffused transistor (PADT) – high speed type of alloy junction transistor, speedier than MAT, a diffused-base transistor. Developed at Philips.
- Tetrode transistor – high speed variant of grown-junction transistor or alloy junction transistor with two connections to base.
- Surface-barrier transistor – high-speed metal barrier junction transistor. Developed at Philco in 1953.
- Drift-field transistor – high speed bipolar junction transistor. Invented by Herbert Kroemer at the Central Bureau of Telecommunications Technology of the German Postal Service, in 1953.
- Spacistor – circa 1957.
- Diffusion transistor – modern type bipolar junction transistor. Prototypes developed at Bell Labs in 1954.
- Diffused-base transistor – first implementation of diffusion transistor.
- Mesa transistor – Developed at Texas Instruments in 1957.
- Planar transistor – the bipolar junction transistor that made mass-produced monolithic integrated circuits possible. Developed by Jean Hoerni at Fairchild in 1959.
- Epitaxial transistor – a bipolar junction transistor made using vapor phase deposition. See epitaxy. Allows very precise control of doping levels and gradients.

THEORY AND MODELING

Transistors can be thought of as two diodes (P–N junctions) sharing a common region that minority carriers can move through. A PNP BJT will function like two diodes that share an N-type cathode region, and the NPN like two diodes sharing a P-type anode region. Connecting two diodes with wires will not make a transistor, since minority carriers will not be able to get from one P–N junction to the other through the wire.

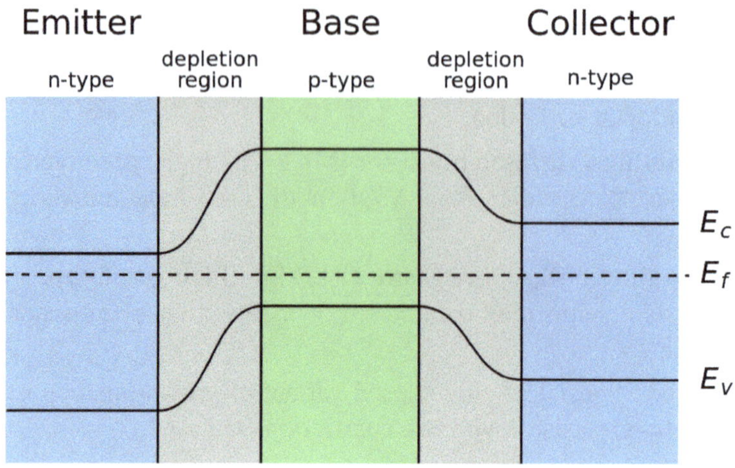

Band diagram for NPN transistor at equilibrium.

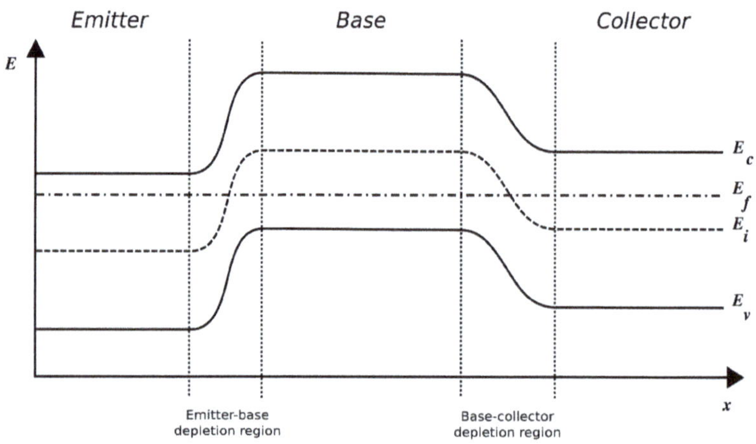

Band diagram for NPN transistor in active mode, showing injection of electrons from emitter to base, and their overshoot into the collector.

Both types of BJT function by letting a small current input to the base control an amplified output from the collector. The result is that the transistor makes a good switch that is controlled by its base input. The BJT also makes a good amplifier, since it can multiply a weak input signal to about 100 times its original strength. Networks of transistors are used to make powerful amplifiers with many different applications. In the discussion below, focus is on the NPN bipolar transistor. In the NPN transistor in what is called active mode, the base–emitter voltage V_{BE} and collector–base voltage V_{CB} are positive, forward biasing the emitter–base junction and reverse-biasing the collector–base junction. In the active mode of operation, electrons are injected from the forward biased n-type emitter region into the p-type base where they diffuse as minority carriers to the reverse-biased n-type collector and are swept away by the electric field in the reverse-biased collector–base junction. For a figure describing forward and reverse bias, see semiconductor diodes.

Large-Signal Models

In 1954, Jew*ell Ja*mes Ebers and John L. Moll introduced their mathematical model of transistor currents:

Ebers–Moll Model

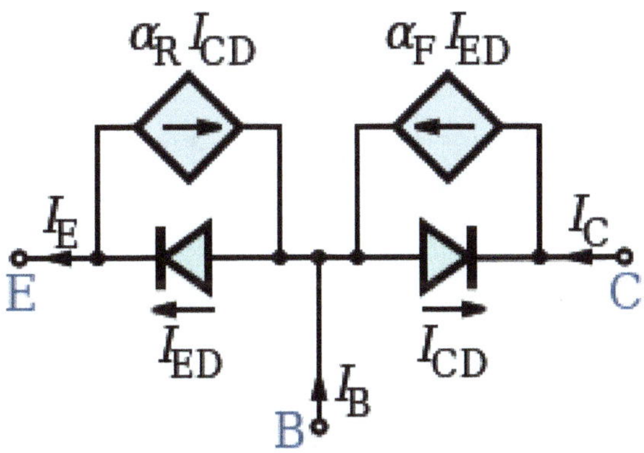

Ebers–Moll model for an NPN transistor * I_B, I_C, I_E: base, collector and emitter currents * I_{CD}, I_{ED}: collector and emitter diode currents * α_F, α_R: forward and reverse common-base current gains.

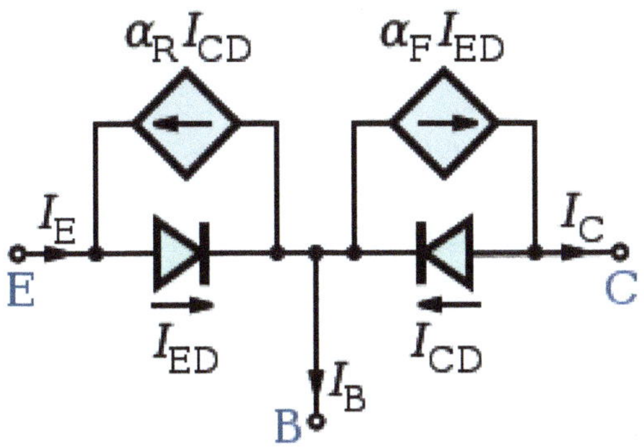

Ebers–Moll model for a PNP transistor.

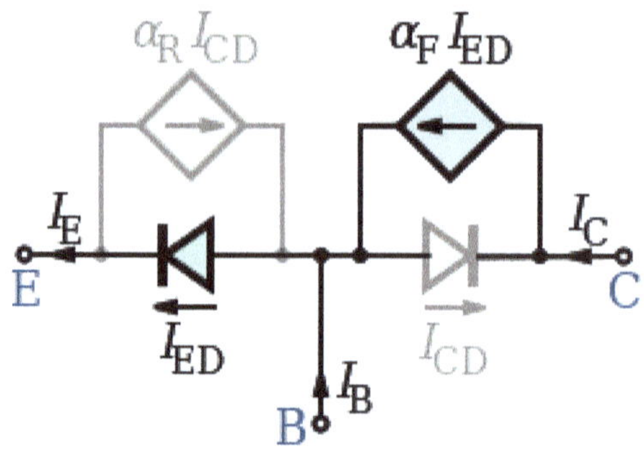

Approximated Ebers–Moll model for an NPN transistor in the forward active mode. The collector diode is reverse-biased so I_{CD} is virtually zero. Most of the emitter diode current (α_F is nearly 1) is drawn from the collector, providing the amplification of the base current.

The DC emitter and collector currents in active mode are well modeled by an approximation to the Ebers–Moll model:

$$I_E = I_{ES}\left(e^{\frac{V_{BE}}{V_T}} - 1\right)$$

$$I_C = \alpha_F I_E$$

$$I_B = (1 - \alpha_F)\, I_E$$

The base internal current is mainly by diffusion (see Fick's law) and

$$J_{n\,(\text{base})} = \frac{1}{W} q D_n n_{bo} e^{\frac{V_{EB}}{V_T}}$$

where

- V_T is the thermal voltage kT/q (approximately 26 mV at 300 K ≈ room temperature).
- I_E is the emitter current
- I_C is the collector current
- α_F is the common base forward short-circuit current gain (0.98 to 0.998)
- I_{ES} is the reverse saturation current of the base–emitter diode (on the order of 10^{-15} to 10^{-12} amperes)
- V_{BE} is the base–emitter voltage
- D_n is the diffusion constant for electrons in the p-type base
- W is the base width

The α and forward β parameters are as described previously. A reverse β is sometimes included in the model.

The unapproximated Ebers–Moll equations used to describe the three currents in any operating region are given below. These equations are based on the transport model for a bipolar junction transistor.

$$i_C = I_S\left[\left(e^{\frac{V_{BE}}{V_T}} - e^{\frac{V_{BC}}{V_T}}\right) - \frac{1}{\beta_R}\left(e^{\frac{V_{BC}}{V_T}} - 1\right)\right]$$

$$i_B = I_S\left[\frac{1}{\beta_F}\left(e^{\frac{V_{BE}}{V_T}} - 1\right) + \frac{1}{\beta_R}\left(e^{\frac{V_{BC}}{V_T}} - 1\right)\right]$$

$$i_E = I_S\left[\left(e^{\frac{V_{BE}}{V_T}} - e^{\frac{V_{BC}}{V_T}}\right) + \frac{1}{\beta_F}\left(e^{\frac{V_{BE}}{V_T}} - 1\right)\right]$$

where

- i_C is the collector current
- i_B is the base current
- i_E is the emitter current
- β_F is the forward common emitter current gain (20 to 500)
- β_R is the reverse common emitter current gain (0 to 20)
- I_S is the reverse saturation current (on the order of 10^{-15} to 10^{-12} amperes)
- V_T is the thermal voltage (approximately 26 mV at 300 K ≈ room temperature).
- V_{BE} is the base–emitter voltage
- V_{BC} is the base–collector voltage

Base-width Modulation

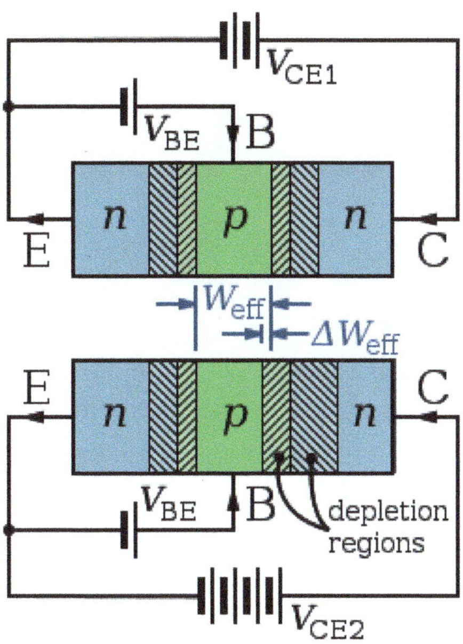

Top: NPN base width for low collector-base reverse bias; Bottom: narrower NPN base width for large collector-base reverse bias. Hashed regions are depleted regions.

As the collector–base voltage $(V_{CB} = V_{CE} - V_{BE})$ varies, the collector–base depletion region varies in size. An increase in the collector–base voltage, for example, causes a greater reverse bias across the collector–base junction, increasing the collector–base depletion region width, and decreasing the width of the base. This variation in base width often is called the "Early effect" after its discoverer James M. Early.

Narrowing of the base width has two consequences:

- There is a lesser chance for recombination within the "smaller" base region.
- The charge gradient is increased across the base, and consequently, the current of minority carriers injected across the emitter junction increases.

Both factors increase the collector or "output" current of the transistor in response to an increase in the collector–base voltage.

In the forward-active region, the Early effect modifies the collector current (i_C) and the forward common emitter current gain (β_F) as given by:

$$i_C = I_S\, e^{\frac{v_{BE}}{V_T}} \left(1 + \frac{V_{CE}}{V_A}\right)$$

$$\beta_F = \beta_{F0} \left(1 + \frac{V_{CB}}{V_A}\right)$$

$$r_o = \frac{V_A}{I_C}$$

where:

V_{CE} is the collector–emitter voltage

V_A is the Early voltage (15 V to 150 V)

β_{F0} is forward common-emitter current gain when V_{CB} = 0 V

r_o is the output impedance

I_C is the collector current

Punchthrough

When the base–collector voltage reaches a certain (device-specific) value, the base–collector depletion region boundary meets the base–emitter depletion region boundary. When in this state the transistor effectively has no base. The device thus loses all gain when in this state.

Gummel–Poon Charge-Control Model

The Gummel–Poon model is a detailed charge-controlled model of BJT dynamics, which has been adopted and elaborated by others to explain transistor dynamics in greater detail than the terminal-based models typically do. This model also includes the dependence of transistor β-values upon the direct current levels in the transistor, which are assumed current-independent in the Ebers–Moll model.

Small-signal Models

Hybrid-pi Model

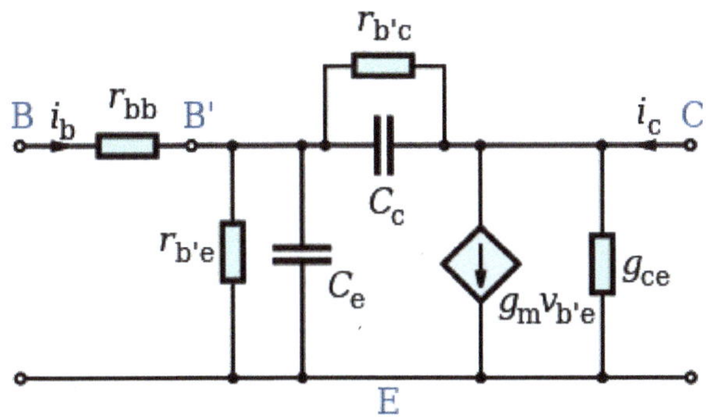

Hybrid-pi model.

The hybrid-pi model is a popular circuit model used for analyzing the small signal behavior of bipolar junction and field effect transistors. Sometimes it is also called *Giacoletto model* because it was introduced by L.J. Giacoletto in 1969. The model can be quite accurate for low-frequency circuits and can easily be adapted for higher-frequency circuits with the addition of appropriate inter-electrode capacitances and other parasitic elements.

h-parameter Model

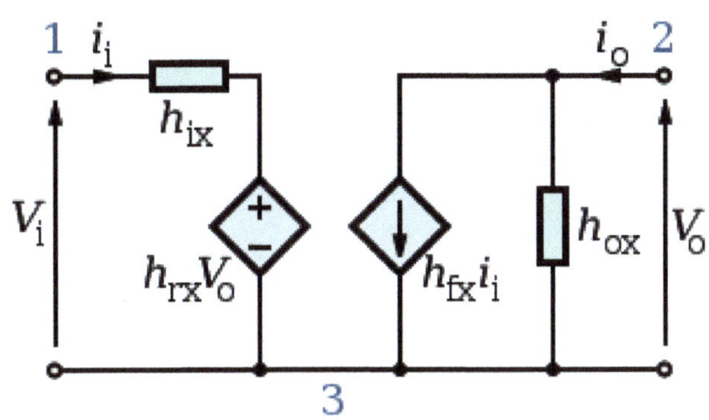

Generalized h-parameter model of an NPN BJT.

Replace ***x*** *with* ***e***, ***b*** *or* ***c*** *for CE, CB and CC topologies respectively.*

Another model commonly used to analyze BJT circuits is the *h-parameter* model, closely related to the hybrid-pi model and the y-parameter two-port, but using input current and output voltage as independent variables, rather than input and output voltages. This two-port network is particularly suited to BJTs as it lends itself easily to the analysis of circuit behaviour, and may be used to develop further accurate models. As shown, the term, x, in the model represents a different BJT lead depending on the topology used. For common-emitter mode the various symbols take on the specific values as:

- Terminal 1, base
- Terminal 2, collector
- Terminal 3 (common), emitter; giving x to be e
- i_i, base current (i_b)
- i_o, collector current (i_c)
- V_{in}, base-to-emitter voltage (V_{BE})
- V_o, collector-to-emitter voltage (V_{CE})

and the h-parameters are given by:

- $h_{ix} = h_{ie}$, the input impedance of the transistor (corresponding to the base resistance r_{pi}).
- $h_{rx} = h_{re}$, represents the dependence of the transistor's I_B–V_{BE} curve on the value of V_{CE}. It is usually very small and is often neglected (assumed to be zero).
- $h_{fx} = h_{fe}$, the current-gain of the transistor. This parameter is often specified as h_{FE} or the DC current-gain (β_{DC}) in datasheets.
- $h_{ox} = 1/h_{oe}$, the output impedance of transistor. The parameter h_{oe} usually corresponds to the output admittance of the bipolar transistor and has to be inverted to convert it to an impedance.

As shown, the h-parameters have lower-case subscripts and hence signify AC conditions or analyses. For DC conditions they are specified in upper-case. For the CE topology, an approximate h-parameter model is commonly used which further simplifies the circuit analysis. For this the h_{oe} and h_{re} parameters are neglected (that is, they are set to infinity and zero, respectively). The h-parameter model as shown is suited to low-frequency, small-signal analysis. For high-frequency analyses the inter-electrode capacitances that are important at high frequencies must be added.

Etymology of h_{FE}

The *h* refers to its being an h-parameter, a set of parameters named for their origin in a *hybrid equivalent circuit* model. *F* is from *forward current amplification* also called the current gain. *E* refers to the transistor operating in a *common emitter* (CE) configuration. Capital letters used in the subscript indicate that h_{FE} refers to a direct current circuit.

Industry Models

The Gummel–Poon SPICE model is often used, but it suffers from several limitations. These have been addressed in various more advanced models: Mextram, VBIC, HICUM, Modella.

APPLICATIONS

The BJT remains a device that excels in some applications, such as discrete circuit design, due to the very wide selection of BJT types available, and because of its high transconductance and output resistance compared to MOSFETs.

The BJT is also the choice for demanding analog circuits, especially for very-high-frequency applications, such as radio-frequency circuits for wireless systems.

High Speed Digital Logic

Emitter-coupled logic (ECL) use BJTs.

Bipolar transistors can be combined with MOSFETs in an integrated circuit by using a BiCMOS process of wafer fabrication to create circuits that take advantage of the application strengths of both types of transistor.

Amplifiers

Electronic amplifier

The transistor parameters α and β characterizes the current gain of the BJT. It is this gain that allows BJTs to be used as the building blocks of electronic amplifiers. The three main BJT amplifier topologies are:

- Common emitter
- Common base
- Common collector

Temperature sensors

Silicon bandgap temperature sensor

Because of the known temperature and current dependence of the forward-biased base–emitter junction voltage, the BJT can be used to measure temperature by subtracting two voltages at two different bias currents in a known ratio.

Logarithmic converters

Because base–emitter voltage varies as the logarithm of the base–emitter and collector–emitter currents, a BJT can also be used to compute logarithms and anti-logarithms. A diode can also perform these nonlinear functions but the transistor provides more circuit flexibility.

VULNERABILITIES

Exposure of the transistor to ionizing radiation causes radiation damage. Radiation causes a buildup of 'defects' in the base region that act as recombination centers. The resulting reduction in minority carrier lifetime causes gradual loss of gain of the transistor.

Power BJTs are subject to a failure mode called secondary breakdown, in which excessive current and normal imperfections in the silicon die cause portions of the silicon inside the device to become disproportionately hotter than the others. The electrical resistivity of doped silicon, like other semiconductors, has a negative temperature coefficient, meaning that it conducts more current at higher temperatures. Thus, the hottest part of the die conducts the most current, causing its conductivity to increase, which then causes it to become progressively hotter again, until the device fails internally. The thermal runaway process associated with secondary breakdown, once triggered, occurs almost instantly and may catastrophically damage the transistor package.

If the emitter-base junction is reverse biased into avalanche or Zener mode and charge flows for a short period of time, the current gain of the BJT will be permanently degraded.

REFERENCES

1. Paul Horowitz and Winfield Hill (1989). The Art of Electronics (2nd ed.). Cambridge University Press. ISBN 978-0-521-37095-0.
2. Juin Jei Liou and Jiann S. Yuan (1998). Semiconductor Device Physics and Simulation. Springer. ISBN 0-306-45724-5.
3. General Electric (1962). Transistor Manual (6th ed.). p. 12.If the principle of space charge neutrality is used in the analysis of the transistor, it is evident that the collector current is controlled by means of the positive charge (hole concentration) in the base region. ... When a transistor is used at higher frequencies, the fundamental limitation is the time it takes the carriers to diffuse across the base region... (same in 4th and 5th editions)
4. Paolo Antognetti and Giuseppe Massobrio (1993). Semiconductor Device Modeling with Spice. McGraw–Hill Professional. ISBN 0-07-134955-3.
5. Alphonse J. Sistino (1996). Essentials of electronic circuitry. CRC Press. p. 64. ISBN 978-0-8247-9693-8.
6. Alphonse J. Sistino (1996). Essentials of electronic circuitry. CRC Press. p. 102. ISBN 978-0-8247-9693-8.
7. D.V. Morgan, Robin H. Williams (Editors) (1991). Physics and Technology of Heterojunction Devices. London: Institution of Electrical Engineers (Peter Peregrinus Ltd.). ISBN 0-86341-204-1.
8. Peter Ashburn (2003). SiGe Heterojunction Bipolar Transistors. New York: Wiley. Chapter 10. ISBN 0-470-84838-3.
9. Paul Horowitz and Winfield Hill (1989). The Art of Electronics (2nd ed.). Cambridge University Press. pp. 62–66. ISBN 978-0-521-37095-0.

10. 1947: Invention of the Point-Contact Transistor - The Silicon Engine - Computer History Museum. Retrieved August 10, 2016.

11. 1948: Conception of the Junction Transistor - The Silicon Engine - Computer History Museum. Retrieved August 10, 2016.

12. Third case study – the solid state advent ArchivedSeptember 27, 2007, at the Wayback Machine. (PDF)

13. Transistor Museum, Historic Transistor Photo Gallery, Bell Labs Type M1752

14. Morris, Peter Robin (1990).4.2. A History of the World Semiconductor Industry. IEE History of Technology Series 12. London: Peter Peregrinus Ltd. p. 29. ISBN 0-86341-227-0.

15. Transistor Museum Photo Gallery RCA TA153. Retrieved August 10, 2016.

16. High Speed Switching Transistor Handbook (2nd ed.). Motorola. 1963. p. 17.

17. Transistor Museum, Historic Transistor Photo Gallery, Western Electric 3N22

18. The Tetrode Power Transistor PDF

19. Transistor Museum Photo Gallery Philco A01 Germanium Surface Barrier Transistor. Retrieved August 10, 2016.

20. Transistor Museum Photo Gallery Germanium Surface Barrier Transistor. Retrieved August 10, 2016.

21. Herb's Bipolar Transistors IEEE Transactions on Electron Devices, vol. 48, no. 11, November 2001 PDF

22. Influence of Mobility and Lifetime Variations on Drift-Field Effects in Silicon-Junction Devices PDF

23. Transistor Museum Photo Gallery Bell Labs Prototype Diffused Base Germanium Silicon Transistor. Retrieved August 10, 2016.

24. Transistor Museum Photo Gallery Fairchild 2N1613 Early Silicon Planar Transistor. Retrieved August 10, 2016.

25. 1960: Epitaxial Deposition Process Enhances Transistor Performance - The Silicon Engine - Computer History Museum. Retrieved August 10, 2016.

26. J.J. Ebers and J.L Moll (1954)Large-signal behavior of junction transistors, Proceedings of the Institute of Radio Engineers, **42** (12) : 1761–1772.

27. Adel S. Sedra and Kenneth C. Smith (1987). Microelectronic Circuits, second ed. p. 903. ISBN 0-03-007328-6.

28. A.S. Sedra and K.C. Smith (2004). Microelectronic Circuits (5th ed.). New York: Oxford. Eqs. 4.103–4.110, p. 305. ISBN 0-19-514251-9.

29. H. K. Gummel and R. C. Poon,An integral charge control model of bipolar transistors, Bell Syst. Tech. J., vol. 49, pp. 827–852, May–June 1970

30. Bipolar Junction Transistors. Retrieved August 10, 2016.

31. A.S. Sedra and K.C. Smith (2004). Microelectronic Circuits (5th ed.). New York: Oxford. p. 509. ISBN 0-19-514251-9.

32. http://www.silvaco.com/content/kbase/smartspice_device_models.pdf

33. Gennady Gildenblat, ed. (2010). Compact Modeling: Principles, Techniques and Applications. Springer Science & Business Media. Part II: Compact Models of Bipolar Junction Transistors, pp. 167-267 cover Mextram and HiCuM in-depth. ISBN 978-90-481-8614-3.
34. Michael Schröter (2010). Compact Hierarchical Bipolar Transistor Modeling with Hicum. World Scientific. ISBN 978-981-4273-21-3.
35. http://joerg-berkner.de/Fachartikel/pdf/2002_ICCAP_UM_Berkner_Compact_Models_4_BJTs.pdf
36. IC Temperature Sensors Find the Hot Spots - Application Note - Maxim. maxim-ic.com. February 21, 2002. Retrieved August 10, 2016.

CHAPTER 8

JFET and MOSFET

Chapter Contents

- **Introduction**
- **History**
- **Structure**
- **Function**
- **Mathematical Model**
- **Mosfet**

A field-effect transistor (FET) is a three-terminal device that can control the amount of electricity allowed to pass through. FET transistors are able to switch or control electricity very rapidly. Some field-effect transistors are able to handle large amounts of electric power, making it useful in a wide variety of electronic and electrical equipment used in consumer, commercial, and military applications.

The FET transistor usually has source, drain, and gate terminals. Electrical resistance between the source and the drain can be manipulated by changing the electrical potential, or voltage, between the gate and the source. Given this effect, the FET transistor can be used in any electronic circuit, such as power supplies, amplifiers, and receivers.

A MOSFET (Metal Oxide Semiconductor Field Effect Transistor) is a semiconductor device. A MOSFET is most commonly used in the field of power electronics. A semiconductor is made of manufactured material that acts neither like a insulator nor a conductor. An insulator is a natural material that will not conduct electricity, such as a dry piece of wood. A conductor is a natural material that conducts or passes electricity. Metals are the most common examples of conductors. Semiconductor material from which devices like a MOSFET are made exhibit both insulation like properties and conduction like properties. Most importantly, semiconductors are designed such that the conduction or insulation properties can be controlled.

INTRODUCTION

The **junction gate field-effect transistor** (**JFET** or **JUGFET**) is the simplest type of field-effect transistor. They are three-terminal semiconductor devices that can be used as electronically-controlled switches, amplifiers, or voltage-controlled resistors.

Unlike bipolar transistors, JFETs are exclusively voltage-controlled in that they do not need a biasing current. Electric charge flows through a semiconducting channel between *source* and *drain* terminals. By applying a reverse bias voltage to a *gate* terminal, the channel is "pinched", so that the electric current is impeded or switched off completely. A JFET is usually on when there is no potential difference between its gate and source terminals. If a potential difference of the proper polarity is applied between its gate and source terminals, the JFET will be more resistive to current flow, which means less current would flow in the channel between the source and drain terminals. Thus, JFETs are sometimes referred to as depletion-mode devices.

JFETs can have an n-type or p-type channel. In the n-type, if the voltage applied to the gate is less than that applied to the source, the current will be reduced (similarly in the p-type, if the voltage applied to the gate is *greater* than that applied to the source). A JFET has a large input impedance (sometimes on the order of 10^{10} ohms), which means that it has a negligible effect on external components or circuits connected to its gate.

HISTORY

A succession of FET-like devices were patented by Julius Lilienfeld in the 1920s and 1930s.However, materials science and fabrication technology would require decades of advances before FETs could actually be made. In 1947, researchers John Bardeen, Walter Houser Brattain, and William Shockley failed in their repeated attempts to make a FET. They discovered the point-contact transistor in the course of trying to diagnose the reasons for their failures. The first practical JFETs were made a decade later.

STRUCTURE

The JFET is a long channel of semiconductor material, doped to contain an abundance of positive charge carriers or holes (*p-type*), or of negative carriers or electrons (*n-type*). Ohmic contacts at each end form the source (S) and the drain (D). A pn-junction is formed on one or both sides of the channel, or surrounding it, using a region with doping opposite to that of the channel, and biased using an ohmic gate contact (G).

FUNCTION

JFET operation can be compared to that of a garden hose. The flow of water through a hose can be controlled by squeezing it to reduce the cross section and the flow of electric charge through a JFET is controlled by constricting the current-carrying channel. The current also depends on the electric field between source and drain (analogous to the difference in pressure on either end of the hose).

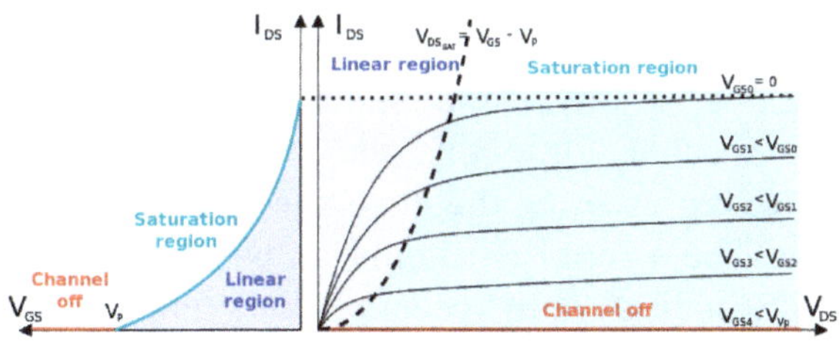

I–V characteristics and output plot of an n-channel JFET.

Constriction of the conducting channel is accomplished using the field effect: a voltage between the gate and the source is applied to reverse bias the gate-source pn-junction, thereby widening the depletion layer of this junction (see top figure), encroaching upon the conducting channel and restricting its cross-sectional area. The depletion layer is so-called because it is depleted of mobile carriers and so is electrically non-conducting for practical purposes.

When the depletion layer spans the width of the conduction channel, *pinch-off* is achieved and drain-to-source conduction stops. Pinch-off occurs at a particular reverse bias (V_{GS}) of the gate-source junction. The pinch-off voltage (V_p) varies considerably, even among devices of the same type. For example, $V_{GS(off)}$ for the Temic J202 device varies from −0.8 V to −4 V. Typical values vary from −0.3 V to −10 V.

To switch off an **n**-channel device requires a **n**egative gate-source voltage (V_{GS}). Conversely, to switch off a **p**-channel device requires **p**ositive V_{GS}.

In normal operation, the electric field developed by the gate blocks source-drain conduction to some extent.

Some JFET devices are symmetrical with respect to the source and drain.

Schematic symbols

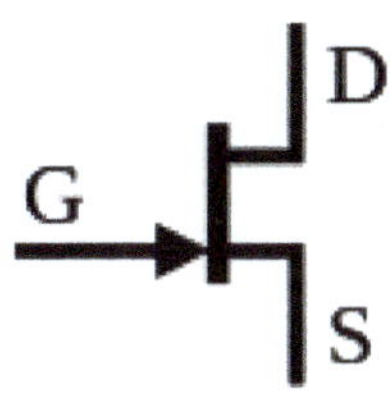

Circuit symbol for an n-Channel JFET.

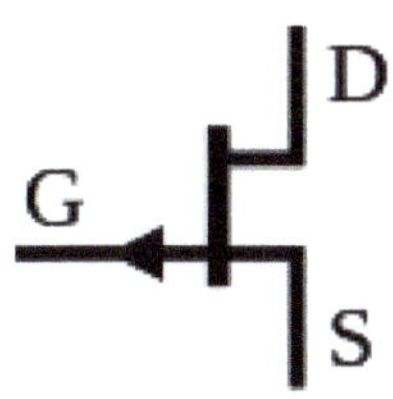

Circuit symbol for a p-Channel JFET.

The JFET gate is sometimes drawn in the middle of the channel (instead of at the drain or source electrode as in these examples). This symmetry suggests that “drain” and “source” are interchangeable, so the symbol should be used only for those JFETs where they are indeed interchangeable. Officially, the style of the symbol should show the component inside a circle (representing the envelope of a discrete device). This is true in both the US and Europe. The symbol is usually drawn without the circle when drawing schematics of integrated circuits. More recently, the symbol is often drawn without its circle even for discrete devices.

In every case the arrow head shows the polarity of the P-N junction formed between the channel and the gate. As with an ordinary diode, the arrow points from P to N, the direction of conventional current when forward-biased. An English mnemonic is that the arrow of an N-channel device “points **in**”.

Comparison with other transistors

At room temperature, JFET gate current (the reverse leakage of the gate-to-channel junction) is comparable to that of a MOSFET (which has insulating oxide between gate and channel), but much less than the base current of a bipolar junction transistor. The JFET has higher gain (transconductance) than the MOSFET, as well as lower flicker noise, and is therefore used in some low-noise, high input-impedance op-amps.

MATHEMATICAL MODEL

The current in N-JFET due to a small voltage V_{DS} (that is, in the linear ohmic region) is given by treating the channel as a rectangular bar of material of electrical conductivity $qN_d\mu_n$:

$$I_D = \frac{bW}{L} qN_d\mu_n V_{DS}$$

where

I_D = drain–source current

b = channel thickness for a given gate voltage

W = channel width

L = channel length

q = electron charge = 1.6 x 10^{-19} C

μ_n = electron mobility

N_d = n-type doping (donor) concentration.

Linear Region

Then the drain current in the *linear region* can be expressed as:

$$I_D = \frac{bW}{L} qN_d\mu_n V_{DS} = \frac{aW}{L} qN_d\mu_n \left(1 - \sqrt{\frac{V_{GS}}{V_P}}\right) V_{DS}$$

In terms of I_{DSS}, the drain current can also be:

$$I_D = \frac{2I_{DSS}}{V_P^2}\left(V_{GS} - V_P - \frac{V_{DS}}{2}\right)V_{DS}$$

Saturation Region

The drain current in the *saturation region* is often approximated in terms of gate bias as:

$$I_{DS} = I_{DSS}\left(1 - \frac{V_{GS}}{V_P}\right)^2$$

where

I_{DSS} is the saturation current at zero gate–source voltage, i.e. the maximum current which can flow through the FET from drain to source at any (permissible) drain-to-source voltage (see, e. g., the I-V characteristics diagram above).

In the *saturation region*, the JFET drain current is most significantly affected by the gate–source voltage and barely affected by the drain–source voltage.

If the channel doping is uniform, such that the depletion region thickness will grow in proportion to the square root of the absolute value of the gate–source voltage, then the channel thickness b can be expressed in terms of the zero-bias channel thickness a as:

$$b = a\left(1 - \sqrt{\frac{V_{GS}}{V_P}}\right)$$

where

V_P is the pinch-off voltage, the gate–source voltage at which the channel thickness goes to zero

a is the channel thickness at zero gate–source voltage.

MOSFET

The **metal–oxide–semiconductor field-effect transistor** (**MOSFET**, **MOS-FET**, or **MOS FET**) is a type of transistor used for amplifying or switching electronic signals.

Although the MOSFET is a four-terminal device with source (S), gate (G), drain (D), and body (B) terminals, the body (or substrate) of the MOSFET is often connected to the source terminal, making it a three-terminal device like other field-effect transistors. Because these two terminals are normally connected to each other (short-circuited) internally, only three terminals appear in electrical diagrams.

The basic principle of the field-effect transistor was first patented by Julius Edgar Lilienfeld in 1925. The main advantage of a MOSFET over a regular transistor is that it requires very little current to turn on (less than 1mA), while delivering a much higher current to a load (10 to 50A or more).

In *enhancement mode* MOSFETs, a voltage drop across the oxide induces a conducting channel between the source and drain contacts *via* the field effect. The term "enhancement mode" refers to the increase of conductivity with increase in oxide field that adds carriers to the channel, also referred to as the *inversion layer*. The channel can contain electrons (called an nMOSFET or nMOS), or holes (called a pMOSFET or pMOS), opposite in type to the substrate, so nMOS is made with a p-type substrate, and pMOS with an n-type substrate (see article on semiconductor devices). In the less common *depletion mode* MOSFET, detailed later on, the channel consists of carriers in a surface impurity layer of opposite type to the substrate, and conductivity is decreased by application of a field that depletes carriers from this surface layer.

The "metal" in the name MOSFET is now often a misnomer because the previously metal gate material is now often a layer of polysilicon (polycrystalline silicon). Aluminium had been the gate material until the mid-1970s, when polysilicon became dominant, due to its capability to form self-aligned gates. Metallic gates are regaining popularity, since it is difficult to increase the speed of operation of transistors without metal gates. Likewise, the "oxide" in the name can be a misnomer, as different dielectric materials are used with the aim of obtaining strong channels with smaller applied voltages.

An insulated-gate field-effect transistor or IGFET is a related term almost synonymous with MOSFET. The term may be more inclusive, since many "MOSFETs" use a gate that is not metal, and a gate insulator that is not oxide. Another synonym is MISFET for metal–insulator–semiconductor FET.

The MOSFET is by far the most common transistor in both digital and analog circuits, though the bipolar junction transistor was at one time much more common.

The basic principle of this kind of transistor was first patented by Julius Edgar Lilienfeld in 1925. Twenty five years later, when Bell Telephone attempted to patent the junction transistor, they found Lilienfeld already holding a patent, worded in a way that would include all types of transistors. Bell Labs was able to work out an agreement with Lilienfeld, who was still alive at that time (it is not known if they paid him money or not). It was at that time the Bell Labs version was given the name bipolar junction transistor, or simply junction transistor, and Lilienfeld's design took the name field effect transistor.

In 1959, Dawon Kahng and Martin M. (John) Atalla at Bell Labs invented the metal–oxide–semiconductor field-effect transistor (MOSFET) as an offshoot to the patented FET design. Operationally and structurally different from the bipolar junction transistor, the MOSFET was made by putting an insulating layer on the surface of the semiconductor and then placing a metallic gate electrode on that. It used crystalline silicon for the semiconductor and a thermally oxidized layer of silicon dioxide for the insulator. The silicon MOSFET did not generate localized electron traps at the interface between the silicon and its native oxide layer, and thus was inherently free from the trapping and scattering of carriers that had impeded the performance of earlier field-effect transistors.

Discrete power MOSFETs are currently widely used as low voltage switches.

Composition

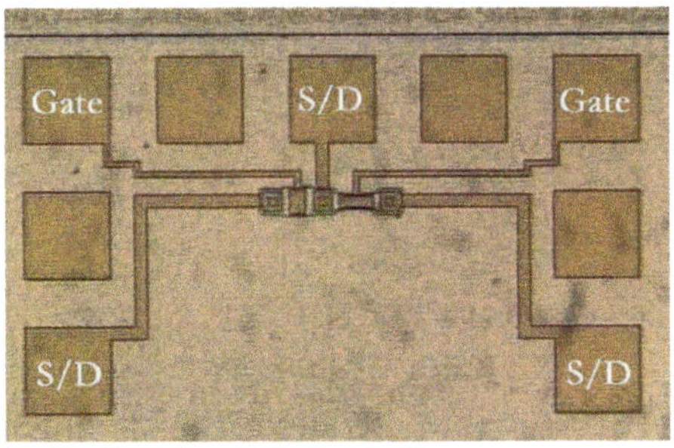

Photomicrograph of two metal-gate MOSFETs in a test pattern. Probe pads for two gates and three source/drain nodes are labeled.

Usually the semiconductor of choice is silicon, but some chip manufacturers, most notably IBM and Intel, recently started using a chemical compound of silicon and germanium (SiGe) in MOSFET channels. Unfortunately, many semiconductors with better electrical properties than silicon, such as gallium arsenide, do not form good semiconductor-to-insulator interfaces, and thus are not suitable for MOSFETs. Research continues on creating insulators with acceptable electrical characteristics on other semiconductor material.

In order to overcome the increase in power consumption due to gate current leakage, a high-κ dielectric is used instead of silicon dioxide for the gate insulator, while polysilicon is replaced by metal gates (see Intel announcement).

The gate is separated from the channel by a thin insulating layer, traditionally of silicon dioxide and later of silicon oxynitride. Some companies have started to introduce a high-κ dielectric + metal gate combination in the 45 nanometer node.

When a voltage is applied between the gate and body terminals, the electric field generated penetrates through the oxide and creates an "inversion layer" or "channel" at the semiconductor-insulator interface. The inversion channel is of the same type, p-type or n-type, as the source and drain, and thus it provides a channel through which current can pass. Varying the voltage between the gate and body modulates the conductivity of this layer and thereby controls the current flow between drain and source. This is known as enhancement mode.

Operation

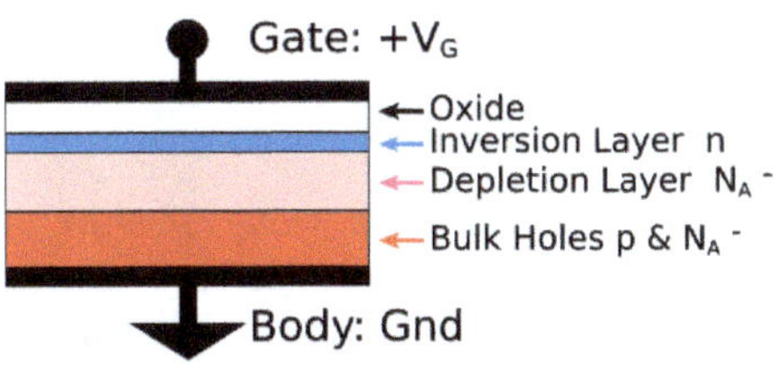

Metal–oxide–semiconductor structure on p-type silicon.

Metal–oxide–Semiconductor Structure

The traditional metal–oxide–semiconductor (MOS) structure is obtained by growing a layer of silicon dioxide (SiO_2) on top of a silicon substrate and depositing a layer of metal or polycrystalline silicon (the latter is commonly used). As the silicon dioxide is a dielectric material, its structure is equivalent to a planar capacitor, with one of the electrodes replaced by a semiconductor.

When a voltage is applied across a MOS structure, it modifies the distribution of charges in the semiconductor. If we consider a p-type semiconductor (with N_A the density of acceptors, p the density of holes; $p = N_A$ in neutral bulk), a positive voltage, V_{GB}, from gate to body (see figure) creates a depletion layer by forcing the positively charged holes away from the gate-insulator/semiconductor interface, leaving exposed a carrier-free region of immobile, negatively charged acceptor ions (see doping (semiconductor)). If V_{GB} is high enough, a high concentration of negative charge carriers forms in an *inversion layer* located in a thin layer next to the interface between the semiconductor and the insulator. Unlike the MOSFET, where the inversion layer electrons are supplied rapidly from the source/drain electrodes, in the MOS capacitor they are produced much more slowly by thermal generation through carrier generation and recombination centers in the depletion region. Conventionally, the gate voltage at which the volume density of electrons in the inversion layer is the same as the volume density of holes in the body is called the threshold voltage. When the voltage between transistor gate and source (V_{GS}) exceeds the threshold voltage (V_{th}), it is known as overdrive voltage.

This structure with p-type body is the basis of the n-type MOSFET, which requires the addition of n-type source and drain regions.

Structure and Channel Formation

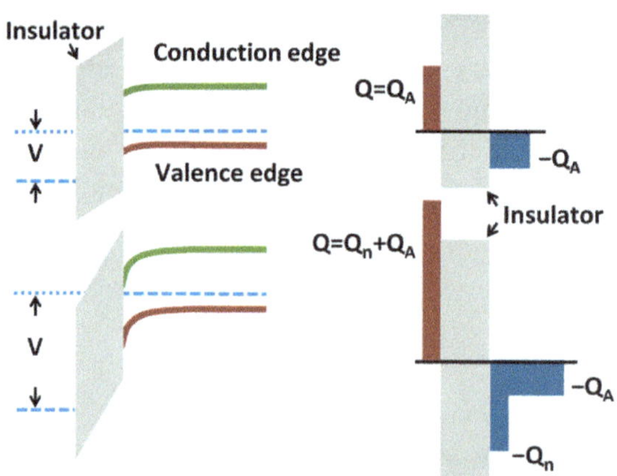

Channel formation in nMOS MOSFET shown as band diagram: Top panels: An applied gate voltage bends bands, depleting holes from surface (left). The charge inducing the bending is balanced by a layer of negative acceptor-ion charge (right). Bottom panel: A larger applied voltage further depletes holes but conduction band lowers enough in energy to populate a conducting channel.

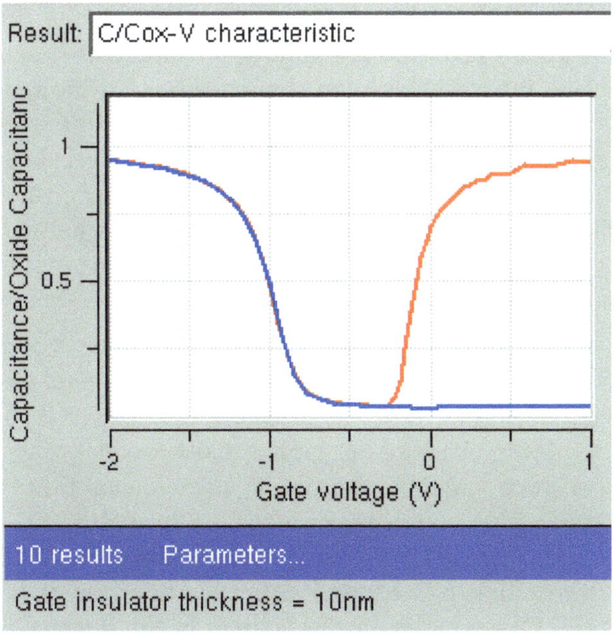

C–V profile for a bulk MOSFET with different oxide thickness. The leftmost part of the curve corresponds to accumulation. The valley in the middle corresponds to depletion. The curve on the right corresponds to inversion.

A MOSFET is based on the modulation of charge concentration by a MOS capacitance between a *body* electrode and a *gate* electrode located above the body and insulated from all other device regions by a gate dielectric layer which in the case of a MOSFET is an oxide, such as silicon dioxide. If dielectrics other than an oxide such as silicon dioxide (often referred to as oxide) are employed the device may be referred to as a metal–insulator–semiconductor FET (MISFET). Compared to the MOS capacitor, the MOSFET includes two additional terminals (*source* and *drain*), each connected to individual highly doped regions that are separated by the body region. These regions can be either p or n type, but they must both be of the same type, and of opposite type to the body region. The source and drain (unlike the body) are highly doped as signified by a "+" sign after the type of doping.

If the MOSFET is an n-channel or nMOS FET, then the source and drain are "n+" regions and the body is a "p" region. If the MOSFET is a p-channel or pMOS FET, then the source and drain are "p+" regions and the body is a "n" region. The source is so named because it is the source of the charge carriers (electrons for n-channel, holes for p-channel) that flow through the channel; similarly, the drain is where the charge carriers leave the channel.

The occupancy of the energy bands in a semiconductor is set by the position of the Fermi level relative to the semiconductor energy-band edges.

With sufficient gate voltage, the valence band edge is driven far from the Fermi level, and holes from the body are driven away from the gate.

At larger gate bias still, near the semiconductor surface the conduction band edge is brought close to the Fermi level, populating the surface with electrons in an *inversion layer* or *n-channel* at the interface between the p region and the oxide. This conducting channel extends between the source and the drain, and current is conducted through it when a voltage is applied between the two electrodes. Increasing the voltage on the gate leads to a higher electron density in the inversion layer and therefore increases the current flow between the source and drain. For gate voltages below the threshold value, the channel is lightly populated, and only a very small subthreshold leakage current can flow between the source and the drain.

When a negative gate-source voltage (positive source-gate) is applied, it creates a *p-channel* at the surface of the n region, analogous to the n-channel case, but with opposite polarities of charges and voltages. When a voltage less negative than the threshold value (a negative voltage for the p-channel) is applied between gate and source, the channel disappears and only a very small subthreshold current can flow between the source and the drain. The device may comprise a silicon on insulator (SOI) device in which a buried oxide (BOX) is formed below a thin semiconductor layer. If the channel region between the gate dielectric and a BOX region is very thin, the channel is referred to as an ultrathin channel (UTC) region with the source and drain regions formed on either side thereof in and/or above the thin semiconductor layer. Alternatively, the device may comprise a semiconductor on insulator (SEMOI) device in which semiconductors other than silicon are employed. Many alternative semiconductor materials may be employed. When the source and drain regions are formed above the channel in whole or in part, they are referred to as raised source/drain (RSD) regions.

Modes of Operation

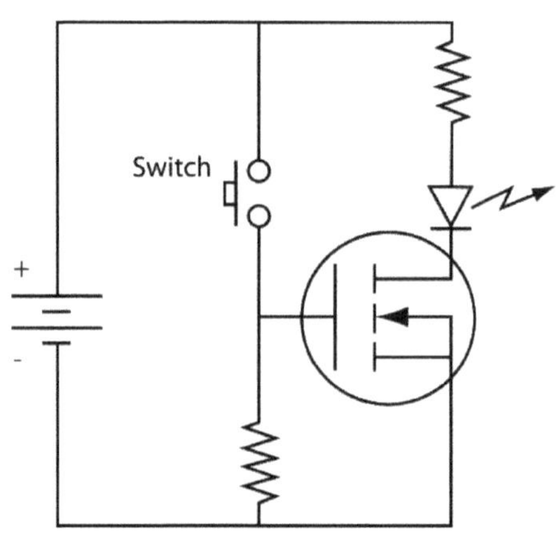

Example application of an N-Channel MOSFET. When the switch is pushed the LED lights up.

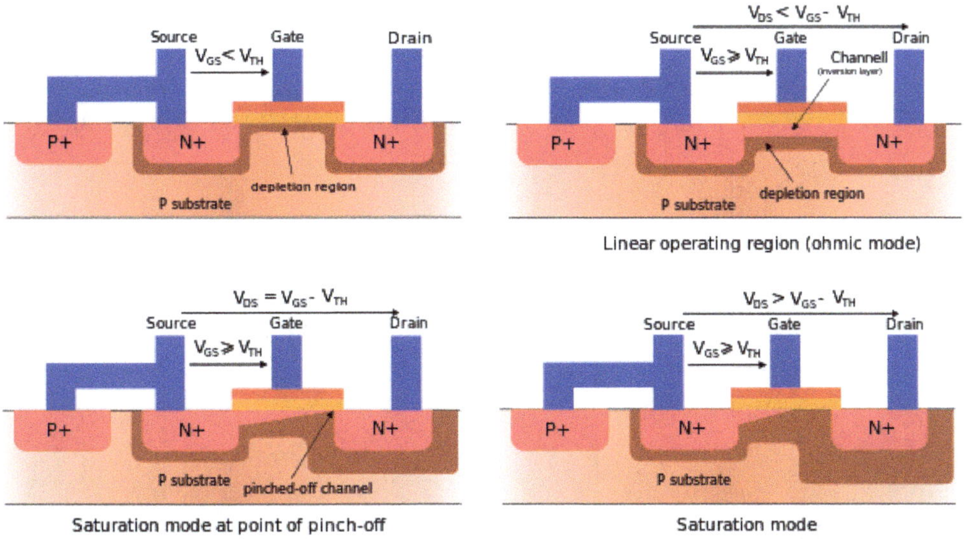

Ohmic contact to body to ensure no body bias; top left:subthreshold, top right:Ohmic mode, bottom left:Active mode at onset of pinch-off, bottom right: Active mode well into pinch-off – channel length modulation evident.

The operation of a MOSFET can be separated into three different modes, depending on the voltages at the terminals. In the following discussion, a simplified algebraic model is used. Modern MOSFET characteristics are more complex than the algebraic model presented here.

For an *enhancement-mode, n-channel MOSFET*, the three operational modes are:

Cutoff, subthreshold, or weak-inversion mode

When $V_{GS} < V_{\text{th}}$:

where V_{GS} is gate-to-source bias and V_{th} is the threshold voltage of the device.

According to the basic threshold model, the transistor is turned off, and there is no conduction between drain and source. A more accurate model considers the effect of thermal energy on the Fermi–Dirac distribution of electron energies which allow some of the more energetic electrons at the source to enter the channel and flow to the drain. This results in a subthreshold current that is an exponential function of gate–source voltage. While the current between drain and source should ideally be zero when the transistor is being used as a turned-off switch, there is a weak-inversion current, sometimes called subthreshold leakage.

In weak inversion where the source is tied to bulk, the current varies exponentially with V_{GS} as given approximately by:

$$I_D \approx I_{D0} e^{\frac{V_{GS}-V_{th}}{nV_T}},$$

where I_{D0} = current at $V_{GS} = V_{th}$, the thermal voltage $V_T = kT/q$ and the slope factor n is given by:

$$n = 1 + C_D/C_{OX},$$

with C_D = capacitance of the depletion layer and C_{OX}= capacitance of the oxide layer. This equation is generally used, but is only an adequate approximation for the source tied to the bulk. For the source not tied to the bulk, the subthreshold equation for drain current in saturation is

$$I_D \approx I_{D0} e^{\frac{\kappa(V_G-V_{th})-V_S}{V_T}},$$

where the K is the channel divider that is given by:

$$\kappa = \frac{C_{OX}}{C_{OX} + C_D},$$

with C_D= capacitance of the depletion layer and C_{OX}= capacitance of the oxide layer. In a long-channel device, there is no drain voltage dependence of the current once $V_{DS} \gg V_T$, but as channel length is reduced drain-induced barrier lowering introduces drain voltage dependence that depends in a complex way upon the device geometry (for example, the channel doping, the junction doping and so on). Frequently, threshold voltage V_{th} for this mode is defined as the gate voltage at which a selected value of current I_{D0} occurs, for example, I_{D0} = 1 μA, which may not be the same V_{th}-value used in the equations for the following modes.

Some micropower analog circuits are designed to take advantage of subthreshold conduction. By working in the weak-inversion region, the MOSFETs in these circuits deliver the highest possible transconductance-to-current ratio, namely: $g_m/I_D = 1/(nV_T)$, almost that of a bipolar transistor.

The subthreshold *I–V curve* depends exponentially upon threshold voltage, introducing a strong dependence on any manufacturing variation that affects threshold voltage; for example: variations in oxide thickness, junction depth, or body doping that change the degree of drain-induced barrier lowering. The resulting sensitivity to fabricational variations complicates optimization for leakage and performance.\

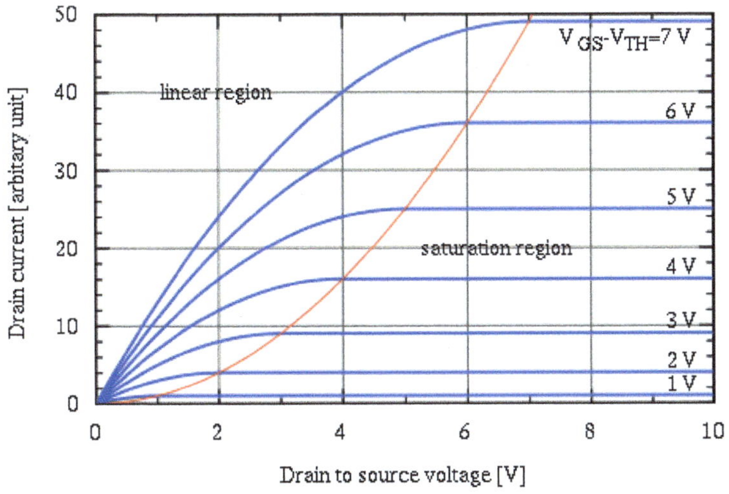

MOSFET drain current vs. drain-to-source voltage for several values of $V_{GS} - V_{th}$; the boundary between *linear* (*Ohmic*) and *saturation* (*active*) modes is indicated by the upward curving parabola

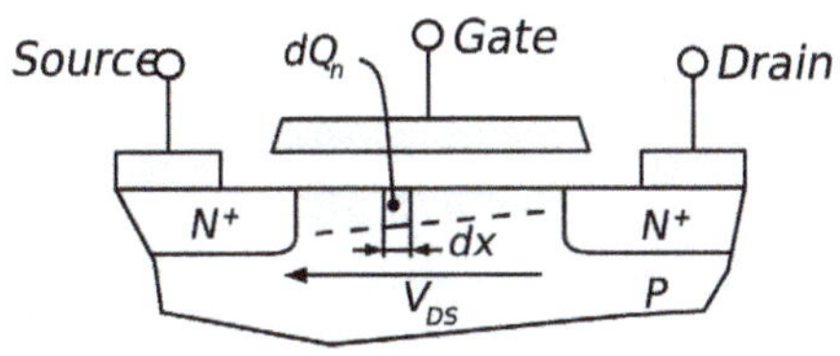

Cross section of a MOSFET operating in the linear (Ohmic) region; strong inversion region present even near drain.

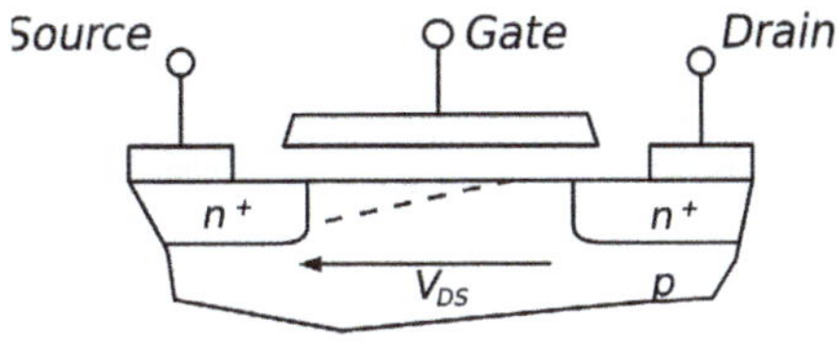

Cross section of a MOSFET operating in the saturation (active) region; channel exhibits channel pinching near drain.

Triode Mode or Linear Region (also known as the Ohmic Mode)

When $V_{GS} > V_{th}$ and $V_{DS} < V_{GS} - V_{th}$:

The transistor is turned on, and a channel has been created which allows current to flow between the drain and the source. The MOSFET operates like a resistor, controlled by the gate voltage relative to both the source and drain voltages. The current from drain to source is modeled as:

$$I_D = \mu_n C_{ox} \frac{W}{L} \left((V_{GS} - V_{th}) V_{DS} - \frac{V_{DS}^2}{2} \right)$$

where μ_n is the charge-carrier effective mobility, W is the gate width, L is the gate length and C_{ox} is the gate oxide capacitance per unit area. The transition from the exponential subthreshold region to the triode region is not as sharp as the equations suggest.

Saturation or Active Mode

When $V_{GS} > V_{th}$ and $V_{DS} \geq (V_{GS} - V_{th})$:

The switch is turned on, and a channel has been created, which allows current to flow between the drain and source. Since the drain voltage is higher than the source voltage, the electrons spread out, and conduction is not through a narrow channel but through a broader, two- or three-dimensional current distribution extending away from the interface and deeper in the substrate. The onset of this region is also known as pinch-off to indicate the lack of channel region near the drain. Although the channel does not extend the full length of the device, the electric field between the drain and the channel is very high, and conduction continues. The drain current is now weakly dependent upon drain voltage and controlled primarily by the gate–source voltage, and modeled approximately as:

$$I_D = \frac{\mu_n C_{ox}}{2} \frac{W}{L} (V_{GS} - V_{th})^2 \left[1 + \lambda (V_{DS} - V_{DSsat})\right].$$

The additional factor involving λ, the channel-length modulation parameter, models current dependence on drain voltage due to the Early effect, or channel length modulation. According to this equation, a key design parameter, the MOSFET transconductance is:

$$g_m = \frac{\partial I_D}{\partial V_{GS}} = \frac{2I_D}{V_{GS} - V_{th}} = \frac{2I_D}{V_{ov}},$$

where the combination $V_{ov} = V_{GS} - V_{th}$ is called the overdrive voltage, and where $V_{DSsat} = V_{GS} - V_{th}$ accounts for a small discontinuity in I_D which would otherwise appear at the transition between the triode and saturation regions.

Another key design parameter is the MOSFET output resistance r_{out} given by:

$$r_{out} = \frac{1}{\lambda I_D}.$$

r_{out} is the inverse of g_{DS} where $g_{DS} = \frac{\partial I_{DS}}{\partial V_{DS}}$. I_D is the expression in saturation region. If λ is taken as zero, an infinite output resistance of the device results that leads to unrealistic circuit predictions, particularly in analog circuits. As the channel length becomes very short, these equations become quite inaccurate. New physical effects arise. For example, carrier transport in the active mode may become limited by velocity saturation. When velocity saturation dominates, the saturation drain current is more nearly linear than quadratic in V_{GS}. At even shorter lengths, carriers transport with near zero scattering, known as quasi-ballistic transport. In the ballistic regime, the carriers travel at an injection velocity that may exceed the saturation velocity and approaches the Fermi velocity at high inversion charge density. In addition, drain-induced barrier lowering increases off-state (cutoff) current and requires an increase in threshold voltage to compensate, which in turn reduces the saturation current.

Body Effect

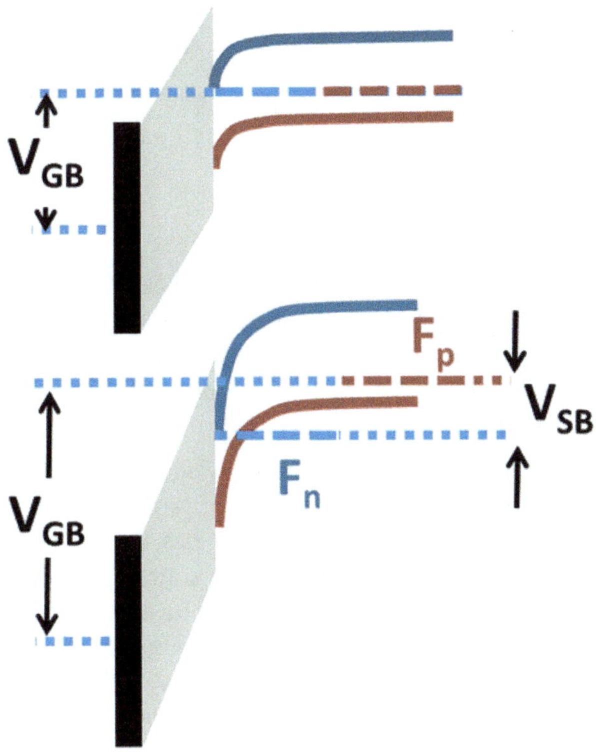

Band diagram showing body effect. V_{SB} splits Fermi levels F_n for electrons and F_p for holes, requiring larger V_{GB} to populate the conduction band in an nMOS MOSFET.

The occupancy of the energy bands in a semiconductor is set by the position of the Fermi level relative to the semiconductor energy-band edges. Application of a source-to-substrate reverse bias of the source-body pn-junction introduces a split between the

Fermi levels for electrons and holes, moving the Fermi level for the channel further from the band edge, lowering the occupancy of the channel. The effect is to increase the gate voltage necessary to establish the channel, as seen in the figure. This change in channel strength by application of reverse bias is called the 'body effect'.

Simply put, using an nMOS example, the gate-to-body bias V_{GB} positions the conduction-band energy levels, while the source-to-body bias V_{SB} positions the electron Fermi level near the interface, deciding occupancy of these levels near the interface, and hence the strength of the inversion layer or channel.

The body effect upon the channel can be described using a modification of the threshold voltage, approximated by the following equation:

$$V_{TB} = V_{T0} + \gamma \left(\sqrt{V_{SB} + 2\varphi_B} - \sqrt{2\varphi_B} \right),$$

where V_{TB} is the threshold voltage with substrate bias present, and V_{T0} is the zero-V_{SB} value of threshold voltage, γ is the body effect parameter, and $2\varphi_B$ is the approximate potential drop between surface and bulk across the depletion layer when $V_{SB} = 0$ and gate bias is sufficient to insure that a channel is present. As this equation shows, a reverse bias $V_{SB} > 0$ causes an increase in threshold voltage V_{TB} and therefore demands a larger gate voltage before the channel populates.

The body can be operated as a second gate, and is sometimes referred to as the "back gate"; the body effect is sometimes called the "back-gate effect".

Circuit Symbols

A variety of symbols are used for the MOSFET. The basic design is generally a line for the channel with the source and drain leaving it at right angles and then bending back at right angles into the same direction as the channel. Sometimes three line segments are used for enhancement mode and a solid line for depletion mode (see depletion and enhancement modes). Another line is drawn parallel to the channel for the gate.

The "bulk" or "body" connection, if shown, is shown connected to the back of the channel with an arrow indicating pMOS or nMOS. Arrows always point from P to N, so an NMOS (N-channel in P-well or P-substrate) has the arrow pointing in (from the bulk to the channel). If the bulk is connected to the source (as is generally the case with discrete devices) it is sometimes angled to meet up with the source leaving the transistor. If the bulk is not shown (as is often the case in IC design as they are generally common bulk) an inversion symbol is sometimes used to indicate PMOS, alternatively an arrow on the source may be used in the same way as for bipolar transistors (out for nMOS, in for pMOS).

Comparison of enhancement-mode and depletion-mode MOSFET symbols, along with JFET symbols. The orientation of the symbols, (most significantly the position of source relative to drain) is such that more positive voltages appear higher on the page than less positive voltages, implying current flowing "down" the page:

Comparison of n- and p-type MOSFETs		
Parameter	**nMOSFET**	**pMOSFET**
Source/drain type	n-type	p-type
Channel type (MOS capacitor)	n-type	p-type
Gate type (poly Si)	n+ poly-Si	p+ poly-Si
Gate type (metal)	φ_m ~ Si CB	φ_m ~ Si VB
Well type	p-type	n-type
Threshold voltage, V_{th}	positive (enhancement) negative (depletion)	negative (enhancement) positive (depletion)
Band-bending	Downwards	Upwards
Inversion layer carriers	electrons	holes
Substrate type	p-type	n-type

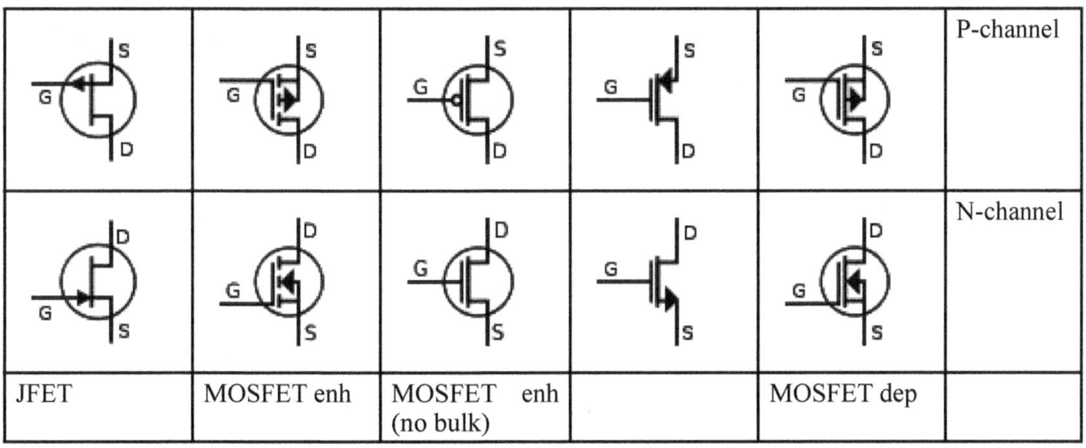

In schematics where G, S, D are not labeled, the detailed features of the symbol indicate which terminal is source and which is drain. For enhancement-mode and depletion-mode MOSFET symbols (in columns two and five), the source terminal is the one connected to the triangle. Additionally, in this diagram, the gate is shown as an "L" shape, whose input leg is closer to S than D, also indicating which is which. However, these symbols are often drawn with a "T" shaped gate (as elsewhere on this page), so it is the triangle which must be relied upon to indicate the source terminal.

For the symbols in which the bulk, or body, terminal is shown, it is here shown internally connected to the source (i.e., the black triangles in the diagrams in columns 2 and 5). This is a typical configuration, but by no means the only important configuration. In general, the MOSFET is a four-terminal device, and in integrated circuits many of the MOSFETs share a body connection, not necessarily connected to the source terminals of all the transistors.

Applications

Digital integrated circuits such as microprocessors and memory devices contain thousands to millions of integrated MOSFET transistors on each device, providing the basic switching functions required to implement logic gates and data storage. Discrete devices are widely used in applications such as switch mode power supplies, variable-frequency drives and other power electronics applications where each device may be switching hundreds or thousands of watts. Radio-frequency amplifiers up to the UHF spectrum use MOSFET transistors as analog signal and power amplifiers. Radio systems also use MOSFETs as oscillators, or mixers to convert frequencies. MOSFET devices are also applied in audio-frequency power amplifiers for public address systems, sound reinforcement and home and automobile sound systems

MOS Integrated Circuits

Following the development of clean rooms to reduce contamination to levels never before thought necessary, and of photolithography and the planar process to allow circuits to be made in very few steps, the Si–SiO_2 system possessed such technical attractions as low cost of production (on a per circuit basis) and ease of integration. Largely because of these two factors, the MOSFET has become the most widely used type of transistor in integrated circuits.

General Microelectronics introduced the first commercial MOS integrated circuit in 1964.

Additionally, the method of coupling two complementary MOSFETS (P-channel and N-channel) into one high/low switch, known as CMOS, means that digital circuits dissipate very little power except when actually switched.

The earliest microprocessors starting in 1970 were all "MOS microprocessors"—i.e., fabricated entirely from PMOS logic or fabricated entirely from NMOS logic. In the 1970s, "MOS microprocessors" were often contrasted with "CMOS microprocessors" and "bipolar bit-slice processors".

CMOS circuits

The MOSFET is used in digital complementary metal–oxide–semiconductor (CMOS) logic, which uses p- and n-channel MOSFETs as building blocks. Overheating is a major concern in integrated circuits since ever more transistors are packed into ever smaller chips. CMOS logic reduces power consumption because no current flows (ideally), and thus no power is consumed, except when the inputs to logic gates are being switched. CMOS accomplishes this current reduction by complementing every nMOSFET with a pMOSFET and connecting both gates and both drains together. A high voltage on the gates will cause the nMOSFET to conduct and the pMOSFET not to conduct and a low voltage on the gates causes the reverse. During the switching time as the voltage goes from one state to another, both MOSFETs will conduct briefly. This arrangement greatly reduces power consumption and heat generation.

Digital

The growth of digital technologies like the microprocessor has provided the motivation to advance MOSFET technology faster than any other type of silicon-based transistor. A big advantage of MOSFETs for digital switching is that the oxide layer between the gate and the channel prevents DC current from flowing through the gate, further reducing power consumption and giving a very large input impedance. The insulating oxide between the gate and channel effectively isolates a MOSFET in one logic stage from earlier and later stages, which allows a single MOSFET output to drive a considerable number of MOSFET inputs. Bipolar transistor-based logic (such as TTL) does not have such a high fanout capacity. This isolation also makes it easier for the designers to ignore to some extent loading effects between logic stages independently. That extent is defined by the operating frequency: as frequencies increase, the input impedance of the MOSFETs decreases.

Analog

The MOSFET's advantages in digital circuits do not translate into supremacy in all analog circuits. The two types of circuit draw upon different features of transistor behavior. Digital circuits switch, spending most of their time outside the switching region, while analog circuits depend on the linearity of response when the MOSFET is held precisely in the switching region. The bipolar junction transistor (BJT) has traditionally been the analog designer's transistor of choice, due largely to its higher transconductance and its lower output impedance (drain-voltage independence) in the switching region.

Nevertheless, MOSFETs are widely used in many types of analog circuits because of certain advantages. The characteristics and performance of many analog circuits can be scaled up or down by changing the sizes (length and width) of the MOSFETs used. By comparison, in most bipolar transistors the size of the device does not significantly affect its performance. MOSFETs' ideal characteristics regarding gate current (zero) and drain-source offset voltage (zero) also make them nearly ideal switch elements, and also make switched capacitor analog circuits practical. In their linear region, MOSFETs can be used as precision resistors, which can have a much higher controlled resistance than BJTs. In high power circuits, MOSFETs sometimes have the advantage of not suffering from thermal runaway as BJTs do. Also, MOSFETs can be configured to perform as capacitors and gyrator circuits which allow op-amps made from them to appear as inductors, thereby allowing all of the normal analog devices on a chip (except for diodes, which can be made smaller than a MOSFET anyway) to be built entirely out of MOSFETs. This means that complete analog circuits can be made on a silicon chip in a much smaller space and with simpler fabrication techniques. MOSFETS are ideally suited to switch inductive loads because of tolerance to inductive kickback.

Some ICs combine analog and digital MOSFET circuitry on a single mixed-signal integrated circuit, making the needed board space even smaller. This creates a need to isolate the analog circuits from the digital circuits on a chip level, leading to the use of isolation rings and Silicon-On-Insulator (SOI). Since MOSFETs require more space to handle a given amount of power than a BJT, fabrication processes can incorporate BJTs and MOSFETs into a single device. Mixed-transistor devices are called Bi-FETs (bipolar FETs) if they contain just one BJT-FET and BiCMOS (bipolar-CMOS) if they contain complementary BJT-FETs. Such devices have the advantages of both insulated gates and higher current density.

Analog Switches

MOSFET analog switches use the MOSFET to pass analog signals when on, and as a high impedance when off. Signals flow in both directions across a MOSFET switch. In this application, the drain and source of a MOSFET exchange places depending on the relative voltages of the source/drain electrodes. The source is the more negative side for an N-MOS or the more positive side for a P-MOS. All of these switches are limited on what signals they can pass or stop by their gate–source, gate–drain and source–drain voltages; exceeding the voltage, current, or power limits will potentially damage the switch.

Single-Type

This analog switch uses a four-terminal simple MOSFET of either P or N type.

In the case of an n-type switch, the body is connected to the most negative supply (usually GND) and the gate is used as the switch control. Whenever the gate voltage exceeds the source voltage by at least a threshold voltage, the MOSFET conducts. The higher the voltage, the more the MOSFET can conduct. An N-MOS switch passes all voltages less than $V_{gate}-V_{tn}$. When the switch is conducting, it typically operates in the linear (or ohmic) mode of operation, since the source and drain voltages will typically be nearly equal.

In the case of a P-MOS, the body is connected to the most positive voltage, and the gate is brought to a lower potential to turn the switch on. The P-MOS switch passes all voltages higher than $V_{gate}-V_{tp}$ (threshold voltage V_{tp} is negative in the case of enhancement-mode P-MOS).

Dual-type (CMOS)

This "complementary" or CMOS type of switch uses one P-MOS and one N-MOS FET to counteract the limitations of the single-type switch. The FETs have their drains and sources connected in parallel, the body of the P-MOS is connected to the high potential (V_{DD}) and the body of the N-MOS is connected to the low potential (Gnd). To turn the switch on, the gate of the P-MOS is driven to the low potential and the gate of the N-MOS is driven to the high potential. For voltages between $V_{DD}-V_{tn}$ and Gnd–V_{tp}, both FETs conduct the signal; for voltages less than Gnd–V_{tp}, the N-MOS conducts alone; and for voltages greater than $V_{DD}-V_{tn}$, the P-MOS conducts alone.

The voltage limits for this switch are the gate–source, gate–drain and source–drain voltage limits for both FETs. Also, the P-MOS is typically two to three times wider than the N-MOS, so the switch will be balanced for speed in the two directions.

Tri-state circuitry sometimes incorporates a CMOS MOSFET switch on its output to provide for a low-ohmic, full-range output when on, and a high-ohmic, mid-level signal when off.

Construction

Gate Material

The primary criterion for the gate material is that it is a good conductor. Highly doped polycrystalline silicon is an acceptable but certainly not ideal conductor, and also suffers from some more technical deficiencies in its role as the standard gate material. Nevertheless, there are several reasons favoring use of polysilicon:

- The threshold voltage (and consequently the drain to source on-current) is modified by the work function difference between the gate material and channel material. Because polysilicon is a semiconductor, its work function can be modulated by adjusting the type and level of doping. Furthermore, because polysilicon has the same bandgap as the underlying silicon channel, it is quite straightforward to tune the work function to achieve low threshold voltages for both NMOS and PMOS devices. By contrast, the work functions of metals are not easily modulated, so tuning the work function to obtain low threshold voltages becomes a significant challenge. Additionally, obtaining low-threshold devices on both PMOS and NMOS devices sometimes requires the use of different metals for each device type. While bimetallic integrated circuits (i.e., one type of metal for gate electrodes of NFETS and a second type of metal for gate electrodes of PFETS) are not common, they are known in patent literature and provide some benefit in terms of tuning electrical circuits' overall electrical perforamnce.
- The silicon-SiO_2 interface has been well studied and is known to have relatively few defects. By contrast many metal–insulator interfaces contain significant levels of defects which can lead to Fermi level pinning, charging, or other phenomena that ultimately degrade device performance.
- In the MOSFET IC fabrication process, it is preferable to deposit the gate material prior to certain high-temperature steps in order to make better-performing transistors. Such high temperature steps would melt some metals, limiting the types of metal that can be used in a metal-gate-based process.

While polysilicon gates have been the de facto standard for the last twenty years, they do have some disadvantages which have led to their likely future replacement by metal gates. These disadvantages include:

- Polysilicon is not a great conductor (approximately 1000 times more resistive than metals) which reduces the signal propagation speed through the material. The resistivity can be lowered by increasing the level of doping, but even highly doped polysilicon is not as conductive as most metals. To improve conductivity further, sometimes a high-temperature metal such as tungsten, titanium, cobalt, and more recently nickel is alloyed with the top layers of the polysilicon. Such a blended material is called silicide. The silicide-polysilicon combination has better electrical properties than polysilicon alone and still does not melt in subsequent processing. Also the threshold voltage is not significantly higher than with polysilicon alone, because the silicide material is not near the channel. The process in which silicide is formed on both the gate electrode and the source and drain regions is sometimes called salicide, self-aligned silicide.

- When the transistors are extremely scaled down, it is necessary to make the gate dielectric layer very thin, around 1 nm in state-of-the-art technologies. A phenomenon observed here is the so-called poly depletion, where a depletion layer is formed in the gate polysilicon layer next to the gate dielectric when the transistor is in the inversion. To avoid this problem, a metal gate is desired. A variety of metal gates such as tantalum, tungsten, tantalum nitride, and titanium nitride are used, usually in conjunction with high-k dielectrics. An alternative is to use fully silicided polysilicon gates, a process known as FUSI.

Present high performance CPUs use metal gate technology, together with high-k dielectrics, a combination known as HKMG (High-K, Metal Gate). The disadvantages of metal gates are overcome by a few techniques:

- The threshold voltage is tuned by including a thin "work function metal" layer between the high-K dielectric and the main metal. This layer is thin enough that the total work function of the gate is influenced by both the main metal and thin metal work functions (either due to alloying during annealing, or simply due to the incomplete screening by the thin metal). The threshold voltage thus can be tuned by the thickness of the thin metal layer.
- High-K dielectrics are now well studied, and their defects are understood.
- HKMG processes exist that do not require the metals to experience high temperature anneals; other processes select metals that can survive the annealing step.

Insulator

As devices are made smaller, insulating layers are made thinner, and at some point tunneling of carriers through the insulator from the channel to the gate electrode takes place. To reduce the resulting leakage current, the insulator can be made thicker by choosing a material with a higher dielectric constant. To see how thickness and dielectric constant are related, note that Gauss's law connects field to charge as:

$$Q = \kappa\epsilon_0 \, E,$$

with Q = charge density, κ = dielectric constant, ε_0 = permittivity of empty space and E = electric field. From this law it appears the same charge can be maintained in the channel at a lower field provided κ is increased. The voltage on the gate is given by:

$$V_G = V_{ch} + E\, t_{ins} = V_{ch} + \frac{Q t_{ins}}{\kappa\epsilon_0},$$

with V_G = gate voltage, V_{ch} = voltage at channel side of insulator, and t_{ins} = insulator thickness. This equation shows the gate voltage will not increase when the insulator thickness increases, provided κ increases to keep t_{ins}/κ = *constant* (see the article on high-κ dielectrics for more detail, and the section in this article on gate-oxide leakage).

The insulator in a MOSFET is a dielectric which can in any event be silicon oxide, but many other dielectric materials are employed. The generic term for the dielectric is gate dielectric since the dielectric lies directly below the gate electrode and above the channel of the MOSFET.

Junction Design

The source-to-body and drain-to-body junctions are the object of much attention because of three major factors: their design affects the current-voltage (*I-V*) characteristics of the device, lowering output resistance, and also the speed of the device through the loading effect of the junction capacitances, and finally, the component of stand-by power dissipation due to junction leakage.

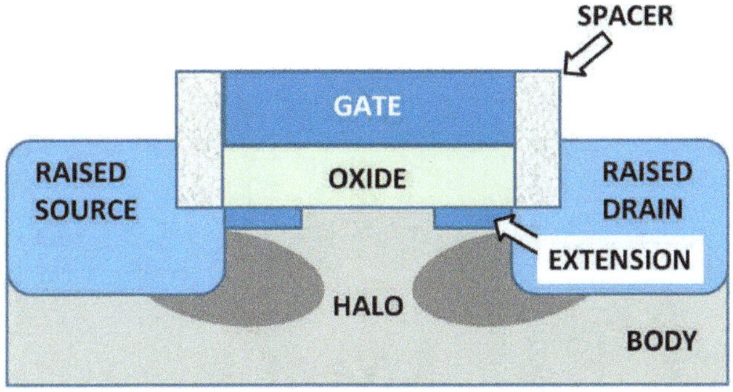

MOSFET showing shallow junction extensions, raised source and drain and halo implant. Raised source and drain separated from gate by oxide spacers.

The drain induced barrier lowering of the threshold voltage and channel length modulation effects upon *I-V* curves are reduced by using shallow junction extensions. In addition, *halo* doping can be used, that is, the addition of very thin heavily doped regions of the same doping type as the body tight against the junction walls to limit the extent of depletion regions.

The capacitive effects are limited by using raised source and drain geometries that make most of the contact area border thick dielectric instead of silicon.

These various features of junction design are shown (with artistic license) in the figure.

Scaling

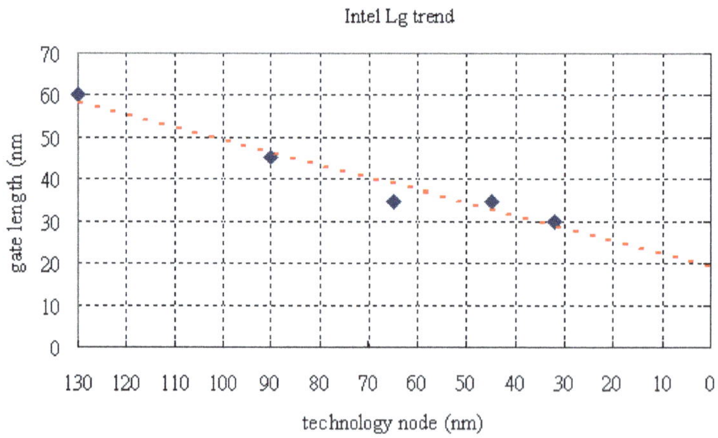

Trend of Intel CPU transistor gate length.

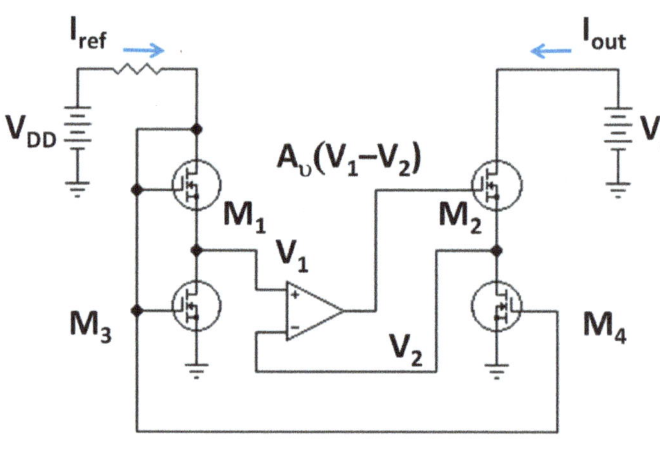

MOSFET version of gain-boosted current mirror; M_1 and M_2 are in active mode, while M_3 and M_4 are in Ohmic mode, and act like resistors. The operational amplifier provides feedback that maintains a high output resistance.

Over the past decades, the MOSFET has continually been scaled down in size; typical MOSFET channel lengths were once several micrometres, but modern integrated circuits are incorporating MOSFETs with channel lengths of tens of nanometers. Robert Dennard's work on scaling theory was pivotal in recognising that this ongoing reduction was possible. Intel began production of a process featuring a 32 nm feature size (with the channel being even shorter) in late 2009. The semiconductor industry maintains a "roadmap", the ITRS, which sets the pace for MOSFET development. Historically, the difficulties with decreasing the size of the MOSFET have been associated with the semiconductor device fabrication process, the need to use very low voltages, and with poorer electrical performance necessitating circuit redesign and innovation (small MOSFETs exhibit higher leakage currents and lower output resistance).

Smaller MOSFETs are desirable for several reasons. The main reason to make transistors smaller is to pack more and more devices in a given chip area. This results in a chip with the same functionality in a smaller area, or chips with more functionality in the same area. Since fabrication costs for a semiconductor wafer are relatively fixed, the cost per integrated circuits is mainly related to the number of chips that can be produced per wafer. Hence, smaller ICs allow more chips per wafer, reducing the price per chip. In fact, over the past 30 years the number of transistors per chip has been doubled every 2–3 years once a new technology node is introduced. For example, the number of MOSFETs in a microprocessor fabricated in a 45 nm technology can well be twice as many as in a 65 nm chip. This doubling of transistor density was first observed by Gordon Moore in 1965 and is commonly referred to as Moore's law. It is also expected that smaller transistors switch faster. For example, one approach to size reduction is a scaling of the MOSFET that requires all device dimensions to reduce proportionally. The main device dimensions are the channel length, channel width, and oxide thickness. When they are scaled down by equal factors, the transistor channel resistance does not change, while gate capacitance is cut by that factor. Hence, the RC delay of the transistor scales with a similar factor. While this has been traditionally the case for the older technologies, for the state-of-the-art MOSFETs reduction of the transistor dimensions does not necessarily translate to higher chip speed because the delay due to interconnections is more significant.

Producing MOSFETs with channel lengths much smaller than a micrometre is a challenge, and the difficulties of semiconductor device fabrication are always a limiting factor in advancing integrated circuit technology. Though processes such as ALD have improved fabrication for small components, the small size of the MOSFET (less than a few tens of nanometers) has created operational problems:

Higher Subthreshold Conduction

As MOSFET geometries shrink, the voltage that can be applied to the gate must be reduced to maintain reliability. To maintain performance, the threshold voltage of the MOSFET has to be reduced as well. As threshold voltage is reduced, the transistor cannot be switched from complete turn-off to complete turn-on with the limited voltage swing available; the circuit design is a compromise between strong current in the "on" case and low current in the "off" case, and the application determines whether to favor one over the other. Subthreshold leakage (including subthreshold conduction, gate-oxide leakage and reverse-biased junction leakage), which was ignored in the past, now can consume upwards of half of the total power consumption of modern high-performance VLSI chips.

Increased Gate-Oxide Leakage

The gate oxide, which serves as insulator between the gate and channel, should be made as thin as possible to increase the channel conductivity and performance when the transistor is on and to reduce subthreshold leakage when the transistor is off. However, with current gate oxides with a thickness of around 1.2 nm (which in silicon is ~5 atoms thick) the quantum mechanical phenomenon of electron tunneling occurs between the gate and channel, leading to increased power consumption. Silicon dioxide has traditionally been used as the gate insulator. Silicon dioxide however has a modest dielectric constant. Increasing the dielectric constant of the gate dielectric allows a thicker layer while maintaining a high capacitance (capacitance is proportional to dielectric constant and inversely proportional to dielectric thickness). All else equal, a higher dielectric thickness reduces the quantum tunneling current through the dielectric between the gate and the channel. Insulators that have a larger dielectric constant than silicon dioxide (referred to as high-k dielectrics), such as group IVb metal silicates e.g. hafnium and zirconium silicates and oxides are being used to reduce the gate leakage from the 45 nanometer technology node onwards. On the other hand, the barrier height of the new gate insulator is an important consideration; the difference in conduction band energy between the semiconductor and the dielectric (and the corresponding difference in valence band energy) also affects leakage current level. For the traditional gate oxide, silicon dioxide, the former barrier is approximately 8 eV. For many alternative dielectrics the value is significantly lower, tending to increase the tunneling current, somewhat negating the advantage of higher dielectric constant. The maximum gate-source voltage is determined by the strength of the electric field able to be sustained by the gate dielectric before significant leakage occurs. As the insulating dielectric is made thinner, the electric field strength within it goes up for a fixed voltage. This necessitates using lower voltages with the thinner dielectric.

Increased Junction Leakage

To make devices smaller, junction design has become more complex, leading to higher doping levels, shallower junctions, "halo" doping and so forth, all to decrease drain-induced barrier lowering (see the section on junction design). To keep these complex junctions in place, the annealing steps formerly used to remove damage and electrically active defects must be curtailed increasing junction leakage. Heavier doping is also associated with thinner depletion layers and more recombination centers that result in increased leakage current, even without lattice damage.

DIBL and V_T roll off

Because of the short-channel effect, channel formation is not entirely done by the gate, but now the drain and source also affect the channel formation. As the channel length decreases, the depletion regions of the source and drain come closer together and make the threshold voltage (V_T) a function of the length of the channel. This is called V_T roll-off. V_T also becomes function of drain to source voltage V_{DS}. As we increase the V_{DS}, the depletion regions increase in size, and a considerable amount of charge is depleted by the V_{DS}. The gate voltage required to form the channel is then lowered, and thus, the V_T decreases with an increase in V_{DS}. This effect is called drain induced barrier lowering (DIBL).

Lower Output Resistance

For analog operation, good gain requires a high MOSFET output impedance, which is to say, the MOSFET current should vary only slightly with the applied drain-to-source voltage. As devices are made smaller, the influence of the drain competes more successfully with that of the gate due to the growing proximity of these two electrodes, increasing the sensitivity of the MOSFET current to the drain voltage. To counteract the resulting decrease in output resistance, circuits are made more complex, either by requiring more devices, for example the cascode and cascade amplifiers, or by feedback circuitry using operational amplifiers, for example a circuit like that in the adjacent figure.

Lower Transconductance

The transconductance of the MOSFET decides its gain and is proportional to hole or electron mobility (depending on device type), at least for low drain voltages. As MOSFET size is reduced, the fields in the channel increase and the dopant impurity levels increase. Both changes reduce the carrier mobility, and hence the transconductance. As channel lengths are reduced without proportional reduction in drain voltage, raising the electric field in the channel, the result is velocity saturation of the carriers, limiting the current and the transconductance.

Interconnect Capacitance

Traditionally, switching time was roughly proportional to the gate capacitance of gates. However, with transistors becoming smaller and more transistors being placed on the chip, interconnect capacitance (the capacitance of the metal-layer connections between different parts of the chip) is becoming a large percentage of capacitance. Signals have to travel through the interconnect, which leads to increased delay and lower performance.

Heat Production

The ever-increasing density of MOSFETs on an integrated circuit creates problems of substantial localized heat generation that can impair circuit operation. Circuits operate more slowly at high temperatures, and have reduced reliability and shorter lifetimes. Heat sinks and other cooling devices and methods are now required for many integrated circuits including microprocessors. Power MOSFETs are at risk of thermal runaway. As their on-state resistance rises with temperature, if the load is approximately a constant-current load then the power loss rises correspondingly, generating further heat. When the heatsink is not able to keep the temperature low enough, the junction temperature may rise quickly and uncontrollably, resulting in destruction of the device.

Process Variations

With MOSFETs becoming smaller, the number of atoms in the silicon that produce many of the transistor's properties is becoming fewer, with the result that control of dopant numbers and placement is more erratic. During chip manufacturing, random process variations affect all transistor dimensions: length, width, junction depths, oxide thickness *etc.*, and become a greater percentage of overall transistor size as the transistor shrinks. The transistor characteristics become less certain, more statistical. The random nature of manufacture means we do not know which particular example MOSFETs actually will end up in a particular instance of the circuit. This uncertainty forces a less optimal design because the design must work for a great variety of possible component MOSFETs. See process variation, design for manufacturability, reliability engineering, and statistical process control.

Modeling Challenges

Modern ICs are computer-simulated with the goal of obtaining working circuits from the very first manufactured lot. As devices are miniaturized, the complexity of the processing makes it difficult to predict exactly what the final devices look like, and modeling of physical processes becomes more challenging as well. In addition, microscopic variations in structure due simply to the probabilistic nature of atomic processes require statistical (not just deterministic) predictions. These factors combine to make adequate simulation and "right the first time" manufacture difficult.

Other types

Dual-gate

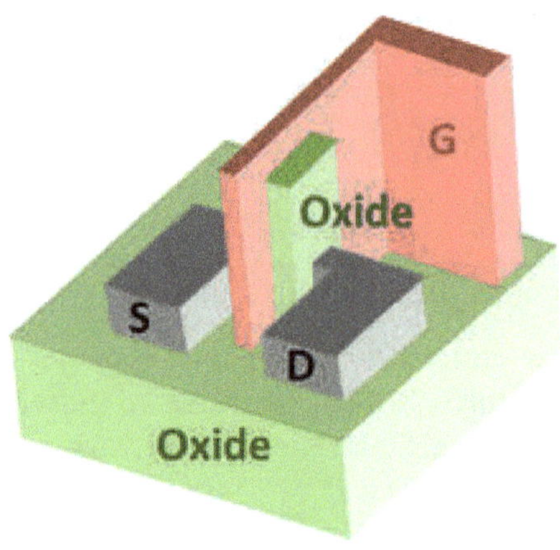

A FinFET.

The dual-gate MOSFET has a tetrode configuration, where both gates control the current in the device. It is commonly used for small-signal devices in radio frequency applications where biasing the drain-side gate at constant potential reduces the gain loss caused by Miller effect, replacing two separate transistors in cascode configuration. Other common uses in RF circuits include gain control and mixing (frequency conversion). The "tetrode" description, though accurate, does not replicate the vacuum-tube tetrode.

Vacuum-tube tetrodes, using a screen grid, exhibit much lower grid-plate capacitance and much higher output impedance and voltage gains than triode vacuum tubes. These improvements are commonly an order of magnitude (10 times) or considerably more. Tetrode transistors (whether bipolar junction or field-effect) do not exhibit improvements of such a great degree.

The FinFET is a double-gate silicon-on-insulator device, one of a number of geometries being introduced to mitigate the effects of short channels and reduce drain-induced barrier lowering. The "fin" refers to the narrow channel between source and drain. A thin insulating oxide layer on either side of the fin separates it from the gate. SOI FinFETs with a thick oxide on top of the fin are called *double-gate* and those with a thin oxide on top as well as on the sides are called *triple-gate* FinFETs.

Depletion-mode

There are *depletion-mode* MOSFET devices, which are less commonly used than the standard *enhancement-mode* devices already described. These are MOSFET devices that are doped so that a channel exists even with zero voltage from gate to source. To control the channel, a negative voltage is applied to the gate (for an n-channel device), depleting the channel, which reduces the current flow through the device. In essence, the depletion-mode device is equivalent to a normally closed (on) switch, while the enhancement-mode device is equivalent to a normally open (off) switch. Due to their low noise figure in the RF region, and better gain, these devices are often preferred to bipolars in RF front-ends such as in TV sets. Depletion-mode MOSFET families include BF 960 by Siemens and BF 980 by Philips (dated 1980s), whose derivatives are still used in AGC and RF mixer front-ends.

NMOS logic

For devices of equal current driving capability, n-channel MOSFETs can be made smaller than p-channel MOSFETs, due to p-channel charge carriers (holes) having lower mobility than do n-channel charge carriers (electrons), and producing only one type of MOSFET on a silicon substrate is cheaper and technically simpler. These were the driving principles in the design of NMOS logic which uses n-channel MOSFETs exclusively. However, neglecting leakage current, unlike CMOS logic, NMOS logic consumes power even when no switching is taking place. With advances in technology, CMOS logic displaced NMOS logic in the mid-1980s to become the preferred process for digital chips.

Power MOSFET

Power MOSFETs have a different structure. As with most power devices, the structure is vertical and not planar. Using a vertical structure, it is possible for the transistor to sustain both high blocking voltage and high current. The voltage rating of the transistor is a function of the doping and thickness of the N-epitaxial layer (see cross section), while the current rating is a function of the channel width (the wider the channel, the higher the current). In a planar structure, the current and breakdown voltage ratings are both a function of the channel dimensions (respectively width and length of the channel), resulting in inefficient use of the "silicon estate". With the vertical structure, the component area is roughly proportional to the current it can sustain, and the component thickness (actually the N-epitaxial layer thickness) is proportional to the breakdown voltage.

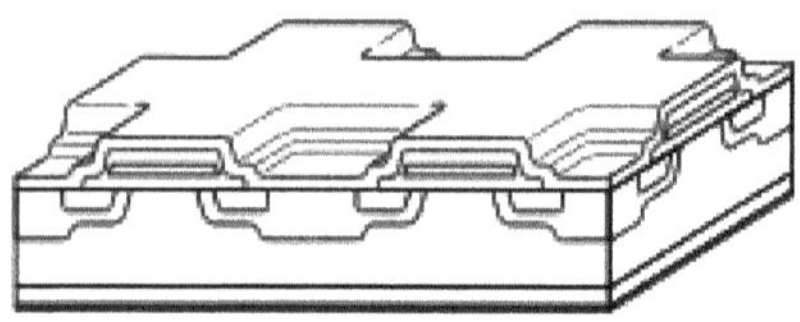

Cross section of a power MOSFET, with square cells. A typical transistor is constituted of several thousand cells.

Power MOSFETs with lateral structure are mainly used in high-end audio amplifiers and high-power PA systems. Their advantage is a better behaviour in the saturated region (corresponding to the linear region of a bipolar transistor) than the vertical MOSFETs. Vertical MOSFETs are designed for switching applications.

DMOS

DMOS stands for double-diffused metal–oxide–semiconductor. There are *LDMOS* (Lateral Double-diffused Metal Oxide Semiconductor) and *VDMOS* (Vertical Double-diffused Metal Oxide Semiconductor). Most power MOSFETs are made using this technology.

RHBD

Semiconductor sub-micrometer and nanometer electronic circuits are the primary concern for operating within the normal tolerance in harsh radiation environments like outer space. One of the design approaches for making a radiation-hardened-by-design (RHBD) device is Enclosed-Layout-Transistor (ELT). Normally, the gate of the MOSFET surrounds the drain, which is placed in the center of the ELT. The source of the MOSFET surrounds the gate. Another RHBD MOSFET is called H-Gate. Both of these transistors have very low leakage current with respect to radiation. However, they are large in size and take more space on silicon than a standard MOSFET. In older STI (shallow trench isolation) designs, radiation strikes near the silicon oxide region cause the channel inversion at the corners of the standard MOSFET due to accumulation of radiation induced trapped charges. If the charges are large enough, the accumulated charges affect STI surface edges along the channel near the channel interface (gate) of the standard MOSFET. Thus the device channel inversion occurs along the channel edges and the device creates off-state leakage path, causing device to turn on. So the reliability of circuits degrades severely. The ELT offers many advantages. These advantages include improvement of reliability by reducing unwanted surface inversion at the gate edges that occurs in the standard MOSFET. Since the gate edges are enclosed in ELT, there is no gate oxide edge (STI at gate interface), and thus the transistor off-state leakage is reduced very much. Low-power microelectronic circuits including computers, communication devices and monitoring systems in space shuttle and satellites are very different from what we use on earth. They are radiation (high-speed atomic particles like proton and neutron, solar flare magnetic energy dissipation in Earth's space, energetic cosmic rays like X-ray, gamma ray etc.) tolerant circuits. These special electronics are designed by applying very different techniques using RHBD MOSFETs to ensure the safe space journey and also space-walk of astronauts.

REFERENCES

1. Yuhua Cheng; Chenming Hu (1999). "§2.1 MOSFET classification and operation". MOSFET modeling & BSIM3 user's guide. Springer. p. 13. ISBN 0-7923-8575-6.
2. U. A. Bakshi; A. P. Godse (2007). "§8.2 The depletion mode MOSFET". Electronic Circuits. Technical Publications. pp. 8–2. ISBN 978-81-8431-284-3.
3. Lilienfeld Julius Edgar U.S. Patent 1,745,175 "Method and apparatus for controlling electric currents". Priority date October 22, 1925
4. 1960 – Metal Oxide Semiconductor (MOS) Transistor Demonstrated: John Atalla and Dawon Kahng fabricate working transistors and demonstrate the first successful MOS field-effect amplifier (U.S. Patent 3,102,230 filed in 1960, issued in 1963)". Computer History Museum.
5. The Silicon Engine | 1948 – Conception of the Junction Transistor. Computer History Museum. 2007.
6. Intel 45nm Hi-k Silicon Technology. Archived from the original on October 6, 2009.
7. Using a MOSFET as a Switch. 090507 brunningsoftware.co.uk
8. H Shichman & DA Hodges (1968). "Modeling and simulation of insulated-gate field-effect transistor switching circuits". IEEE Journal of Solid-state circuits. **SC–3** *(3): 285–289. doi:10.1109/JSSC.1968.1049902. Archived from the original on June 10, 2013.* and often is called the Shichman-Hodges model.
9. For example, see Yuhua Cheng; Chenming Hu (1999). MOSFET modeling & BSIM3 user's guide. Springer. ISBN 0-7923-8575-6.. The most recent version of the BSIM model is described in Sriramkumar V.; Navid Paydavosi; Darsen Lu; Chung-Hsun Lin; Mohan Dunga; Shijing Yao; Tanvir Morshed; Ali Niknejad & Chenming Hu (2012). "BSIM-CMG 106.1.0beta Multi-Gate MOSFET Compact Model" (PDF). Department of EE and CS, UC Berkeley. Retrieved 2012-04-01.
10. =P R Gray; P J Hurst; S H Lewis & R G Meyer (2001). Analysis and Design of Analog Integrated Circuits (Fourth ed.). New York: Wiley. pp. 66–67. ISBN 0-471-32168-0.
11. P. R. van der Meer, A. van Staveren, A. H. M. van Roermund (2004). Low-Power Deep Sub-Micron CMOS Logic: Subthreshold Current Reduction. Dordrecht:

Springer. p. 78. ISBN 1-4020-2848-2.

12. Degnan, Brian. "Wikipedia fails subvt".
13. Mead, Carver (1989). Analog VLSI and Neural Systems. Reading, MA: Addison-Wesley. p. 370. ISBN 9780201059922.
14. =Leslie S. Smith; Alister Hamilton (1998). Neuromorphic Systems: Engineering Silicon from Neurobiology. World Scientific. pp. 52–56. ISBN 981-02-3377-9.
15. Satish Kumar (2004). Neural Networks: A Classroom Approach. Tata McGraw-Hill. p. 688. ISBN 0-07-048292-6.
16. =Manfred Glesner; Peter Zipf; Michel Renovell (2002). Field-programmable Logic and Applications: 12th International Conference. Dordrecht: Springer. p. 425. ISBN 3-540-44108-5.
17. Eric A. Vittoz (1996). "The Fundamentals of Analog Micropower Design". In Chris Toumazou; Nicholas C. Battersby; Sonia Porta. Circuits and systems tutorials. John Wiley and Sons. pp. 365–372. ISBN 978-0-7803-1170-1.
18. Sandeep K. Shukla; R. Iris Bahar (2004). Nano, Quantum and Molecular Computing. Springer. p. 10 and Fig. 1.4, p. 11. ISBN 1-4020-8067-0.
19. Ashish Srivastava; Dennis Sylvester; David Blaauw (2005). Statistical Analysis and Optimization For VLSI: Timing and Power. Springer. p. 135. ISBN 0-387-25738-1.
20. C Galup-Montoro & Schneider MC (2007). MOSFET modeling for circuit analysis and design. London/Singapore: World Scientific. p. 83. ISBN 981-256-810-7.
21. Norbert R Malik (1995). Electronic circuits: analysis, simulation, and design. Englewood Cliffs, NJ: Prentice Hall. pp. 315–316. ISBN 0-02-374910-5.
22. PR Gray, PJ Hurst, SH Lewis & RG Meyer. §1.5.2 p. 45. ISBN 0-471-32168-0.
23. A. S. Sedra & K.C. Smith (2004). Microelectronic circuits (Fifth ed.). New York: Oxford. p. 552. ISBN 0-19-514251-9.
24. A. S. Sedra & K.C. Smith. p. 250, Eq. 4.14. ISBN 0-19-514251-9.
25. For a uniformly doped p-type substrate with bulk acceptor doping of N_A per unit volume, the intrinsic mobile carrier density per unit volume in the bulk. See, for example, Narain Arora (2007). "Equation 5.12". Mosfet modeling for VLSI simulation: theory and practice. World Scientific. p. 173. ISBN 981-256-862-X.
26. Body effect". Equars.com. Retrieved 2012-06-02.
27. "Electronic Circuit Symbols". circuitstoday.com. 9 November 2011
28. IEEE Std 315-1975 — Graphic Symbols for Electrical and Electronics Diagrams (Including Reference Designation Letters)
29. Microelectronic Circuit Design, Jaeger, Blalock — Figure 4.15 IEEE Standard MOS transistor circuit symbols.
30. memory components data book" (PDF). memory components data book. Intel. pp. 2–1. Retrieved 30 August 2015.
31. Computer History Museum – The Silicon Engine | 1955 – Photolithography

Techniques Are Used to Make Silicon Devices". Computerhistory.org. Retrieved 2012-06-02.

32. "1964 - First Commercial MOS IC Introduced".
33. Robert H. Cushman. "2-1/2-generation μP's-$10 parts that perform like low-end mini's". EDN September 20, 1975.
34. Computer History Museum – The Silicon Engine | 1963 – Complementary MOS Circuit Configuration is Invented". Computerhistory.org. Retrieved 2012-06-02.
35. Computer History Museum – Exhibits – Microprocessors". Computerhistory.org. Retrieved 2012-06-02.
36. "ReVera's FinFET Control. revera.com.
37. Jean-Pierre Colinge; Cynthia A. Colinge (2002). Physics of Semiconductor Devices. Dordrecht: Springer. p. 233, Figure 7.46. ISBN 1-4020-7018-7.
38. Eicke R. Weber, Jarek Dabrowski (Editors) (2004). Predictive Simulation of Semiconductor Processing: Status and Challenges. Dordrecht: Springer. p. 5, Figure 1.2. ISBN 3-540-20481-4.
39. International Technology Roadmap for Semiconductors".
40. 1965 – "Moore's Law" Predicts the Future of Integrated Circuits". Computer History Museum.
41. Kaushik Roy; Kiat Seng Yeo (2004). Low Voltage, Low Power VLSI Subsystems. McGraw-Hill Professional. Fig. 2.1, p. 44. ISBN 0-07-143786-X.
42. Kaushik Roy; Kiat Seng Yeo (2004). Low Voltage, Low Power VLSI Subsystems. McGraw-Hill Professional. Fig. 1.1, p. 4. ISBN 0-07-143786-X.
43. Dragica Vasileska; Stephen Goodnick (2006). Computational Electronics. Morgan & Claypool. p. 103. ISBN 1-59829-056-8.
44. Frontier Semiconductor Paper" (PDF). Archived from the original (PDF) on February 27, 2012. Retrieved 2012-06-02.
45. Wai-Kai Chen (2006). The VLSI Handbook. CRC Press. Fig. 2.28, p. 2–22. ISBN 0-8493-4199-X.
46. R. Lindsay; Pawlak; Kittl; Henson; Torregiani; Giangrandi; Surdeanu; Vandervorst; Mayur; Ross; McCoy; Gelpey; Elliott; Pages; Satta; Lauwers; Stolk; Maex (2011). "A Comparison of Spike, Flash, SPER and Laser Annealing for 45nm CMOS". MRS Proceedings. **765**. doi:10.1557/PROC-765-D7.4.
47. VLSI wiring capacitance. IBM Journal of Research and Development.
48. D Soudris, P Pirsch, E Barke (Editors) (2000). Integrated Circuit Design: Power and Timing Modeling, Optimization, and Simulation (10th Int. Workshop). Springer. p. 38. ISBN 3-540-41068-6.
49. Michael Orshansky; Sani Nassif; Duane Boning (2007). Design for Manufacturability And Statistical Design: A Constructive Approach. New York 309284: Springer.
50. PM Zeitzoff; JA Hutchby; HR Huff (2002). "Figure 12: Simplified cross section

of FinFET double-gate MOSFET.". In Yoon-Soo Park; Michael Shur; William Tang. Frontiers in electronics: future chips : proceedings of the 2002 Workshop on Frontiers in Electronics (WOFE-02), St Croix, Virgin Islands, USA, 6–11 January 2002. World Scientific. p. 82. ISBN 981-238-222-4.

51. J-H Lee, J-W Lee, H-A-R Jung and B-K Choi (2009). "Comparison of SOI FinFETs and bulk FinFETs: Figure 2". Silicon-on-Insulator Technology and Devices. The Electrochemical Society. p. 102. ISBN 1-56677-712-7.
52. Techweb. Techweb (2010-01-29). Retrieved 2010-11-27.
53. Power Semiconductor Devices, B. Jayant Baliga, PWS publishing Company, Boston. ISBN 0-534-94098-6
54. Power MOSFET Basics: Understanding MOSFET Characteristics Associated With The Figure of Merit. element14. Retrieved 2010-11-27. Archived April 5, 2015, at the Wayback Machine.
55. Power MOSFET Basics: Understanding Gate Charge and Using It To Assess Switching Performance. element14. Retrieved 2010-11-27. Archived June 30, 2014, at the Wayback Machine.

Index